the Guru's Guide to the Universe...

A Practical Guide to Modern Physics

Rick Tavares, Ph.D.

The Guru's Guide to the Universe...
A Practical Guide to Modern Physics

Published by Kiwibird Press of Kiwibird, LLC

Printed in the United States of America

First Printing, 2013

ISBN 978-0-9885401-0-1

For information or inquiries please contact:

Kiwibird, LLC.
P.O. Box 488
Ramona, CA 92065

www.kiwibirdpress.com

~ Contents ~

Introduction

Physics Anyone?

Just look around at today's world! Misconceptions, fallacy, and fad-science have become so prevalent in our modern technologically advanced society that there is often more *pseudo-science* tossed about during the course of a day, than the genuine article. With so much simplistic, incomplete, and even outright incorrect information about science being promulgated, is it any wonder how real scientific ideas can get so muddled up with science fiction in the popular thought.

We are exposed through television and the internet to a daily diet of *one-sided* hubris-driven science, *science-by-consensus*, or flat-out *junk-science and techno-gibberish.* It is commonplace to see 'scientific arguments' made whereby certain physical laws are suspended to favor others, or a physical phenomenon is ignored just because it does not fit a specific notion a commentator might have of how things are.

Wherever there is a lack of information on a particular topic, human beings begin to ascribe inaccurate assumptions and predictions {guesses}, bad cause-and-effect {urban legends}, or curious non-scientific notions {magic and superstition} to describe a physical process or phenomenon, and the context and value of the problem are completely lost.

There are many great books out there on physics and mechanics, light and sound propagation, electromagnetism, quantum mechanics and the like. Strangely enough, this vast quantity of literary material does not directly translate into our society being well versed in science. Sadly, it is quite to the contrary.

The books out there are either too technically complex or mathematically rigorous for the average audience (a good thing for a graduate level course on the subject, but a bad thing when you're trying to learn the subject from the beginning), or are watered-down to the point of being completely useless. There needs to be a middle-ground, a place where good science comes together in an informative and illustrative fashion, with technical clarity in a diverse range of subject matter, and presented in such a way that the layperson or novice can understand, enjoy, and more importantly apply each and every day.

So, what's a person to do? The answer is simple: *read this book!*

The *Guru's Guide to the Universe* is a <u>practical</u> book on modern physics. Note the emphasis on the word 'practical'. It is designed to be a fun, informative, and educational book on how the world, and universe, around you works. Think of this book as a science fiction novel – *minus the 'fiction' part.* We're going to tell you a fantastic story, stranger than any fiction, and when it is done, not only will you understand how things work, you will have gained an enlightening background proficiency in physical science, that you can enjoy and apply in everyday life.

From learning what 'stuff' the universe is made of, to being able to calculate the forces in nature, to understanding why some types of time travel might not be a very good idea, this book will take you there and back – *all from the comfort of your arm chair.*

The *Guru's Guide to the Universe* provides a thorough treatment on the universality of physical science, while not burying the reader in rigorous and complex mathematics. This book is written for anyone with a keen interest in learning applied physics. The only real prerequisites for the Guru's course on physics is that you know how to read and have an imagination and curiosity about the unknown.

Ideal applications for the *Guru's Guide to the Universe* would include an entertaining brush-up and reference for individuals interested in improving or refreshing their understanding of the topic, an introductory or supplemental book for freshman/sophomore college students having to take their 'dreaded' physical science courses, or even as a high school primer on physics.

The Natural Scientists Amongst Us

The first few years of a baby's life are principally spent sleeping, eating, and 'playing', where "playing" is essentially comprised of a vigorous study of language, sociology, and science.

Babies love learning science, especially physical science. They have a great zeal for exploring and investigating every little detail about the physical properties and behavior of the world around them. It is new, amazingly varied, and absolutely fascinating!

Since at first, everything is unknown to them, babies must start by examining the basics. So they begin by setting up little science experiments (such as throwing or dropping a pacifier), taking data with any sensing devices they have at their disposal (all five built-in senses), and observing and noting anything that seems relevant.

Then, they think about their acquired data/observations, and reformulate their understanding of how things are. Usually, they will repeat the experiments, to see if the same thing happens again. Babies apparently recognize what all true scientists know: that repeatability of results is a very important part of the process of forming a scientific theory.

But, more study is still needed. So next, the babies begin making slight modifications to their experiments, whenever opportunities to do so are available. Thus, later the pacifier may be thrown in different directions. The baby observes, *"Hmm, curiously enough, it always ends up down on the floor, somehow."* Or, the object being thrown might be changed and the baby observes, *"Hmm, the chunk of banana and sippy-cup made completely different sounds when they hit the floor."*

Of course, since we all were babies once, it therefore must follow that we have all been eager scientists in the past. Thus, there is no reason to not to continue our zealous study of science as a life long hobby. Unfortunately, as we grow up we become aware of some societal biases and misconceptions about science (such as "it's too difficult", "it's too boring", or "we already know everything now, so there's nothing more to learn"), which can serve to needlessly dissuade or discourage us from further pursuit of the topic. But the babies have the right idea. To them learning physics is fun.

For many, science seems to have originated from some really thick books, but it really comes from human curiosity. When we do not understand something, our natural inclination is a desire to find out the answer, to learn about it. This is why as babies, we take matters into our own hands and start experimenting. Then, as we gain a command of language, we ask questions of those who might already know the answer to a question. When we can go to school and read, we can learn more from books. But where did that information come from, and what happens when there is still more that is unknown and must be discovered?

This is when we must revert to the method the babies use. Only now, since we are more mature, we can be more organized, and using fancy experiments and gadgets in laboratories, we can embark on a more systematic course of scientific investigation, which we refer to as the *scientific method*.

The Order of the Guru's Universe

The specific branch of science known as *Physics* is concerned with what makes the universe 'tick'. It is the field of study of the *nature and properties of matter and energy* (i.e., everything around us). Physics, which was originally called *Natural Philosophy*, is one of the oldest academic disciplines, and probably <u>the oldest</u> depending on how you look at the historical record of humans.

Just as the actual universe follows an ordered direction, so too does the Guru's Guide, in the method and order that new topics in physics are presented. Each new topic is designed to build upon the previous, through the development of the scientific principles, new technical vocabulary, and examples of the principles applied to a wide variety of topics. The goal is to not only have the reader become conversant in physics at a level far beyond that of a layperson, but also to have the necessary tools to apply the learned physics to everyday problems encountered in life that are outside the scope of this book.

Given this, we're going to start at the basics. The Guru assumes that the reader has some basic algebra skills and knows some very elementary aspects of trigonometry (or geometry). The derivations in this book are designed to be *Calculus free*, to the maximum extent possible (which is not an easy thing to do in a physics book), so some artistic-license has been taken without any loss in scientific generality for the technical purists out there.

Our path through the universe will start out by developing an understanding of how we measure physical phenomena and what units we should use. We'll then touch upon some elementary forces and develop the necessary mathematical tools to describe any physical process, whether it occurs here on Earth, or on the other side of the universe.

We'll continue our exploration by seeing what this 'stuff' called *matter* is made of, and how it moves about. Once we can quantify its motion, we can move right into the forces that produce that motion, just as early physicists did. Along the way we'll look at forces produced by collisions, the concepts of work and energy, and how things rotate freely in space. In totality, we'll cover the area of physics known as *mechanics*.

Being mechanical 'experts' at this point, we will shift gears and look at how energy is moved throughout the universe by means of fluid and wave motion. Discovering that the concepts of work, energy, and heat seem to keep popping up everywhere, we'll wrap up this part of our journey with the *Laws of Thermodynamics*, which describe the physical mechanisms that actually provide the power to keep the universe moving right along. We'll also learn which universal laws can never be violated (not even with a time machine on a good day), and discover that out of all the forces we can imagine in the universe, there are really only four of them that are unique to everything.

The last part of our journey will take us into some of the intangibles of the universe starting with electricity and magnetism, and what strange behavior light seems to possess. Finally, we'll dare to venture into the world of special relativity and see that the actual universe is stranger than any work of science fiction.

Upon completion of this book, the reader should be able to pick up any college level text on physics and have a good general understanding of the material minus, of course, the '*fear factor*' some associate with reading such books.

So get ready for a fast-paced, fun, and informative romp through the inner-workings of what makes the universe tick. All the tools are here at your disposal to help you understand the fundamentals of most topics in modern physics, and give you the necessary vocabulary to be able to do further research in your own personal areas of interest.

About the Guru and the Guide

The *'Guru'* is in real-life Dr. Rick Tavares. Although he looks nothing like the guru professor who will guide you through this book, Dr. Tavares' role is to animate this enthusiastic scientist character on your exploration through the universe. Dr. Tavares has a Bachelor of Science degree in aerospace engineering, Master of Science degrees in mechanical engineering and structural engineering, and a Doctorate in civil engineering. As an engineer and a scientific consultant, he has made it his business to solve the world's problems using applied physics, and has taught numerous engineering courses at the university level. All of this knowledge has been imparted to our cartoon professor friend, so you are in good hands with the Guru.

In researching the *Guru's Guide to the Universe*, Dr. Tavares combed through dozens of physics, mechanics, optics, thermodynamics, fluid and solid mechanics, astrodynamics, and college level mathematics books in his personal library, searching for the unifying theme that holds these seemingly diverse topics together. The result is the book you hold in your hands, a summary and introduction of the notable aspects of modern physics gleaned from his more than 20 years of work in the field.

Section I – Back to the Basics

Chapter 1: Measuring the Universe

Because we really do need to know how fast, how tall, how slow, and how small

A good place to start our exploration of the universe is to work out some of the rudimentary details of how we measure things. The Guru assumes that the reader has an understanding of arithmetic and basic Algebra, so this section should be more of a review than anything else. If there is something that looks new or strange, don't worry, as you continue reading the fog will lift.

Numbers, Numbers, Everywhere…

In the world of physics, we work with two different quantities: numbers and units. We might have some truly baffling mathematical expressions along the way, but in the long run we are only interested in a numerical answer and the associated units.

The best way to arrive at a good functional test of any hypothesis is to compare *numbers-to-numbers* using a consistent set of units and coordinates that everyone agrees upon.

Due to the wide range of values we can see in physics, and the need to keep our accuracy as high as possible, a system, known as *scientific notation,* was developed to express big and small numbers accurately and concisely. Scientific notation expresses any number as a coefficient times a power of 10. Thus, all numbers within the study of physics can be written in a similar numerical format.

For example, in a single gram of hydrogen gas, there are roughly 602,200,000,000,000,000,000,000 atoms, each of which has a mass defined as *one atomic mass unit*. This numerical constant is called *Avogadro's number* and is the conversion constant between the units of grams and atomic mass units (or amu's) of any substance, enabling us to determine how many atoms are present in a measurable sample.

If we had to write this number out in longhand each and every time we used it, we would, 1) get writer's cramp, and, 2) invariably make a mistake somewhere in the arithmetic by losing track of some of those zeros. The solution to this dilemma is to write this number in a different way, in scientific notation as,

$$602{,}200{,}000{,}000{,}000{,}000{,}000{,}000 = 6.022x10^{23}$$

Which would be spoken to another scientist as, *"6.022 times the number 10 to the 23 power"* or simply *"6.022 times 10 to the 23"*. This more compact scientific notational form is just as accurate as the long-winded version of the number and actually allows us to perform arithmetic on a number this large with greater ease. Generally speaking, any number can be written in scientific notation format as follows,

$$\text{any number} = \text{coefficient} \times 10^{\text{exponent}}$$

The coefficient's job is to maintain the numerical accuracy of the number, while the exponent keeps track of all those annoying zeros that keep getting in the way. If we need more precision in the number, we simply add more digits to the coefficient. It's as simple as that.

Let's now look at a really small number. The mass of an electron (the negatively charged part of an atom) is roughly $9.10938215 \times 10^{-31}$ kilograms (abbreviated as 'kg'). This is actually an incredibly precise number, since if we were to expand it out in longhand notation it would look like,

$$9.10938215x10^{-31} = 0.000000000000000000000000000000910938215$$

There isn't a handheld calculator that can handle this number outside the realm of scientific notation (even Microsoft Excel stops at 30 digits; our example here requires 39). One can easily see that using scientific notation is more of a requirement, than a nicety, when exploring the universe.

So, if we are given a value for some physical phenomenon, say 12.081 meters, and we want to convert this value into scientific notation, we would merely move the decimal point to the left until there was only one number remaining to the left of it, and count the number of times we had to do this – which, by default becomes our exponent. In the above simple example, we would move the decimal point one place to the left, and then rewrite this value in scientific notation as 1.2081×10^{1} meters.

Problem Solving Tip...

All modern calculators include a button labeled 'E', 'EE', or 'EXP' allowing entry of scientific notation. The procedure is usually the same, in that you enter the coefficient, then the 'EE' button, and then the exponent (with the sign).

This is, by and large, one of the most used buttons on a scientific calculator – *ask any student of science.*

Likewise, if someone gave us a distance in scientific notation as 1.2081×10^{1} meters, and we wanted to convert it to ordinary decimal longhand notation, we would use the exponent to tell us how many places to shift the decimal point to the right (or, to the left in the case of a negative exponent).

The Guru Says...

Humans have been using base 10 as a number system for a very long time – most likely because we have 10 fingers (or toes) to count with...

The Guru hypothesizes that real aliens near the star Proxima Centauri would have a little trouble learning their scientific notation as well, so you're not alone, and if they have only eight fingers and toes, you can imagine what their numbers must look like!

Confused? Many beginning science students are at first, mostly due to a matter of perception, rather than understanding. We have been taught to look at numbers in a wholly decimal fashion, and so the concept of representing them, as a 'power of ten' is somewhat alien. Not to worry, the Guru has created a simple scientific notation flowchart in Figure 1.1 to help you figure all this out.

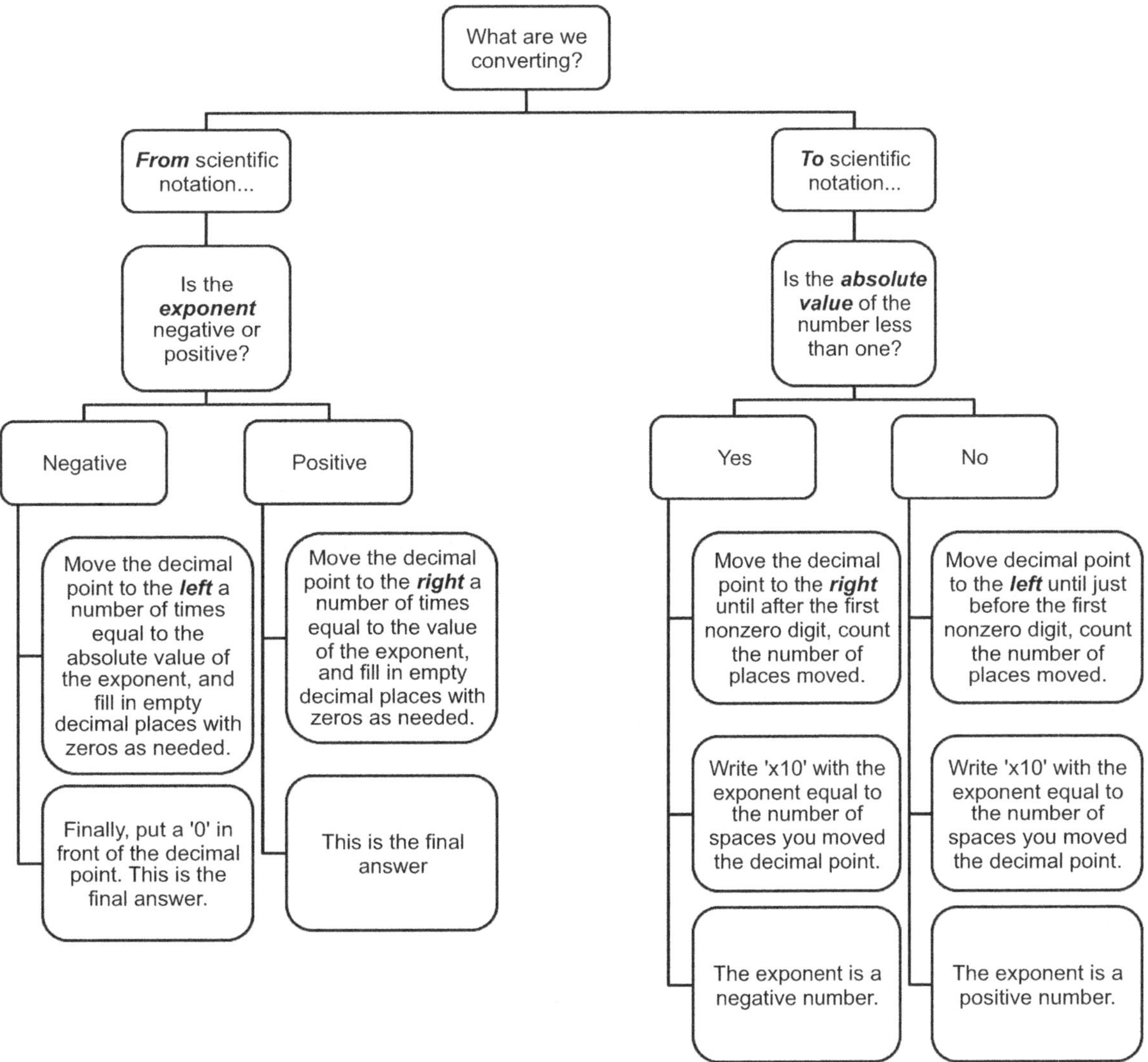

Figure 1.1: The Guru's Legendary Scientific Notation *'Convert-o-Matic'* Flowchart

We will see many examples of the use of scientific notation in our travels through the field of physics. For the time being, it is only important to have a feel for the magnitude of a number when it is expressed as an exponent of the number 10 since most mistakes are made by not recognizing a large from a small number. Let's look at an example.

Example 1.1: *The Guru asks you to add, subtract, multiply and divide the following two distance measurements in scientific notation together and write the result to four significant digits: $1.234x10^2$ and $4.321x10^{-2}$. What is the answer?*

When we have two numbers that we want to add together, we simply *convert them to the same exponent and add the coefficients together*. Thus we can write,

$$1.234x10^2 + 4.321x10^{-2} = ?$$
$$= 1.234x10^2 + 0.0004321x10^2$$
$$= (1.234 + 0.0004321)x10^2 = 1.2344321x10^2$$
$$\approx \mathbf{1.234x10^2}$$

Throughout this book, the Guru will show the results of a problem or example in **bold text** so you can see what the final answer looks like.

Thus, the addition of the two numbers equates to $1.2344321x10^2$, which is approximately '$\approx$' equal to $1.234x10^2$, when we round the answer to four digits. Since we are adding a large number to a small number, and then rounding the result, it is mathematically acceptable that the smaller number gets 'lost' in the rounding.

The Guru Says...

There is an old saying in mathematics that goes something like, "**P**lease **E**xcuse **M**y **D**ear **A**unt **S**ally". This goofy quote is actually a memory trick used to remember the order of operation of mathematical equations – **PEMDAS**, or **P**arenthesis, **E**xponents, **M**ultiplication, **D**ivision, **A**ddition, and **S**ubtraction.

Remember these rules and Aunt Sally won't get mad at you...

The same logic applies if we were to subtract the numbers. We would typically adjust the smaller value to have the same exponent as the larger one, and subtract the coefficients from each other.

We can write,

$$1.234x10^2 - 4.321x10^{-2} = ?$$
$$= 1.234x10^2 - 0.0004321x10^2$$
$$= (1.234 - 0.0004321)x10^2 = 1.2335679x10^2$$
$$\approx \mathbf{1.234x10^2}$$

Again this gives a value of $1.234x10^2$ when we round the answer to four digits. The Guru asks the reader, why does this occur?

Now, when we multiply two numbers in scientific notation, the procedure is a little bit different. In this case, we *multiply the coefficients together and add the exponents*. You can prove this to yourself on a calculator that multiplying two 'powers of 10' together is the same as adding their exponents.

We would get for our two sample numbers,

$$1.234x10^{2} \cdot 4.321x10^{-2} = \ ?$$

$$= (1.234x10^{2}\ ) \cdot (4.321x10^{-2}) = (1.234) \cdot (4.321)x10^{(2-2)}$$

$$= (5.332)x10^{(0)} = 5.332x1 \quad Note\ :\ 10^{0} = 1$$

$$= \mathbf{5.332}$$

So, the trick when multiplying two numbers in scientific notation is to *separate out the coefficients and multiply them together; then add the exponents to achieve the new exponent* (which can be positive, negative, or zero as we have seen above).

Finally, dividing two numbers in scientific notation works the same way, with the exception being that *we divide the coefficients and subtract the exponents*. Let's look at the same problem one more time, and instead of multiplication, we will divide the second number into the first.

$$1.234x10^{2} \div 4.321x10^{-2} = ?$$

$$= \frac{1.234x10^{2}}{4.321x10^{-2}} = \frac{1.234}{4.321}x10^{(2-(-2))}$$

$$= \frac{1.234}{4.321}x10^{(2+2)} \qquad Note: 2 - (-2) = 2 + 2$$

$$= \frac{1.234}{4.321}x10^{4} = 0.2856x10^{4} = \mathbf{2.856x10^{3} = 2{,}856}$$

Now that we know how to perform arithmetic on numbers in scientific notation, we can start to approach practical problems in physics in a quantitative way – *no more guessing for us, we're in the 'big leagues'.* Let's do another example and toss in some units for good measure.

***Example 1.2:** Suppose a hypothetical scientist measures a room as a future home for a time machine and finds out that its length is $50x10^{2}$ meters (m) and its width is $100x10^{3}$ m. The floor to ceiling height is 3x10 m.*

She then learns that a 2x10 m by 2x10 m section of the floor of the room cannot be used in the calculation. What is the final usable volume of the room?

Let's break this word problem down into its pieces, if only to see the operations on the scientific notation and the use of PEMDAS. Mathematically, we have the following expression describing the volume of the room.

$$\text{Room Volume} = (length \cdot width \cdot height)_{whole\ room} - (length \cdot width \cdot height)_{part\ we\ can't\ use}$$

So, the solution would be,

$$\begin{aligned}\text{Room Volume} &= \left[(50x10^2)(100x10^3)(3x10^1)\right]-\left[(2x10^2)(2x10^2)(3x10^1)\right] \\ &= (50)(100)(3)x10^{(2+3+1)} - (2)(2)(3)x10^{(2+2+1)} \\ &= (50)(100)(3)x10^6 - (2)(2)(3)x10^5 \\ &= 15000x10^6 - 12x10^5 = 15000x10^6 - 1.2x10^6 = 14998.8x10^6 \\ &= \mathbf{1.4999x10^{10}\ meters^3}\end{aligned}$$

...a very large room indeed, which would have a footprint the size of a small city!

So, the moral to the story is that a time machine requires a lot of space to work properly, and we need to be extremely careful when working with exponents since they represent *powers of 10*, and can create very large or very small numbers easily.

Problem Solving Tip...

Unless you're stranded on a desert island, with nothing but pencil and paper, you'll probably want to do this calculation using a scientific calculator.

In this case, you can perform the entire calculation in one-shot on a TI Calculator by typing: (50EE2*100EE3*3EE1) - (2EE2*2EE2*3EE1) and pressing ENTER.

Now that we have the quantification part of our understanding of physics under control, it's time to take a closer look at the other part of the equation – *the physical units*.

The History of Physical Measurements

By mid-1700, there were a lot of different units floating around (notably the English system of units, consisting of the foot, the pint, the slug, degrees Fahrenheit, etc.), all of which required experimenters to know how to convert from one particular set of units to their own in order to be able to reproduce a particular experiment. Then there was another really big problem – *exactly how long was the unit of a foot anyway?* One measuring stick might be a little different than one somewhere else since there was no standardization of anything during these times.

As the required accuracy of physical experiments increased through the centuries, the need for a more unified approach to measurement was required. Thus, on June 22, 1799, right at the end of the French Revolution no less, a group of notable scientists created two platinum bars, one representing the length of the unit of the meter, and the second the mass of the kilogram, and stored them safely in the *Archives de la République*, or Archives of the Republic, under glass and in a vacuum to prevent

Table 1.1: Length of Various Things in the Universe

Object of Interest	Length in Meters
Diameter of a Hydrogen Atom	1.0×10^{-10}
Thickness of a Sheet of Paper	1.0×10^{-4}
Average Height of a Human	1.8×10^{0}
Height of Mt. Everest	8.9×10^{3}
Diameter of Earth	1.2×10^{7}
Diameter of the Solar System	1.2×10^{13}
Distance to Nearest Star (Alpha Centauri)	4.3×10^{16}
Distance to Nearest Galaxy (Andromeda)	2.1×10^{22}

environmental contamination. All physicists then agreed that this would be the standard against which all rulers and scale mass weights would be compared.

Later in 1889, due to storage difficulties associated with the original platinum meter standard, a new 'X' shaped 90% platinum 10% iridium bar was created since this metal alloy does not oxidize or appreciably change with respect to temperature.

This standard was in place until 1960, when the meter was again redefined in terms of the wavelength of light of a specific isotope of the noble gas Krypton (Kr-86 to be specific).

Table 1.2: Time Intervals of Events in the Universe

OBJECT OF INTEREST	TIME IN SECONDS
One Cycle of a Cesium Atomic Clock	1.10×10^{-10}
One Cycle of an FM Radio Wave	1.00×10^{-08}
One Tick on a Clock's Second Hand	1.00×10^{0}
One Solar Day on Earth	8.60×10^{4}
One Solar Year on Earth	3.10×10^{7}
Lifespan of a Human	2.00×10^{9}
Age of Earth	1.30×10^{17}
Age of the Universe	6.30×10^{17}

Finally, in 1983 all this artificial measurement jazz was dispensed with, and the meter is now currently defined by the length a beam of light travels in a vacuum during a specified time interval.

Knowing that there was now a place where you can go to mark off a really good ruler, or compare your laboratory mass weights, provided a boost to the physical sciences, by allowing all experimenters to know whether a new finding was indeed new, or due to a bent ruler or dropped mass. Even with these improvements, there was still a need for a unification of units in order to get everyone's work on the 'same page', as it were.

The Basic Units of Physics

In 1889 it was internationally agreed that the meter, the kilogram, and the second were to be the *officially approved units* for all physical work. This simple step provided a gigantic leap in providing consistency amongst scientific measurements. Using these adopted units, Tables 1.1 through 1.3, which quantify various objects in the universe, now take on the familiar form that we are used to seeing today.

Later in 1954, the units of the ampere (for electrical current), the Kelvin (for temperature), and the candela (for light intensity, to make astronomers happy) were added to account for all the new findings in physics since the turn of the century.

In 1960, the codified system of units started in 1889 was officially named the *International System of Units,* or SI units. The last addition to the SI system was in 1971 when chemists added their two-cents, by including the unit of the *mole* as the basic unit of substance.

Table 1.3: Masses of Objects in the Universe

OBJECT OF INTEREST	MASS IN KILOGRAMS
An Electron	9.10×10^{-31}
A Proton	1.70×10^{-27}
A Speck of Dust	6.70×10^{-10}
An Average Human	6.00×10^{1}
The Moon	7.40×10^{22}
Earth	6.00×10^{24}
Our Sun	2.00×10^{30}
Milky Way Galaxy	2.20×10^{41}

Now that we know where science has been, quantifying the physical nature of the universe is not all that hard. As it turns out, there are actually only seven (7) different units to know, which describe the *'whole shebang'*. These seven units are collectively called the *'base units'* of physics in that any other units you might encounter are derived from these seven basic physical quantities. These physical quantities are: *length, mass, time, electric current, temperature, the amount of a substance, and how bright something is*. They are shown with their 'official SI definition' in Table 1.4. Don't worry if it doesn't all make sense now, it will later on.

Table 1.4: The Basic Units of the SI System

PHYSICAL MEASUREMENT	SI UNIT NAME	ABBREVIATION	HOW IT'S *OFFICIALLY* DEFINED BY THE SI...
Length	meter	m	The meter is the length of the path travelled by light in vacuum during a time interval of 1/299,792,458 of a second.
Mass	kilogram	kg	The kilogram is the unit of mass; it is equal to the mass of the international prototype of the kilogram {the Platinum-Iridium bar we talked about earlier}.
Time	second	s	The second is the duration of 9,192,631,770 periods of the radiation corresponding to the transition between the two-hyperfine levels of the ground state of the Cesium 133 atom.
Electric Current	ampere	A	The ampere is that constant current which, if maintained in two straight parallel conductors of infinite length, of negligible circular cross-section, and placed 1 meter apart in vacuum, would produce between these conductors a force equal to 2 x 10^{-7} Newton per meter of length.
Temperature	Kelvin	K	The Kelvin, unit of thermodynamic temperature, is the fraction 1/273.16 of the thermodynamic temperature of the triple point of water.
Amount of Substance	mole	mol	The mole is the amount of substance of a system, which contains as many elementary entities as there are atoms in 0.012 kilogram of Carbon 12; its symbol is "mol." When the mole is used, the elementary entities must be specified and may be atoms, molecules, ions, electrons, other particles, or specified groups of such particles.
Light Intensity	candela	cd	The candela is the luminous intensity, in a given direction, of a source that emits monochromatic radiation of frequency 540 x 1012 hertz and that has a radiant intensity in that direction of 1/683 watt per steradian.

That's it! If you know these basic units, you can tackle any problem in modern physics, chemistry and science, no matter how difficult or obtuse. As we can see, the physical definitions of the various unit standards employed by the SI system are extremely straightforward and can be created in any modern physical science laboratory having the minimum of scientific instrumentation. In fact, the only standard that would require us to leave the lab would be for mass, since it is compared to the platinum-iridium bar in France, but we could always order out for a cheaper calibrated copy from a laboratory supply house.

Deriving Units in the SI System

We can now take our base SI units and start constructing units of higher complexity. For example, we know that length is measured in the unit of the meter. If we have a large block of concrete ½-meter long on each side, we can calculate the area of the block, as well as its volume, since all we need to do is add another dimension to the basic meter to get meter-squared (m^2) for area, and meter-cubed (m^3) for volume.

If for some strange reason we were desirous to shoot our ½-meter long concrete block out of a cannon and wanted to know how fast it travels, well that's easy too. The basic SI unit of time is the second, so how far something travels in a given period of time is simply the quantity of length per measured time, or meters per second (m/s). Playing around with the various base SI units allows us to create some of the new derived physical units as shown in Table 1.5.

Table 1.5: Some Derived Units of the SI System

DERIVED SI QUANTITY	UNITS NAME	ITS UNITS...	WHAT IT'S USED FOR...
Area	square meter	m^2	How big of a footprint is something.
Volume	cubic meter	m^3	How much 'space' does something take.
Speed or Velocity	meter per second	m/s	How fast is something moving.
Acceleration	meter per second squared	m/s^2	How fast is something changing speed.
Mass Density	kilogram per cubic meter	kg/m^3	How compact is something.
Current Density	ampere per square meter	A/m^2	How much electric charge is passing through in a certain area.
Magnetic Field Strength	ampere per meter	A/m	How strong is a magnetic field.
Concentration Amount	mole per cubic meter	mol/m^3	How concentrated is the 'stuff' you're looking at.
Luminance	candela per square meter	cd/m^2	How 'bright' is the object you're looking at.

As you can see, it would start of become pretty cumbersome to only work in the base SI units or their derivations. After a while, our units would get very long indeed, and the possibility of introducing error into our work by missing a squared term on the meter or second would become greater.

Think about how confusing it would be to tell someone what the voltage is in your home using just the base SI units. Instead of saying 120 Volts (a derived working unit as we'll see in a moment), we would have to say something along the lines that the electromotive force in a house is 120 $m^2 \cdot kg/(s^3 \cdot A)$ or *120 kilogram-square-meters per seconds-cubed per ampere – yuck!*

The Working Units of Physics

To solve the problem of base- and derived-unit overload, we need to define a set of *working units* for modern physics that will become the new starting point in our physical calculations. These more complex units will be the system that we will always work in, and try to reduce our work into. The question then becomes as to what would be a good collection of working units. Luckily, physicists have already figured out what the best possible set of working units would look like. Let's take a look at them in Table 1.6.

Table 1.6: The Working Units of the SI System

WORKING UNIT (DERIVED)	WHAT IT MEASURES...	SI SYMBOL	ITS DERIVED UNITS...	IN TERMS OF SI BASE UNITS...
Radian	Plane Angle (2D)	rad	-	$m \cdot m^{-1} = 1$
Steradian	Solid Angle (3D)	sr	-	$m^2 \cdot m^{-2} = 1$
Hertz	Frequency	Hz	-	s^{-1}
Newton	Force	N	-	$m \cdot kg \cdot s^{-2}$
Pascal	Pressure, Stress	Pa	N/m^2	$m^{-1} \cdot kg \cdot s^{-2}$
Joule	Energy, Work, Quantity of Heat	J	$N \cdot m$	$m^2 \cdot kg \cdot s^{-2}$
Watt	Power, Radiant Flux	W	J/s	$m^2 \cdot kg \cdot s^{-3}$
Coulomb	Quantity of Electric Charge	C	-	$s \cdot A$
Volt	Electric Potential	V	W/A	$m^2 \cdot kg \cdot s^{-3} \cdot A^{-1}$
Farad	Capacitance	F	C/V	$m^{-2} \cdot kg^{-1} \cdot s^4 \cdot A^2$
Ohm	Electric Resistance	Ω	V/A	$m^2 \cdot kg \cdot s^{-3} \cdot A^{-2}$
Weber	Magnetic Flux	Wb	$V \cdot s$	$m^2 \cdot kg \cdot s^{-2} \cdot A^{-1}$
Tesla	Magnetic Flux Density	T	Wb/m^2	$kg \cdot s^{-2} \cdot A^{-1}$
Henry	Inductance	H	Wb/A	$m^2 \cdot kg \cdot s^{-2} \cdot A^{-2}$
Celsius	Temperature	°C	-	°K
Lumen	Luminous Flux	lm	$cd \cdot sr$	$m^2 \cdot m^{-2} \cdot cd = cd$
Lux	Luminance	lx	lm/m^2	$m^2 \cdot m^{-4} \cdot cd = m^{-2} \cdot cd$
Gray	Absorbed Radiation Dose	Gy	J/kg	$m^2 \cdot s^{-2}$

Looking at the table we see, in order from left to right, the name of the new working unit, what the unit measures, its common symbol in the SI, its units derived from other working units, and finally its units as derived from the original base SI units. It should be noted that the mathematical convention of multiplying by a negative exponent is the same as dividing by a positive exponent of the same value (i.e., one Pascal is one Newton per square-meter = $N/m^2 = N \cdot m^{-2}$ and so on).

It is still important to remember that these are derived quantities and, when in doubt, they can always be reduced back to the base SI units. This is an extremely important concept, especially when examining or reducing complex answers down to something more understandable, and to check if your final answer is correct.

With the introduction of these new working units, we now have two new dimensionless (i.e., no units or unit-less) quantities to measure angles in a plane (the radian) and in solids (the steradian). The unit of frequency, the Hertz, which everyone who has ever worked a radio is aware of, is nothing more than the inverse of time (i.e., one divided by seconds or $1/s = s^{-1}$). The unit of force, the Newton, is derived from *Newton's Second Law* (more on that in Chapter 5), and is equal to the mass of an object (in kilograms), times the acceleration due to gravity (in meters per second-squared, or m/s^2). Thus a Newton is equal to a kilogram-meter-per-second-squared ($kg{\cdot}m/s^2$ or $kg{\cdot}m{\cdot}s^{-2}$).

From the Newton, we can derive the unit of pressure (the Pascal) by observing that the phenomenon of pressure is nothing more than a force divided by its area of application. A Pascal therefore has units of Newtons per square meter, or $N{\cdot}m^{-2}$.

Energy and work is physically defined as a force times a distance and can be derived in a manner similar to what was done for pressure. More simply though, we'll just use the unit of the *Joule* (or Newton-meter, N·m) to describe this quantity. Similarly, power is nothing more than how fast work can be performed; so we just take a Joule and divide by time (seconds), to get a new unit – *the Watt.*

The Guru Says...

At this point you should realize that a Kilowatt-Hour, the unit that our household electrical meters use to measure how much electricity we use, is not really measuring power.

Look at the units, a Kilowatt (energy/time) times an hour (time) equals just energy (the time units cancel out).

Four new units are derived as working units to describe anything of an 'electrical nature' in the universe. They are the *Coulomb*, the *Volt*, the *Farad*, and the *Ohm*, which measure, respectively: how much electron charge, how hard the electrons are being pushed-on, how much electron charge can be stored somewhere, and how much resistance does an electron feel when traveling though something. Adding to these units are three quantities describing electricity's *alter ego*, *magnetism*. These magnetic units are the *Weber*, the *Tesla*, and the *Henry,* which describe: how strong a magnetic field is over a certain area, how strong a magnetic field is over a unit area, and how much resistance does a magnetic field feel (the counterpart to the electrical Ohm, which is called inductance).

If temperature is your game, the only unit you need to worry about is the unit of *Celsius.* Celsius has no derived units, it is based solely on the base SI unit of the Kelvin and is shifted in scale so that *water freezes at zero degrees Celsius and boils at 100 degrees Celsius.*

If you're interested in how bright something is, then the units of the *Lumen* and the *Lux* will be occurring often. A Lumen is a measure of how much overall illumination, or perceived brightness, something produces, while a Lux is a measure of the intensity of light emitted from a surface per unit area. Household light bulbs are typically rated in one or the other of these units.

Finally, the unit of the *Gray* is a measure of how much energy in Joules something absorbs per kilogram of mass. It is the working unit for the measurement of radiation dose and an important concept when we start looking at some of the radioactive elements in Chapter 3.

The Scale of Things in the Universe

Things that occur in the universe don't usually occur in a *'onesy-twosey'* fashion; rather, there are typically a *whole lot of them* or a *scant few*, and their scale is either *really, really* large, or *teeny-tiny* small. Thus, we need some way to describe the extreme range of values we will encounter while examining a phenomenon scientifically.

A convenient way of approaching this problem of scale is to break any quantity in question into 'chunks of 1,000', and every time multiply or divide our quantity by this value, we assign a new prefix to the physical SI unit. We've already seen this used in the base unit section. Notice that the base unit for mass is not the gram – rather it is the kilogram. *Kilo* is the Greek prefix for 1,000, so the base SI unit of mass is actually 1,000 grams because a gram mass is actually kind of a small dinky little thing, and pretty easy to lose in a physics lab.

Now take a close look at Table 1.7, which represents the extremes in unit magnitude of everything you'll find while exploring the universe. To read this table, you start in the center of the table at the big bold number '**1**', which is the physical unit you are working with. Generally, for every 1,000-fold increase or decrease in the magnitude of the unit, you modify the prefix of the unit as shown in the table.

Thus, if we use the 'meter' as our unit of interest, 10 meters could be referred to as a *dekameter*. In the Guru's many years in science, however, he's never heard the prefix 'deka' used outside of a textbook setting, so if you're standing in a room full of physicists, it's probably not a good idea to call 10 meters a dekameter – they might look at you funny.

The Guru Says...

You'll notice that the metric prefixes are divided into groups of 1,000's except for areas immediately surrounding the unit itself. In this case, the divisions are a little finer in detail, since most common laboratory work typically never exceeds these bounds.

In fact, most work in physics uses the range of unit prefixes from micro (one-millionth of 'something') to mega (one million 'somethings').

The same applies for the prefix 'deci'. We would technically call one-tenth of a meter a *decimeter*. Real physicists just call this 0.1 meters though.

On the other hand, the prefix 'centi' is used everywhere in science. A 'centi' is one one-hundredth of something. A *centimeter* is one one-hundredth of a meter (or there are 100 centimeters in a meter). In medicine they often refer to volume through the units of cubic-centimeters (or cc's), since this is a good all-purpose unit of measure for this discipline. So what is a volume of 1,000 cubic centimeters? There's a shorthand name for that too – *it's called a Liter*.

Remember the Guru's warning, and never sound silly – solids are measured in grams (or fractions thereof) and liquids are measured in liters (or fractions of liters called milliliters). There are no exceptions (unless of course you're a doctor).

Table 1.7: Various Unit Prefixes and Notation within the SI System

Where the Unit is Used in Science	Metric Prefix	Symbol	Scientific Notation	Decimal Scale	English Name
Astronomical Work	exa	E	10^{18}	1,000,000,000,000,000,000	quintillion
	peta	P	10^{15}	1,000,000,000,000,000	quadrillion
	tera	T	10^{12}	1,000,000,000,000	trillion
	giga	G	10^{9}	1,000,000,000	billion
	mega	M	10^{6}	1,000,000	million
Naked Eye Work / Human Experience	kilo	k	10^{3}	1,000	thousand
	hecto	h	10^{2}	100	hundred
	deka	dk or da	10^{1}	10	ten
			10^{0}	**1**	
	deci	d	10^{-1}	0.1	tenth
	centi	c	10^{-2}	0.01	hundredth
	milli	m	10^{-3}	0.001	thousandth
	micro	μ	10^{-6}	0.000001	millionth
Atomic and Cellular Work	nano	n	10^{-9}	0.000000001	billionth
	pico	p	10^{-12}	0.000000000001	trillionth
	femto	f	10^{-15}	0.000000000000001	quadrillionth
Subatomic Work	atto	a	10^{-18}	0.000000000000000001	quintillionth

In most practical physics problems, large measurements usually work in the *kilo-, mega-,* and sometimes the *giga-* range, while the small stuff is measured in the *milli-, micro-,* and *nano-* range. Really small stuff is sometimes expressed in the *pico-* scale, but at that point we are looking down into the atomic or quantum world, which is only done by a small group of physicists.

We now have at our disposal, in these few short pages, every tool imaginable to measure, weigh, listen to, see, touch and otherwise observe the world of physics. The rest of this book will concentrate on explaining the physical nature of the universe from the simple to the downright weird starting with some fundamental mathematical concepts.

The Guru Says...

It is also worth mentioning that prefix symbols for physical quantities of $1\text{x}10^6$ base units (or larger) are capitalized, while those for quantities of $1\text{x}10^3$ base units (or less) are not. Most people don't care that much, but now you can be a prefix symbol elitist when the need arises.

The Guru's Guide is also about problem solving, and having the reader be able to apply the concepts in this book to their own problems, so let's put our newfound information to the test and work out a couple of examples.

Example 1.3: *The speed of light in a vacuum is 299,792,458 meters per second. What is the speed of light in scientific notation? Roughly how far is a light year (the distance light travels in one year) in meters?*

Following our scientific notation flowchart, we note that the number is greater than one, and thus the exponent would be positive. We move the decimal point to the left eight times until it is after the two and write the result as $2.99792458x10^8$ meters per second. In physics it's generally okay to round the speed of light to $3.0x10^8$ meters per second.

Now, one year is 365 days, one day is 24 hours, one hour is 60 minutes, and one minute is 60 seconds. Since we want all the units of time to cancel out (except for those of seconds), our conversion from years to seconds would look like the following,

$$1.0\ \cancel{\text{year}} \cdot \frac{365\ \cancel{\text{days}}}{1\ \cancel{\text{year}}} \cdot \frac{24\ \cancel{\text{hours}}}{1\ \cancel{\text{day}}} \cdot \frac{60\ \cancel{\text{minutes}}}{1\ \cancel{\text{hour}}} \cdot \frac{60\ \text{seconds}}{1\ \cancel{\text{minute}}} = \mathbf{3.15x10^7\ seconds}$$

So, there are $3.15x10^7$ seconds in one year. Since the speed of light is $3.0x10^8$ meters per second, one light year would be,

$$3.15x10^7\ \cancel{\text{seconds}} \cdot \frac{3.0x10^8\ \text{meters}}{\cancel{\text{second}}} = 9.45x10^{15}\ \text{meters} \approx \mathbf{1.0x10^{16}\ meters}$$

Thus a light year is $9.45x10^{15}$ meters, which astronomers round up to $1.0x10^{16}$ meters, since what are a few meters amongst friends in the universal scale of things.

Example 1.4: *A person is lifting a 50 N weight to a height of 3 m. How much work is being performed?*

Taking a look at the units for work, we see that it is expressed in Joules, and has derived units of the Newton-meter (N·m). Since we are told that the weight is in Newtons, and the height that the weight is lifted is 3 m, the process becomes one of simple multiplication. Thus the work, 'W', to lift the weight is,

$$W = (50\ \text{N}) \cdot (3\ \text{m}) = 150\ \text{N} \cdot \text{m} = \mathbf{150\ J = 150\ Joules}$$

Example 1.5: *The voltage in a typical house is 120 V. If the current drain of an appliance is 10A, what is the power consumption in Watts? If the current is left on for 30 seconds, how many Joules are expended?*

Again, looking at the derived units for voltage we see that it is defined as a watt per ampere (or W/A). Mathematically we therefore have,

$$Volts = Watts \,/\, Amperes, \quad or, \quad V = \frac{W}{A}$$

Rearranging terms and substituting the given values shown in the example we obtain,

$$\mathrm{V} = \frac{\mathrm{W}}{\mathrm{A}}; \quad \mathrm{W} = \mathrm{V} \cdot \mathrm{A}$$

$$W = (120\ \mathrm{V}) \cdot (10\ \mathrm{A}) = \mathbf{1200\ W = 1200\ Watts}$$

Now, looking at the definition of a Watt we can see that is it a derived unit, defined as a Joule per second. Rearranging terms again and substituting the time that the switch was left closed (30 seconds); we obtain the final answer as,

$$W = \frac{J}{s}; \quad J = W \cdot s$$

$$J = (1200\ \mathrm{W}) \cdot (30\ \mathrm{s}) = \mathbf{3.6x10^4\ J = 36{,}000\ Joules}$$

Later, in Chapter 12, we'll discover some very handy relationships regarding electricity that will allow us to tackle any problem like this that comes our way.

Example 1.6: *An object changes its temperature by 30 degrees Centigrade (°C). What is this temperature change expressed in degrees Kelvin (°K)?*

This example is a 'trick question', since we are stating that the change in temperature of the object is 30 degrees, and not the absolute temperature itself. Looking at our derived SI units table, we see that the unit of the Kelvin and the unit of the Centigrade are exactly the same (they are just offset from each other with respect to their starting points). **Thus, a 30° Centigrade temperature change is the same as a 30° change in Kelvin**.

So why have two temperature scales? Convenience mostly. The Kelvin scale is an *absolute thermodynamic temperature scale*. It is defined such that at zero degrees Kelvin, all atomic activity stops. It is the lowest temperature that can ever be reached in the universe – *anywhere*.

Using this absolute scale we find that water freezes at 273.15°K and boils at 373.15°K. Imagine having a laboratory thermometer which starts at 273.15°K and goes to 373.15°K – *what a headache that would be*. Thus, the unit of Centigrade is merely shifted by 273.15°K to set the freezing point of water to 0°C and its boiling point to 100°C. It makes the math a little bit easier, and thermometers just a little bit shorter.

Chapter 2: The Language of the Universe

Why the word 'vector' is not just another pretty name

In Chapter 1, we spent a bit of time talking about numbers and their units, and how they are expressed in physics. Now it's time to *kick it up a notch* by taking a look at some of the more useful mathematical concepts from an everyday standpoint.

In physics it is not only necessary to know how big a quantity is, it's also important to know where this quantity has been, or where it's going. To describe this type of problem, we need to construct a mathematical framework, which allows us to specify not only the size of the physical property under examination, but in what direction it is, or has been, traveling. By using the word 'traveling', we are not only implying movement through three-dimensional space, but also through time. Thus, the expressions we are going to derive here are valid for four-dimensional work (which is a good thing if we're ever going to get that time machine project off the drawing board).

In fact, the mathematics of physics is really not limited to any number of dimensions. Current theories of the universe hypothesize that there could be up to 10 dimensions (something called superstring theory), or 11 dimensions (called M-theory), or better yet, the fabulous *Bosonic string theory* with a total of 26 dimensions that would dazzle even Rod Serling.

Fortunately for us though, everything with which humans can interact works quite nicely in four dimensions (three spatial and one temporal or 'time'), so we'll just stick to these four and be no worse for the experience.

The Mathematics of Scalars

In the beginning, *there were scalars*. Scalars are nothing more than numbers. Numbers that specify the magnitude of some type of physical phenomenon (such as mass, height, speed, radioactive content, etc.) are all classified as scalars. If we say that the speed along a roadway is 50 kilometers-per-hour, we are using scalar terminology. If something weighs 100 Newtons, that's a scalar too. Mathematically, you've been working with scalars all your life. For example, using the scalar unit of the 'apple' we can add or subtract scalars in any fashion, as for example,

$$4 \textit{ apples} + 10 \textit{ apples} - 2 \textit{ apples} = \mathbf{12\ apples}$$

Or, being a little more technical, we can calculate the scalar volume 'V' of a spherical weather balloon of radius 5 meters and see that scalars can be multiplied or divided by any quantity with the same relative ease.

$$V = \frac{4}{3}\pi r^3 = \frac{4}{3}\pi(5\text{ meters})^3 = \frac{4}{3}\pi(5)^3 \cdot (\text{meters})^3$$

$$= \frac{4}{3}\pi(125)\text{ meters}^3 = 166.67\pi\text{ meters}^3 \approx \mathbf{523.6\text{ meters}^3}$$

While scalars tell us the magnitude of something, they do not impart any information as to what the phenomenon will be doing in the future, or has been doing in the past – they are absolutely silent on the topic. If we wanted to tell someone how fast we were going, *and in which direction*, a scalar wouldn't cut it and we would need a more powerful tool – *the vector.*

Cartesian Vectors – Where the Real Fun Begins

The word *'vector'* is ancient Greek for *'carrier'*; so by virtue of its name, a vector provides some type of information as to the path of the phenomenon it represents.

If we now say, *"I'm traveling at 50 kilometers-per-hour towards the east"*, I am using vector terminology. I now know the magnitude of the phenomenon, *"I'm traveling at 50 kilometers-per-hour..."* and in which direction, *"... towards the east"*. Every vector, by definition, has information specifying its *magnitude and its direction.*

From a practical standpoint, a vector is nothing more than a *super-scalar* in that it contains all the information of the scalar, with the added knowledge of how the scalar information is being applied in the universe (such as, 50 kilometers-per-hour easterly, 100 Newtons straight-down, etc.). A vector is your first important tool to describe things around you, and no matter how simple or complex the problem, they are always lurking about somewhere.

Knowing the basic definition of a vector, we now need some way to draw one in space, and ultimately perform some type mathematics on it, like we do with scalars. To do this, we're going to start with something called the *Cartesian coordinate system*.

The Guru Says...

The Cartesian Coordinate System is named after the famous 17th century philosopher René Descartes. His coordinate system revolutionized mathematics by providing a link between Euclidean geometry and modern algebra, allowing geometric shapes to be expressed using mathematical equations.

He's also the guy who made the observation, *"I think, therefore I am..."*

You've seen this system before; it's the 'X-Y' and sometimes 'X-Y-Z' coordinate system for drawing things in space. Let's start by drawing both coordinate systems, as they would be seen on a sheet of paper or in a mathematics or physics book.

Take a look at the two different coordinate systems shown in Figure 2.1. The first shows the point P_1 in two-dimensional space as being represented by the coordinates (x,y), where 'x' and 'y' can be any two scalar numbers. If we assume, for the purposes of our example, that the origin of this graph (the 0,0 point) is as shown in the figure, with the 'x' axis being in a positive sense towards the right and the 'y' axis being increasingly positive in an upward direction, then the point P_1 is therefore defined by two positive numbers. Both the 'x' and 'y' axis are at right angles, or perpendicular, to each other.

Using similar logic, our three-dimensional coordinate system in the second figure pane, would also be shown in a positive sense. Thus, point P_2, which is defined now by three scalars, would also be positive. *So far, so good?*

It is important to note that for all practical purposes, paper is a two-dimensional object. Thus, the three-dimensional coordinate system shown, when drawn on two-dimensional paper, actually has all axes at right angles to each other, even though it doesn't look like it. This is the penalty we pay in the universe for losing a dimension – right angles don't appear to be 90 degrees apart, *or orthogonal*, anymore. Even though no one can correctly draw a three-dimensional object in two dimensions, mathematically the axes are still orthogonal to each other and the vectors themselves couldn't care less whether you draw a coordinate system or not.

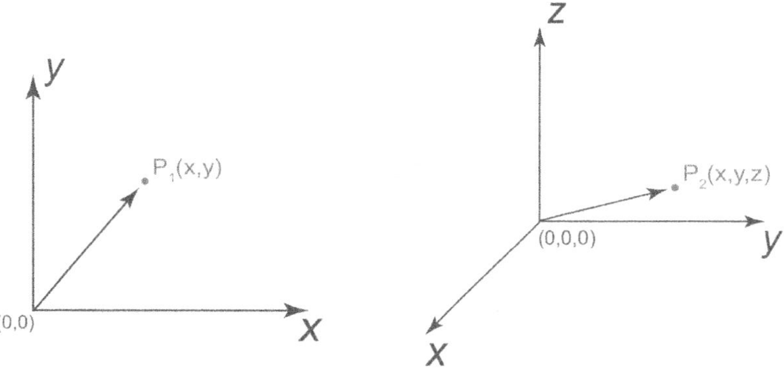

Figure 2.1: Two- and Three-Dimensional Coordinate Systems in Space Showing Positive Sense

Now stare at the arrows that run from the origin of the coordinate system to the points P_1 and P_2. Those arrows are the 'fixed' position vectors defining these two points in space. They have magnitude (as indicated by the length of the arrow), and they have direction (the arrow points in a specific direction). If we move either point, the corresponding position vector will change to redefine its location in Cartesian space.

You'll soon discover that we are not limited to just vectors that are 'bolted' to the origin of the coordinate system. Take a look at Figure 2.2, which shows what a 'free vector' between two different points 'A' and 'B' would look like.

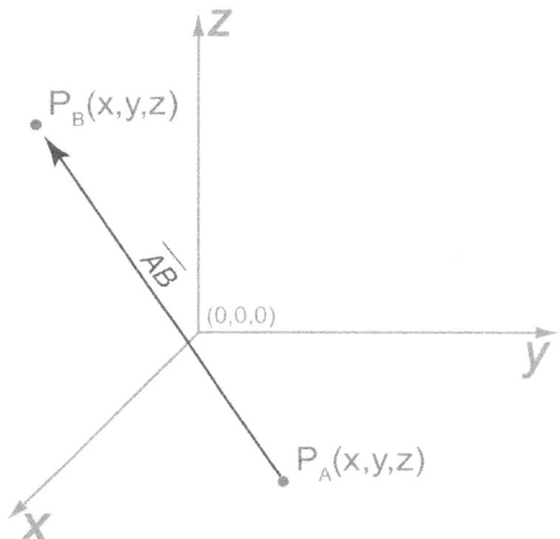

Figure 2.2: Representing a Vector between Points 'A' and 'B'

In this case, we have two points in space denoted as P_A and P_B. Both P_A and P_B would each have position vectors relative to the (0,0,0) coordinate origin, but they also have a vector relative to each other, as we have shown in the figure. We denote this vector as $\overline{AB}$ with the little bar above the letters indicating that this is a vector quantity and not a scalar. It is read literally as, *"the vector from A to B"*. This concept can be extended to any number of points, and any number of dimensions, each having vectors relative to one another, and relative to some fixed point in space, like our (0,0,0) origin point shown in Figure 2.1. In fact, we could add another dimension 't' for 'time', after the 'z' in our figure and technically we would have a four-dimensional vector, which could physically represent three spatial position dimensions varying with time for something like a satellite or rocket. *Pretty cool – huh?*

Cartesian Vector Notation

So, what have we learned about vectors so far? Well, a vector is a scalar that is definitely going somewhere. It has a magnitude and a direction, and appears to be relative to whatever coordinate system we place on it. It could be a *fixed position vector*, or a *free vector* between two different points in space, depending on what aspect of physics we are looking at. Now, pay very close attention to the next paragraph and the example that follows.

We <u>mathematically represent</u> vectors by writing their notation starting from *the tail of the vector and continuing to its tip*. Thus, if we're going from point 'A' to 'B' as shown in Figure 2.2, we denote the vector as $\overline{AB}$. Consequently, we <u>calculate</u> a vector by *subtracting its tail coordinates from its tip* coordinates. This is a convention that once mastered will make operating with vectors as easy as working with scalars and you'll find that you don't even need to draw them or a coordinate system. Probably the best way to illustrate this concept is through an example.

Problem Solving Tip...

Remember, vectors are written from <u>tail to tip</u> and calculated from <u>tip to tail</u>. This convention is followed for all vector notation and mathematical operations.

Doing this incorrectly will produce the negative of the result you are looking for (and the wrong answer).

Example 2.1: Suppose in Figure 2.2 that point P_A lies in the 'x-y' plane at point (5,3,0) and point P_B is in the 'x-z' plane at (2,0,5). What is the position vector that is equivalent to $\overline{AB}$?

To perform this mathematical operation, we are going to subtract, term-by-term, the 'x', 'y', and 'z' values of point P_A from those of point P_B (i.e., we're going to subtract the 'x' value of point P_A from the 'x' value of point P_B, and so on). Thus, we would write the vector as follows,

$$\begin{aligned}\overline{AB} &= P_B - P_A \\ &= (2,0,5) - (5,3,0) = (2-5)\hat{i} + (0-3)\hat{j} + (5-0)\hat{k} \\ &= \mathbf{-3\hat{i} + -3\hat{j} + 5\hat{k} = < -3,-3,5 >}\end{aligned}$$

And this yielded an answer with some unusual notation. At this point, we need to extend our discussion to explain what we just did. You'll notice that determining the scalar portion of the vector is accomplished by subtracting individual coordinate terms as we discussed (which is no big deal), but what's up with the $\hat{i}$, $\hat{j}$, and $\hat{k}$ stuff and those '< >' brackets?

Well, since a vector has to specify its direction in multi-dimensional space, we need to have a way to write down what each component is doing with respect to its coordinate axes. The $\hat{i}$, $\hat{j}$, and $\hat{k}$ values are the bookkeeping notation required to specify the contribution of each part of the vector to the 'x' axis (the $\hat{i}$ part), the 'y' axis (the $\hat{j}$ part), and the 'z' axis (the $\hat{k}$ part). Alternatively, we can write the answer in *vector shorthand notation* as $<-3,-3,5>$, where it is assumed that the answer is a vector since we enclosed it with '< >' symbols and the order of the terms shown is implied. It is really a matter of choice which notation to use, although you will discover that the shorthand approach is preferable to writing all those little $\hat{i}$, $\hat{j}$, and $\hat{k}$'s after a while.

Thus we know that something that moves from point P_A to point P_B, moves along a path defined by $-3\hat{i}+-3\hat{j}+5\hat{k}=<-3,-3,5>$, or, it moves backwards (-3) in the 'x' direction, backwards (-3) in the 'y' direction, and up (+5) in the 'z' direction. Vectors tell you not only how much something has moved, but in which direction in space. Adding more dimensions is as easy as adding more terms to the vector. You can work with a 10-dimentional vector just as easily as you can work with a three-dimensional one.

Resultant Magnitude of a Vector

Suppose now that we want to know how far the scalar distance covered by vector $\overline{AB}$ is. In this case, we're interested in calculating what is known as the *resultant magnitude of the vector*, which gives a scalar quantity equal to the length of the vector. The resultant magnitude 'R' of a Cartesian vector can be calculated by solving as follows,

$$R=\sqrt{x_i^2+y_j^2+z_k^2}$$

Where x_i, y_j, and z_k are the *components of the vector* in the 'x', 'y', and 'z' directions. Throughout this book, the Guru will highlight important mathematical concepts or findings with a box like the one shown above. This will make it easier to find the starting point for a particular problem solution later on.

Example 2.2: For our previous example, what is the magnitude 'R' of the vector $\overline{AB}$?

The magnitude of the vector $\overline{AB}$, which can also be expressed as $|AB|$, is therefore given as,

$$R=|AB|=\sqrt{(-3)^2+(-3)^2+(5)^2}=\sqrt{9+9+25}=\sqrt{43}\approx\mathbf{6.55}$$

Thus, the vector $\overline{AB}$ is 6.55 'units' long. If everything were expressed in meters, then $\overline{AB}$ would be written as $-3\hat{i}+-3\hat{j}+5\hat{k}=<-3,-3,5>$ meters, and the scalar length of $\overline{AB}$ would be 6.55 meters.

Addition of Vector Quantities

Adding vectors together is as simple as adding the components of the $\hat{i}$, $\hat{j}$, and $\hat{k}$ terms together to create a new vector. Let's examine this through an example.

> **Example 2.3:** Suppose we now have two vectors that represent the velocities of two objects as shown in Figure 2.3. If these two velocities are added together, what is the resultant final velocity vector R of the two objects?

To solve this problem we first have to figure out the numerical values of vectors $\overline{AB}$ and $\overline{AC}$, and add them together. Once this is done, we can then determine the resultant of the final vector. So, from a previous example we found out that,

$$\overline{AB}=P_B-P_A=<-3,-3,5>$$

Using the same method, we now find the value of vector $\overline{AC}$ as,

$$\overline{AC}=P_C-P_A=<-5,3,6>$$

Addition of these two vectors would be performed in the following manner,

$$\begin{aligned}\overline{AB}+\overline{AC}&=\overline{AC}+\overline{AB}\\&=<-3,-3,5>+<-5,3,6>\\&=<(-3+(-5)),(-3+3),(5+6)>\\&=\mathbf{<-8,0,11>}\end{aligned}$$

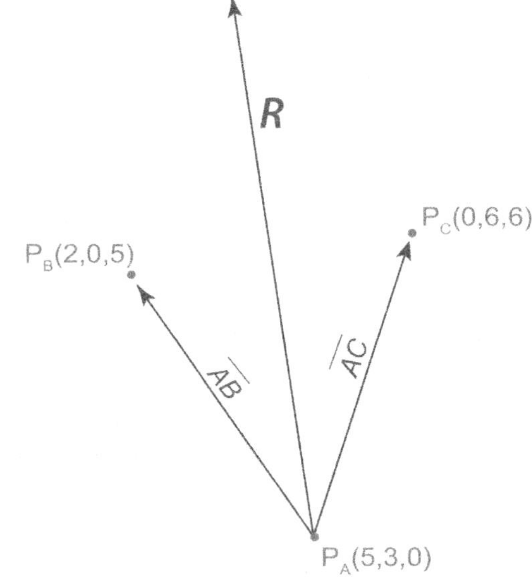

Figure 2.3: Adding Vectors $\overline{AB}$ and $\overline{AC}$ to Obtain the Resultant Vector 'R'

The benefits of using vectors to solve our problems in physics should be obvious now. We can instantly see that the motion imparted to the two objects P_A and P_B is equivalent to a single vector having a backwards (negative) motion in the 'x' direction, no motion in the 'y' direction, and a substantial upwards 'z' motion, or, $<-8,0,11>$. Finally, the scalar size (magnitude) of the resultant vector is calculated as follows,

$$R=\sqrt{(-8)^2+(0)^2+(11)^2}=\sqrt{64+0+121}=\sqrt{185}\approx\mathbf{13.6}$$

Solving this problem in the days before Descartes invented his coordinate system would have required quite a bit of geometric headache by constructing the various triangles between the objects, adding them together, and then finding out the resultant triangle which fit the points. We can now solve problems like this, and even ones of vastly greater complexity, *without ever sketching a single line*.

It is also important to note that if we were to move the vector $\overline{AC}$ such that its tail touched the tip of vector $\overline{AB}$, we would still get the same result. *Why is this so?*

I'm glad you asked. You see, when we solve a problem like the one in Figure 2.4, we can always treat it like the problem shown in Figure 2.3. They are mathematically identical.

The Guru Says...

We should note that if we know the angle between a vector and a coordinate axis, we can find out the component of the vector along that axis by taking the cosine of the angle and multiplying the result by the resultant magnitude. In fact, we can just as easily define a vector in all three directions by specifying its magnitude and three angles from the vector to the coordinate axes known as the 'direction cosines'.

We'll see more of this when we talk about the 'dot product' of two vectors in space.

Vector addition automatically moves all the vectors in a problem to a common starting point and then adds them together. So, if you had a complex motion of a particle to analyze (i.e., each vector given in a tail to tip fashion), geometrically you would move all the vector tails to a common point, and add them up. In either case, you will find that it is a whole lot easier to solve the problem mathematically first, and then sketch the result later to see what the final solution looks like.

Subtraction of Vector Quantities

Subtraction of vectors is just the opposite of adding vectors (in the same way that subtraction of scalars is the opposite operation of addition). In this case though, we have to be careful as to which vector is being subtracted from which, otherwise we will get the opposite vector of the one we are looking for.

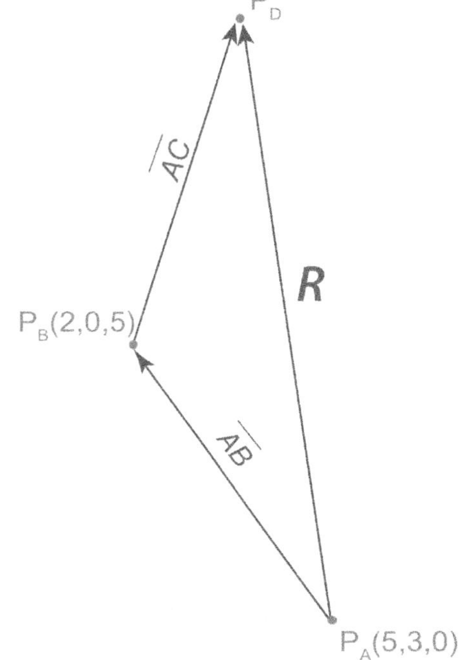

Figure 2.4: Another Way of Looking at the Same Problem in Figure 2.3

Let's take a look at the vectors from the previous example and instead of adding them together, let's subtract $\overline{AB}$ from $\overline{AC}$ and see what we get. Also, because we are now advanced vector experts, we can take the 'training wheels' off the problem and not have to draw the actual sketch of the vectors to understand the problem (although the reader is welcomed to sketch out the problem on graph paper if seeing what the vectors look like helps in the understanding).

Example 2.4: What is the vector quantity $\overline{AC} - \overline{AB}$ equal to?

So, we have the following expression,

$$\overline{AC} - \overline{AB} = < -5,3,6 > - < -3,-3,5 >$$
$$= < (-5) - (-3), (3 - (-3), (6 - 5) > = \mathbf{< -2,6,1 >}$$

And just to prove a point, let's work the subtraction the other way around to show that we would get the negative of the correct result,

$$\overline{AB} - \overline{AC} = < -3,-3,5 > - < -5,3,6 >$$
$$= < (-3) - (-5), (-3 - 3), (5 - 6) >$$
$$= \mathbf{< 2,-6,-1 >}$$

The Guru Says...

Note that the vector quantities $\overline{AB} + \overline{AC} = \overline{AC} + \overline{AB}$ are the same. This is called the commutative property of vectors, in that it does not matter which order we add them together. The same rule applies for scalars as well.

As we can expect, <u>subtraction of vectors is not commutative</u>. If we reverse the order of subtraction we wind up getting the negative of the correct answer.

Multiplication of Vectors and the Venerable "*Right Hand Rule*"

Multiplication of vector quantities is the only really tricky part of working with vectors since there are two different ways of going about it depending on whether or not you want the answer to be a scalar, or a vector. *Which answer you get is entirely dependent on the type of physical phenomena you are examining*, and the operations are not interchangeable. Let's start out with multiplying two vectors in such a way that the result is a scalar quantity – *the so-called scalar dot product.*

The dot product of two vectors is formally written as $\overline{A} \bullet \overline{B}$ and is geometrically equal to the 'amount' of vector $\overline{A}$ that is projected along the line forming vector $\overline{B}$ as shown in Figure 2.5. This is just a fancy way of saying, *How much of vector $\overline{A}$ occurs in the direction of vector $\overline{B}$.*

Why would we care about how much of a vector occurs in the direction of some other vector? Simple, many physical concepts in the universe are the product of some phenomenon occurring in some direction other than the direction of interest.

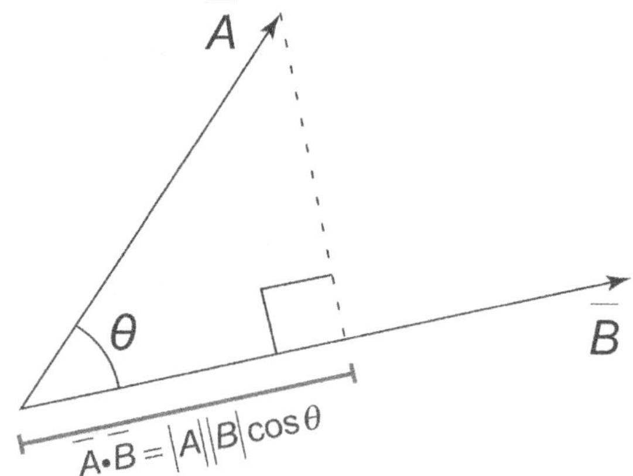

Figure 2.5: The Dot Product of Two Vectors $\overline{A}$ and $\overline{B}$.

Take the concept of 'work' for example; it's defined as the amount of force applied in a certain user-specified direction. As we should suspect by now, things like force and distance can be expressed as vectors. Work, alas, is a scalar quantity, hence the need for the dot product.

Problem Solving Tip...

We denote the dot product of two vectors with a 'big dot' to distinguish it from ordinary scalar multiplication.

Thus the dot product of two vectors is written as $\overline{A} \bullet \overline{B}$ and the scalar multiplication of two numbers would be written as $A \cdot B$.

Mathematically, taking the dot product of a couple of vectors isn't any harder than performing the addition we did earlier. We merely multiply each of the $\hat{i}$, $\hat{j}$, and $\hat{k}$ terms of the vectors together and add up the results. Formally we would write the following expression for the dot product of two three-dimensional vectors $\overline{A}$ and $\overline{B}$ as,

$$\overline{A} \bullet \overline{B} = A_x B_x + A_y B_y + A_z B_z$$

The result of this operation is a scalar that describes the magnitude of the vector $\overline{A}$ in the direction of $\overline{B}$. Probably the best way to show this is through an example.

Example 2.5: Work is defined physically as the product of a force times the distance over which it is applied. If we have a force that is applied in space only along the x direction according to the vector <12,0,0> N, what is the amount of work that is performed along a path defined by the vector <9,3,6> m?

We are told that work is defined as *the force times the dot product of the distance* and is a scalar quantity. Thus we need to use the dot product of vectors to figure this problem out. Mathematically we would write the following solution,

$$\text{Work} = \overline{\text{Force}} \bullet \overline{\text{Distance}}$$

Where the $\overline{\text{Force}} = <12,0,0>\text{N}$ and the $\overline{\text{Distance}} = <9,3,6>\text{m}$. Thus, we have the following expression for work as,

$$\begin{aligned} Work = W &= <12,0,0> \bullet <9,3,6> \\ &= (12\text{ N}) \cdot (9\text{ m}) + (0\text{ N}) \cdot (3\text{ m}) + (0\text{ N}) \cdot (6\text{ m}) \\ &= \mathbf{108\ Newton\text{-}meters = 108\ N \cdot m = 108\ J} \end{aligned}$$

So the work produced by this force is 108 Joules. See, you're already solving difficult problems using some pretty sophisticated tools – *and we haven't even gotten to the real physics yet.*

We should note that while the above definition of a dot product is vector based, the dot product also has a meaning in trigonometry. The projection of $\overline{A}$ in the direction of $\overline{B}$ is nothing more than the resultant magnitude (or scalar length) of $\overline{A}$ times the magnitude of $\overline{B}$ times the cosine of the angle between the two vectors, or,

$$\overline{A} \bullet \overline{B} = |A| \cdot |B| \cdot \cos\theta$$

Although when we work in vectors we are not too terribly interested in the scalar solution method of the dot product, we will use it occasionally to find out the angle 'θ' between two vectors. This can be a real time saver, especially if we have 10-dimensional vectors and no multi-dimensional graph paper lying around to draw a hyper-triangle in space.

Our last vector multiplication trick involves the case where we multiply a vector by another vector and get a vector as the result – *the so-called cross product of two vectors*. This is by far the most common vector operation, with applications in physics ranging from mechanical torque in an automobile, to the rotational spin of a planet, to describing how electric and magnetic fields behave. Yes indeed, nature likes vector cross products, to be certain.

The Guru Says...

The dot product of vectors is also a commutative property since it does not matter which way we perform the operation; we are effectively projecting one vector atop the other.

We should also note that we can multiply or divide a vector by a scalar using the 'distributive property' of vectors, which is sometimes helpful in making the math easier to read.

The vector cross product is the trickiest mathematical operation you'll find in the *Guru's Guide*, so if you can figure this one out, everything else pretty much boils down to simple algebra and careful bookkeeping of the units.

We'll begin by denoting the cross product operation with the symbol 'x'. But, so as not to confuse it with the operation of basic multiplication, we'll put a circle around it like '⊗'. This makes the mathematics look all the more impressive since we're using a cool new symbol that most people have never seen.

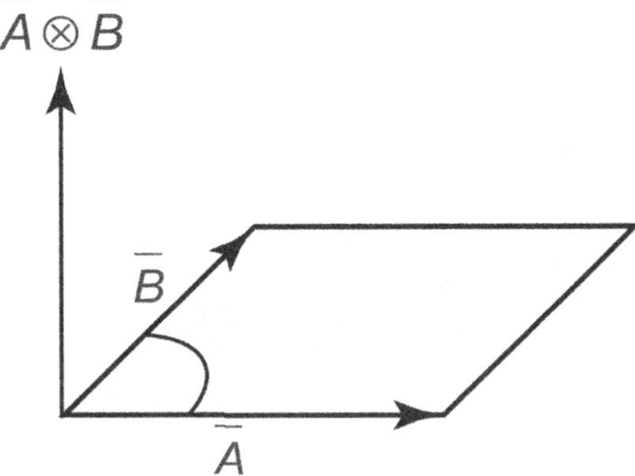

Figure 2.6: The Cross Product of Two Vectors $\overline{A}$ and $\overline{B}$.

Mathematically, we would express the resulting vector cross-product operation shown in Figure 2.6 as,

$$\overline{A} \otimes \overline{B} = <(A_yB_z - A_zB_y),(A_zB_x - A_xB_z),(A_xB_y - A_yB_x)>$$

Wow, what a mess… Fortunately, it's not as bad as it seems. If we look carefully at what is going on, we see that all we are doing is subtracting alternating terms of each vector component from each other such that the contribution to the 'x' terms of the vector are determined by what goes on in the 'y' and 'z' directions, the 'y' terms of the vector are determined by what goes on in the 'x' and 'z' directions, and the 'z' terms of the vector are determined by what goes on in the 'x' and 'y' directions. So it's really not that bad of a thing to figure out if you look at it for a couple of minutes.

What we do notice from the expression is that *each of the components of the final vector is solely a function of the components of the other two directions*. Because we know that all directions are orthogonal (i.e., perpendicular, or at 90-degrees) to each other, this would imply that the cross product has the interesting ability to add another dimension {literally} to the problem. The dot product on the other hand, as we have seen previously, reduces vectors to scalars, which is most certainly a *dimensional killer*.

Take a look at the above equation, and zero-out all the 'z' terms so that you have something that looks like the following,

$$\bar{A}\otimes\bar{B} = <(A_y(0)-(0)B_y),((0)B_x - A_x(0)),(A_xB_y - A_yB_x)> = <0,0,(A_xB_y - A_yB_x)>$$

Even though our original vectors had no 'z' coordinate terms (only 'x' and 'y', meaning that they are two-dimensional and only in the x-y plane, as shown in Figure 2.6), the cross product will generate the third 'z' dimension automatically, and in this case, the resultant answer lies entirely in that newly added dimension. The mathematics governing how the universe operates has *'required' there to be another dimension!!!*

Problem Solving Tip…

There are really clever and elegant ways to calculate the cross product of vectors using methods found in Calculus or Matrix Algebra; however, for the purposes of computing things within this book – <u>just utilize the form of this equation</u>.

Apply a wrench to tighten a bolt and you have a physical example of this exact type of problem. The force applied to the wrench is applied in the 'x-y' plane. The handle of the wrench is also in the 'x-y' plane. The applied mechanical 'torque', which is the physical phenomenon that actually turns the bolt and is defined as *the length of the wrench times the <u>cross product</u> of the force*, however, is actually at 90-degrees to both the wrench and the force, in a totally different dimensional plane. We'll get lots of experience with the concept of torque in future chapters, and find out that the added dimension is what helps keep planets spinning.

This vector cross-product concept works in two dimensions, it works in three-dimensions, and it works in the Bosonic string universe (whatever that looks like). All we have to do is keep adding terms to our vector. If we now had three vectors that were all a function of 'x', 'y', 'z', and say time 't', we could construct the product $\bar{A}\otimes\bar{B}\otimes\bar{C}$ as,

$$\begin{aligned}\bar{A}\otimes\bar{B}\otimes\bar{C} = <&(A_t\cdot(B_yC_z - B_zC_y) + A_y\cdot(B_zC_t - B_tC_z) + A_z\cdot(B_tC_y - B_yC_t)),\\ &(A_t.(B_zC_x - B_xC_z) + A_x\cdot(B_tC_z - B_zC_t) + A_z\cdot(B_xC_t - B_tC_x)),\\ &(A_t\cdot(B_xC_y - B_yC_x) + A_x\cdot(B_yC_t - B_tC_y) + A_y\cdot(B_tC_x - B_xC_t))\ >\end{aligned}$$

We won't be using this expression in any of our work, but it's interesting to note what it would look like and how we can keep adding dimensions to the problem. For everything we will examine in classical and modern physics, sticking to plain old $\overline{A} \otimes \overline{B}$ works just fine and we don't need to go beyond that.

Finally, it is important to note that $\overline{A} \otimes \overline{B}$ is not the same as $\overline{B} \otimes \overline{A}$. Don't believe me; take another look at Figure 2.6. If you flip the operation around, you not only get the negative of the solution, it also points in the opposite direction.

To keep track of the correct orientation of your vectors, some clever fellow discovered that if you point the first two fingers of your right hand in the direction of vectors $\overline{A}$ and $\overline{B}$, your thumb automatically points in the positive direction of the quantity $\overline{A} \otimes \overline{B}$. Thus was born the *so-called right hand rule* as shown in Figure 2.7.

You must have a right hand in order for this to work. If you don't have a right hand, then borrow one from someone else.

Additionally, we should note that like the vector dot product, the vector cross product also has a scalar equivalent. It is simply defined as,

$$\overline{A} \otimes \overline{B} = |A| \cdot |B| \cdot \sin\theta$$

Where, the magnitude of vector $\overline{A}$ and $\overline{B}$ are multiplied together times the sine of the angle between the two vectors. This equation is also very useful when you want to find the angle between vectors 'A' and 'B'.

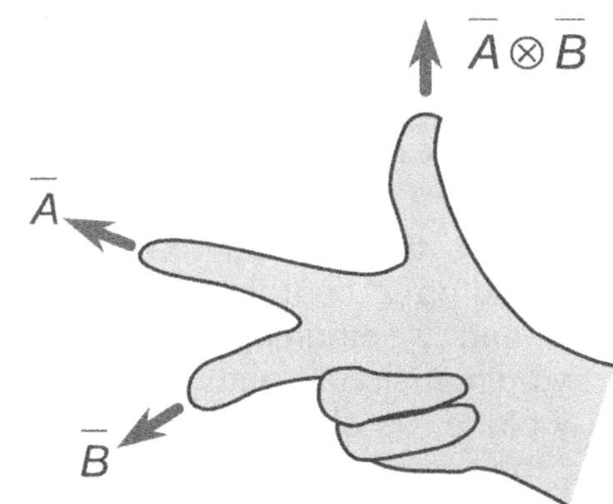

Figure 2.7: A Visual Representation of the Right Hand Rule

Division of Vectors – *It Doesn't Exist*

There is no such mathematical operation as dividing two vectors. It is a non-sensical concept not worthy of discussion by the Guru.

Anyway, do you really want to perform the operation shown in Figure 2.8 and risk tearing a hole in the fabric of space and time with such a careless use of mathematics?

I didn't think so…

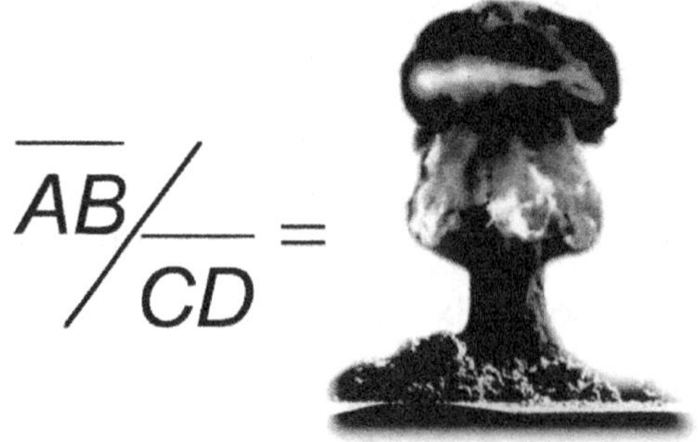

Figure 2.8: We Never Divide Vectors – *Don't Even Think About It !!!*

Examples of 'Vector Awesomeness'

Let's round out this one-and-only chapter devoted entirely to mathematics with a couple of simple examples that encapsulate the different types of things you'll see throughout this book. The concepts we've discussed so far, along with a basic understanding of algebra, some *not-too-terribly-difficult* arithmetic, and some common sense are all a person needs to solve most physics problems encountered in everyday life.

Example 2.6: You are carrying a 90 N bowling ball in the 'y' direction for a distance of 100 m. Express this problem in vector notation and calculate the total amount of work you do.

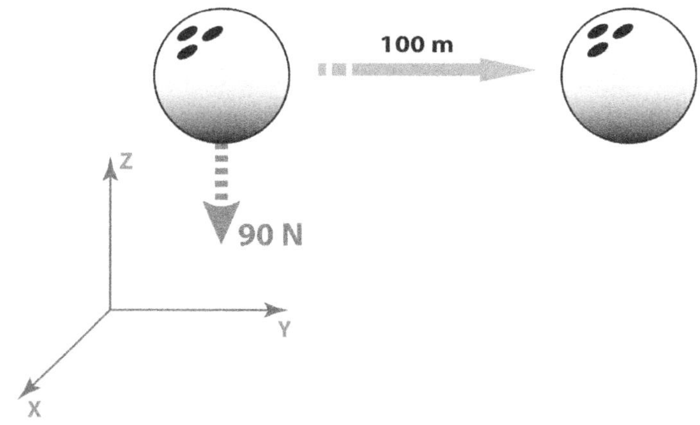

From a previous example, we know that work is equal to a force times a distance. A 90 N bowling ball would have all of its force acting in the '-z' direction since gravity always acts downward. We are moving the bowling ball to the right (in the positive 'y' direction as shown in the figure). Thus, our gravity force on the bowling ball (due to Earth) would be $<0,0,-90>$ N, which we could express using the distributive property of scalars as $90<0,0,-1>$N. Both of these expressions for the force vector are identical as you can quite readily factor, or distribute, scalars through vectors with no loss of generality. Knowing this, our distance over which we apply the force would be written as,

$$<0,100,0>\text{ m, which is the same as } 100<0,1,0>\text{ m}$$

Applying the relationship for determining work, we can express our problem mathematically as,

$$\begin{aligned} Work = W &= <0,0,\text{-}90> \bullet <0,100,0> \\ &= (0\text{ N})\cdot(0\text{ m}) + (0\text{ N})\cdot(100\text{ m}) + (\text{-}90\text{ N})\cdot(0\text{ m}) \\ &= (0+0+0)\text{ N}\cdot\text{m} = \mathbf{0\ Joules} \end{aligned}$$

Wow – what happened? How can we actually be moving an object through space and predict no work?

Problem Solving Tip...

The cosine of zero is equal to one, which implies that two vectors acting along the same 'line-of-action' produce the greatest mechanical work. The cosine of 90 degrees (which is what we have here, since the force {weight} acts downward and we're moving to the right) is zero. So, a force that acts perpendicular to the direction of motion produces no work.

Sure, your arms might be tired after carrying a heavy weight across the length of a football field, but you didn't produce any net work (which is a super-really-important concept when we get into the world of thermodynamics).

Simple, this is a trick question, and a good example of why it is very important to set up your problem correctly and understand what you are doing. In this example, the physical work, as defined, is indeed zero.

If you go back and look at the scalar definition of the dot product you'll see that it is equal to the magnitude of the first vector (the bowling ball's weight in this case, or 90 N) times the magnitude of the second vector (the distance of 100 meters). If we just multiplied this together, we would get -9,000 $\mathrm{N \cdot m}$, but, what about that pesky cosine of the angle between the two vectors?

Remember, that work is defined as the dot product of two vectors, which is the amount of one vector in the direction of the other. Two vectors that do not act (even slightly) in the same direction cannot, and do not, produce any usable work in the universe. Also remember that the Guru is a tricky fellow. Always pay attention to what the problem is asking for, and be suspicious about physics that does not fit the definition of the science being examined as this problem is actually a variant of one solved incorrectly by television 'science experts'.

Example 2.7: The Guru gives you a handful of vectors and asks you to perform the mathematical operation below. Can it be done?

$$\left(\frac{\overline{C}}{\overline{B}\bullet\overline{A}}\right)\bullet\left(\overline{B}\otimes\overline{C}\right)=?$$

This example combines the various vector and scalar operations we have talked about in this chapter, as well as the PEMDAS concepts of Chapter 1. In order for the problem to be solvable, it must be expressed in a form such that the various mathematical operations are valid. For instance, we know that we can multiply a scalar by anything, that the dot product requires two vectors and yields a scalar, and that the cross product requires two vectors and yields another vector. So, let's break apart the expression graphically and see what's going on.

This Yields a Vector

$$\left(\frac{\overline{C}}{\overline{B}\bullet\overline{A}}\right)\bullet\left(\overline{B}\otimes\overline{C}\right)=?$$

This Yields a Scalar

Therefore we wind up having a vector times the dot product of another vector, which is a defined operation. So the answer would be, '***yes***', this mathematical operation on vectors could be performed.

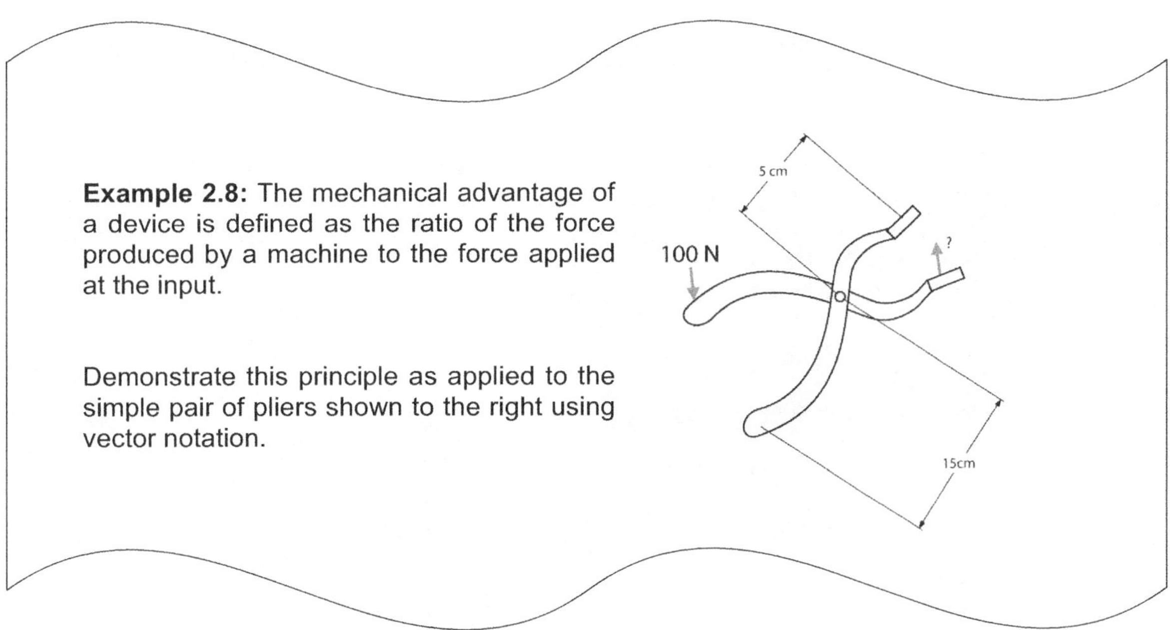

Example 2.8: The mechanical advantage of a device is defined as the ratio of the force produced by a machine to the force applied at the input.

Demonstrate this principle as applied to the simple pair of pliers shown to the right using vector notation.

This is a good example to get you exposed to looking at a real world object and seeing how the vectors are applied. When you have a simple mechanical device, which pivots about a fixed point such as a wrench, pliers, a lever, etc., a *mechanical advantage* is obtained by virtue of the ratio of how long the input arm is, compared to the output length. The longer the input lever compared to the output, the greater the mechanical advantage.

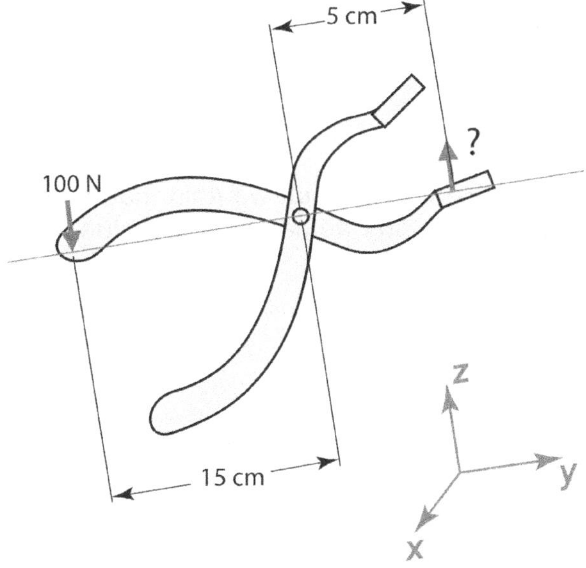

So, looking at the sketch of our pair of pliers, we have a 100 N force that is applied roughly perpendicularly to the 'y' axis as shown and acting in the '-z' direction. The gripping force of the pliers is unknown, but also acts perpendicularly to the 'y' axis in the '+z' direction. The distance from the applied force on the handle to the pivot pin is 15 cm. The distance from the pivot pin to the gripping end of the pliers is 5 cm.

Earlier, we talked about the concept of torque and how it is defined as the cross product of the distance times the force. In this case, we'll simply break the problem into two pieces analyzing the torque produced by the handle about the pivot pin (so watch the minus sign), and then figure out the torque produced by the gripping end of the pliers about the same point.

These two values for torque must be equal, and they sum to zero, since the physics of the problem dictates that <u>anything that can freely rotate cannot have any restraining torque</u>. We'll use this to our advantage to solve this problem.

The force applied to the handle can be represented as <0,0,-100>N. The force applied by the grippers of the pliers is currently unknown, so we'll just call it <0,0,P> to get it into the correct spatial orientation and be able to define it as a vector.

We chose the variable 'P' to represent the unknown force since pliers can **P**inch your fingers if you're not careful, but you can make this vector anything you like just as long as you make sure that all the coordinate directions are consistent.

Thus, the torque produced by the handle on the pivot pin of the pliers can be written as the following cross product expression,

$$\begin{aligned} \text{Torque}_{\text{handle}} &= \text{Distance} \otimes \text{Force} \\ &= <0,15,0> \otimes <0,0,-100> \\ &= <(A_yB_z - A_zB_y),(A_zB_x - A_xB_z),(A_xB_y - A_yB_x)> \\ &= <(15\cdot(-100) - 0\cdot 0),(0\cdot 0 - 0\cdot(-100)),(0\cdot 0 - 15\cdot 0)> \\ &= <-1500,0,0> \ \text{N}\cdot\text{cm} \\ &= \mathbf{-1500\ N\cdot cm\ in\ the\ 'x'\ direction} \end{aligned}$$

Similarly, the torque produced by the gripper end on the pivot pin (our axis origin, so again watch the minus signs) can be written as follows,

$$\begin{aligned} \text{Torque}_{\text{gripper}} &= \text{Distance} \otimes \text{Force} \\ &= <0,-5,0> \otimes <0,0,+P> \\ &= <(-5\cdot P - 0\cdot 0),(0\cdot 0 - 0\cdot P),(0\cdot 0 - (-5)\cdot 0)> \\ &= <-5P,0,0> \ \text{N}\cdot\text{cm} \\ &= \mathbf{-5P\ N\cdot cm\ in\ the\ 'x'\ direction} \end{aligned}$$

Remember that we said that the sum of the torque about any pinned connection is zero, and thus we can find the value of 'P' by equating the two expressions,

$$\begin{aligned} \text{Torque}_{\text{handle}} &= \text{Torque}_{\text{gripper}} \\ -1500\ \text{N}\cdot\text{cm} &= -5P\ \text{N}\cdot\text{cm} \\ -1500 &= -5P \\ \mathbf{300} &= \mathbf{P} \end{aligned}$$

So the gripping force produced by the pliers would be 300 N. Since we know that the mechanical advantage is the ratio of the force produced by a machine to the force applied at the input we can easily determine this value as,

$$\text{Mechanical Advantage} = \left|\frac{\text{Force Produced}}{\text{Force Applied}}\right| = \frac{300}{100} = \mathbf{3}$$

Of course, now that we know how the vectors behave, we can see that the mechanical advantage of a system is merely the ratio of the lever arms, which in our case is 15 cm / 5 cm = 3, but we wouldn't have had all this fun working with vectors.

Congratulations, you successfully made it through the final chapter of prerequisite work in the Guru's course in physics. Armed with your knowledge of numbers, units, vectors, and scalars, we can finally start examining the fundamentals of the universe on a meaningful and intellectual level. We've already started with some fundamental concepts and terms such as force, work, torque and distance; now let's take a look at what actually makes up *mass* and *matter* and discover some interesting facts about the 'stuff' that surrounds us everywhere.

Chapter 3: What's the Matter?

Nature's building blocks are mostly empty space

The Building Blocks of the Universe

Most people know that the universe is composed of something called *space*. Space is that all-encompassing term to describe a vast nothingness, a vacuum, or a general lack of anything solid or tangible. There is no air in space, it's very cold, it's very large (like this chapter), and the odds on you randomly running into something if you were floating in it are pretty darn remote. Yet, if you were to 'bump' into some solid matter, it would probably be an atom of the element Hydrogen.

Hydrogen comprises about 74-percent of the universe by mass with the element helium currently taking up the remaining approximately 26-percent. That means that somewhere in the scientific notation rounding error of the Guru's estimate is a fraction of a percent of *odds-and-ends stuff*, which constitutes all the other elements in the universe.

Think about that for a minute…

As you look around at your surroundings you definitely can see a whole lot of matter. There are items constructed out of wood (which is mostly carbon), and steel (mostly iron with maybe some chromium or nickel), brass objects (copper and zinc), and gold and silver. We breathe atmospheric oxygen, but there's also a lot of nitrogen in our air along with a trace amount of argon as well. If you stop to use your cell phone, there are literally dozens of strange elements in play including silicon, germanium, copper, europium, and phosphorous. There are a lot of elements out there, but they are still only a fraction of a percentage of the total mass of the universe. Most of that mass is still the simplest element out there – hydrogen.

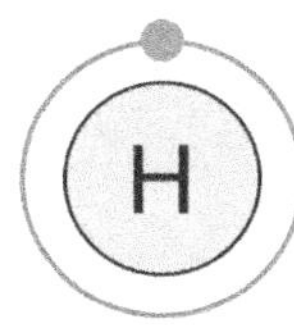

The Fundamental Construction Material of the Universe – The Hydrogen Atom

In fact, at the start of the universe about 20 or so billion years ago (plus or minus a week and a half), the ratio of hydrogen to helium was only 75 to 25 percent, so in the billions of years since the Big Bang, only about one-percent of that original hydrogen matter has been converted into helium and all the other heavier elements that we have around us right now.

So, how did the universe convert this basic material building block into the roughly 90 naturally occurring elements that we see today? This is the role of stars in the universe.

Every other element in the known universe was synthesized deep within the hellish temperature and gravitational extremes of stars. The reason for this is simple; it takes an awful lot of force to *push* (i.e., fuse as in the word *fusion*) a couple of hydrogen atoms together to make a helium (He) atom. Yet, this process occurs as a matter of routine within every star in the universe. Our sun is currently consuming roughly 600 million tons of hydrogen per second and producing 596 million tons of helium. The difference between hydrogen-in and helium-out is pure energy (Professor Einstein will have more to say about that in a later chapter), and the formation of the heavier elements around us. This process has been going on in our sun for the past 4.5 billion years and will continue to do so for at least that long again.

Once a star uses up its hydrogen, it begins to perform the fusion trick on helium until all the helium is exhausted, at which point the process breaks down (i.e., the nuclear process stops) and the star is effectively dead. A dead star isn't a really bad thing in the universe; it's just the process of getting there that's murder.

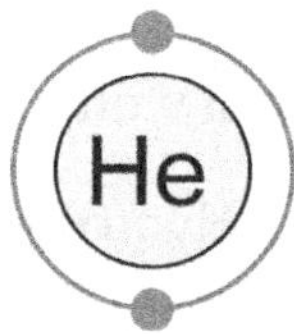

The Fusion Product of Hydrogen in a Star – A Helium Atom

You see, by the time a star starts to fuse helium as part of its daily routine, the hottest part of the star (the core) has already cooled down to the point to where it starts to contract. The contraction supplies energy back to the core of the star (through the crushing force of gravity), which then overheats it, and many of the heavier elements are produced.

At this point one of two things will generally happen, namely:

1. The star started out with a large amount of mass (a Blue Giant type star much larger than our sun) and gravity alone cannot contain the energy being produced by the runaway fusion reaction. The star goes critical and, like a very large atomic bomb, explodes. We call this process a *supernova*. Heavier elements are shot out everywhere, or,

2. Gravity can hold the mass together and the heat from the core is conveyed to the outer layers, which speeds up the hydrogen-to-helium fusion in the outer layers. This in turn causes the exterior surface of the star to cool off and expand enormously. We call this star a Red Giant and it will ultimately collapse into a nebula and/or a small dense cool star called a white dwarf. This is the ultimate fate or our sun (but not for another 5 billion or more years, so relax).

Now that we know how matter is formed in the universe, it's probably time to take a look at some of the various elements synthesized in stars – the veritable *nuts and bolts of matter*.

The Nuts and Bolts of Matter

An element is the fundamental building block of the universe. You can never divide any object such as an apple or a piece of paper into a quantity smaller than the elements that make it up no matter how sharp a knife you have. The best you can do is break down an object into its collection of elements.

Elements can neither be created nor destroyed (at least with the non-nuclear naturally occurring elements) and the totality of the mass of any compound, when divided into its constituent elements, will always be the same (i.e., if a piece of wood is burned to ashes in a closed container containing oxygen, the total mass of the products of the ash and smoke residue remain the same). This is a statement of the *Principle of Conservation of Mass,* and was first postulated by the French chemist Antoine Lavoisier around 1789.

Going back to our old friend hydrogen, we can see that this basic element consists of a single central entity called a *proton*. The proton has a positive electrical charge and constitutes most of the mass of the hydrogen atom. This central part of the atom has a special name; it is called the *nucleus* of the atom.

The Guru Says...

So, why the lecture on stars in a chapter talking about the elements and matter?

Well, the processes described are the universal mechanisms whereby denser elements are dispersed back into the universe and how everything on planet Earth that is denser than hydrogen got here – including the stuff that makes up your own body.

This cosmic matter (or *'Star Stuff'* as Dr. Carl Sagan referred to it) comprises all the known elements in the universe.

Surrounding the proton is a single; electrically negative charged entity, called an *electron*. The electron, for all practical purposes, contributes no mass to the element. It orbits the nucleus of the atom at a predefined distance (i.e., about 10,000 times greater than the diameter of the nucleus) in order to maintain a zero electrical balance between itself and the proton. If we were to strip off the electron, we would have a positively charged hydrogen atom, which if you think about it, is nothing more than a single proton floating in space. Chemists refer to a charged hydrogen atom (H^+) as simply a proton and use it as such in their experiments when they need to add some positive charge to the chemical 'recipe'.

A Friendly Hydrogen Atom

The electron, which is shown stationary in the figure (primarily because no one has figured out how to animate a book yet), would actually be spinning around the nucleus in a hazy cloud with no predictable trajectory. A German physicist by the name of Werner Karl Heisenberg in 1927 looked at the problem of how to figure out the path and speed of the electron. He ultimately concluded that it couldn't be done, at least not at the same time, and explained how if one knows where an electron is, you cannot determine its speed – and vice versa. His theory, which is a fundamental property of all quantum mechanical systems is appropriately known as, the *Uncertainty Principle*. It gave physicists a glimpse into the reality that the universe when examined at the elemental building block level is a very strange place indeed.

Since we know that an atom of hydrogen is really a positively charged 'blob' of material with an extremely small particle called an electron whirling around it at a high rate of speed, the question still becomes – *how do you draw such a thing on paper so you can explain it to someone else?* Enter physicist Niels Bohr, the inventor of the atomic model, and the fellow who gave us the sketch of the hydrogen atom above (minus the smiley face, that was the Guru's handiwork). Bohr models of atoms are never drawn to scale; the Guru however, has drawn them to relative scale in Figure 3.1 so you can see the difference between small atoms and their larger counterparts.

Finally, there is one more constituent to an atom that we should mention, it is the *neutron*. A neutron is a proton and electron smashed together to form a slightly heavier object with no net electrical charge. The total combination of neutrons and protons is what determines the mass of an atom.

Neutrons, when they are present in an element, always reside in the nucleus and account for the difference between an element's *atomic number* (the number of protons) and its *atomic mass* (the number of protons and neutrons). The number of electrons, however, is always the same as the number of protons (elements are very touchy that way; they like electrical balance and will steal or share electrons, if they have to, in order to maintain this balance).

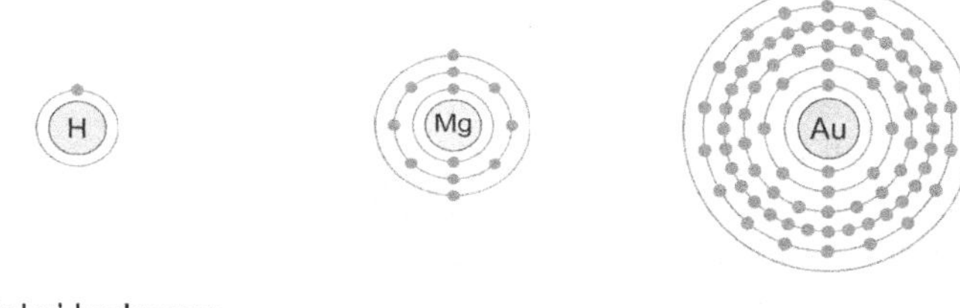

Figure 3.1: Bohr Models of Various Atoms Drawn to Indicate Their Relative Size

Let's take a look at this atomic structure concept in closer detail and see if we can figure out what Dr. Bohr's model says. To do this, the Guru has synthesized an atom of the rare and fictitious element *unobtainium* is his lab as shown in Figure 3.2.

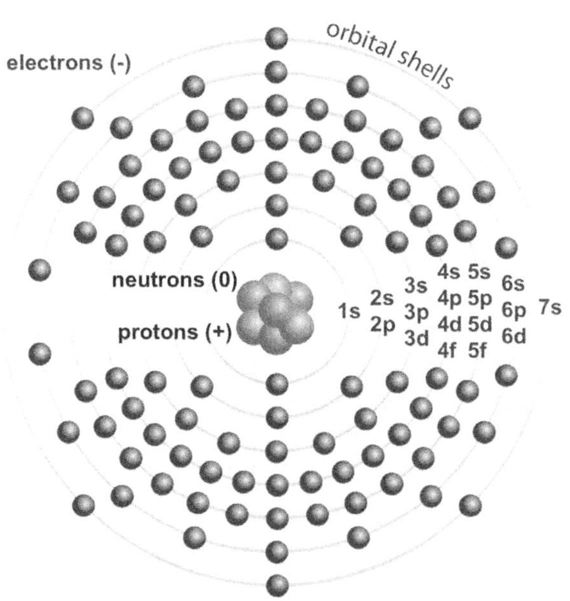

Figure 3.2 : The Fictional Element *'unobtainium'* Showing the Inner-workings of the Atom

The element unobtainium (*un·ob·tain·e·um*) is the rarest of all made-up elements with no known properties or uses except for teaching purposes. Unobtainium should never be confused with its more toxic component, known as *'weapons grade bullonium'*, which is sometimes thrown around with reckless abandon.

Our fictitious two-dimensional atomic model shows circular shells of electrons surrounding a central nucleus containing protons and neutrons for clarity. The shells are sequentially numbered and have the letter designations of 's', 'p', 'd', and 'f' to denote how many electrons can fit into a particular type of shell (more on that in a second).

While the 's' orbital shells are actually spherical in three-dimensions, the 'p', 'd', and 'f' orbits are far from simple geometries. The 'p' orbitals resemble three figure-eight patterns when drawn in two-dimensions, and something akin to six water balloons tied together in three-dimensions. The 'd' and 'f' orbitals are real messes indeed, with truly complex spherical and torroidal combinations of electron movement. For our purposes of understanding the inner-workings of the atom though, the simple orbital rings shown in the Bohr model are quite good enough.

One thing becomes immediately apparent from examining our sample of unobtainium, namely, electrons only reside at certain fixed distances away from the nucleus of the atom. The reasons for this are long and complicated, and found deep in the world of quantum mechanics. For the purposes of understanding within the Guru's universe, it is safe to say that it all has to do with balancing the electrical charge between the nucleus and the electrons in such a manner so as to achieve the lowest possible *electrical potential* (think voltage) difference between any given pairs of electrons (which have opposing negative charges) and the nucleus (which has an attractive positive charge).

The Guru Says...

The orbital shells of atoms have a labeling shorthand developed as an adaptation of the Bohr model and the findings of quantum mechanics as to how electrons actually order themselves within an atom.

They are denoted as '**s**', '**p**', '**d**', and '**f**', based upon an early naming scheme calling them the **s**harp, **p**rincipal, **d**iffuse, and **f**undamental electron shells. Each 's' shell can hold up to 2 electrons, each 'p' shell can have 6 electrons, 'd' shells can have 10 electrons, 'f' shells can have 14, and 'g' shells can have a total of 18 electrons. Thus, the naming scheme refers to, 1) which shell, 2) the shell name, and, 3) how many electrons are in the particular shell.

For our element hydrogen (H), we have one 's-shell' containing one electron; therefore we would write the Bohr electron notation for this as $1s^1$. Since helium (He) can occupy a second electron location in the first 's' orbital, we would write its notation as $1s^2$. Higher ordered elements simply build upon the previous elements by adding more shells and electrons until the assembled collection starts to resemble our unobtainium example (i.e., 1s, then 2s, then 2p, then 3s, then 3p, and so on). It should be noted that by the time we start getting to the really huge elements (like uranium) the atom has a hard time electrically holding itself together and starts to break apart, a term called *fission or radioactive decay*. Radioactive decay also occurs in simpler elements that have too many neutrons. We call atoms with an excess of neutrons *isotopes* of the element.

For example, for our old friend hydrogen, the atomic number is one (one proton and one electron), and the atomic mass is 1.00794 (which rounds to the number one); so we know that by subtracting the two values there are usually no neutrons present in hydrogen. This is not always the case though. Deuterium, which is an isotope of hydrogen, has one extra neutron and is not radioactive. Heavy water is nothing more than deuterium oxide and is a great shield against neutron radiation. A second isotope of hydrogen, with yet another neutron, is called *tritium*. It is very useful for creating thermonuclear weapon triggers, glow in the dark gadgets, and is somewhat radioactive.

This concept of isotopes applies to every other element in the universe as well. For example, the inert gas krypton has an atomic number of 36, so it would normally have 36 protons and 36 electrons. However, its atomic mass is 83.80, which rounds up to 84. Thus, the number of neutrons would be 84 minus 36, or 48. We would then state that under normal circumstances, an atom of krypton contains 48 neutrons. So, krypton-85 (Kr_{85}) would have one additional neutron and would be radioactive. This idea also works the other way around. Naturally occurring uranium has an atomic weight of 238 and is mildly radioactive. However, roughly 0.7-percent of naturally occurring U_{238}

contains the isotope U_{235} (the same element missing three neutrons), and this stuff is called weapons grade uranium used in an atomic bomb.

There's also another small, and almost trivial, behavior that elements exhibit from the perspective of the study of physics dealing with what the electrons are doing in their outermost, *or valence*, electron shells. As we will see from our Bohr models of each element, the outermost electron shell may, *or more commonly may not*, have all of its electrons. It's this imbalance of electrons in the outer shell, and only the outer shell, which gives rise to an element's complete chemical properties and behavior – and there happens to be a distinct pattern to how the valence shell electrons of one element interact with the electrons of another element.

Enter the Russian chemist and inventor Dmitri Mendeleev with very a unique observation that revolutionized science and laid the foundation for the field of science we call *Chemistry*.

Mendeleev's Hike Through a Periodic Wilderness

In 1868, Mendeleev became intrigued with all the new elements that were being discovered at the time (typically a new one was discovered each year), and had noticed that the roughly 60 known elements discovered as of that time, had some common chemical properties associated with them. By this time it was known that elements *chemically bonded* when they exchanged or shared valence electrons, and the resulting chemical compounds were dependent on the elements involved and the type of bonding. Still, there appeared to be some predictability, or *periodicity*, in how the elements themselves behaved and how they reacted with other elements.

Mendeleev wrote a textbook of his observations entitled, *Principles of Chemistry*, where for the first time he cataloged all of the known elements according to their chemical properties, thus creating the first *Periodic Table of the Elements*. His observations were quite remarkable at the time with the two most notable aspects being that he, 1) correctly identified that if the known elements were ordered according to their atomic weight, they exhibited a periodic nature to their chemical behavior, and, 2) he observed that elements of similar ranges of atomic weights have similar chemical properties. This second finding allowed for the properties of unknown elements to be predicted based solely upon atomic weight – a big deal at the time. His work also allowed him to predict using his new periodic table the suspected existence of elements that had yet to be discovered.

The Guru Says...

From an atomic standpoint, we now have a pretty good handle on what this 'star stuff' consists of. The startling thing about it is that there's a whole lot of empty space between the nucleus and the electrons. In fact, there's a whole lot of nothing between the protons and neutrons as well (and if we looked deeper we'd find out that there's not much matter that makes up these particles either). So when you think about it, most of the matter in the universe is empty space.

The current periodic table of the elements is shown in Figure 3.3 and encompasses 118 known or suspected elements with all naturally occurring elements in the universe being accounted for. Elements are organized into what are called 'groups' and 'periods' with the vertical groups (columns) corresponding to elements with similar chemical properties (i.e., valence electron shells), and the horizontal periods (rows) representing the increasing atomic masses within those periods.

Group → / ↓ Period	1	2	3	4	5	6	7	8	9	10	11	12	13	14	15	16	17	18
1	1 H																	2 He
2	3 Li	4 Be											5 B	6 C	7 N	8 O	9 F	10 Ne
3	11 Na	12 Mg											13 Al	14 Si	15 P	16 S	17 Cl	18 Ar
4	19 K	20 Ca	21 Sc	22 Ti	23 V	24 Cr	25 Mn	26 Fe	27 Co	28 Ni	29 Cu	30 Zn	31 Ga	32 Ge	33 As	34 Se	35 Br	36 Kr
5	37 Rb	38 Sr	39 Y	40 Zr	41 Nb	42 Mo	43 Tc	44 Ru	45 Rh	46 Pd	47 Ag	48 Cd	49 In	50 Sn	51 Sb	52 Te	53 I	54 Xe
6	55 Cs	56 Ba	57-71 La-Lu	72 Hf	73 Ta	74 W	75 Re	76 Os	77 Ir	78 Pt	79 Au	80 Hg	81 Tl	82 Pb	83 Bi	84 Po	85 At	86 Rn
7	87 Fr	88 Ra	89-103 Ac-Lr	104 Rf	105 Db	106 Sg	107 Bh	108 Hs	109 Mt	110 Ds	111 Rg	112 Cn	113 Uut	114 Fl	115 Uup	116 Lv	117 Uus	118 Uuo

Lanthanides	57 La	58 Ce	59 Pr	60 Nd	61 Pm	62 Sm	63 Eu	64 Gd	65 Tb	66 Dy	67 Ho	68 Er	69 Tm	70 Yb	71 Lu
Actinides	89 Ac	90 Th	91 Pa	92 U	93 Np	94 Pu	95 Am	96 Cm	97 Bk	98 Cf	99 Es	100 Fm	101 Md	102 No	103 Lr

Figure 3.3: The Current Periodic Table of the Elements as of 2013

One of the things you might notice is that certain elements are shown in different shades. This shading indicates the relative classifications of the various elements thereby giving you a picture on their physical properties and chemical behavior. These classifications are shown in Table 3.1 below.

Table 3.1: Classifications of Elements Within the Periodic Table

3 Li	**Alkali metals**. The alkali metals make up Group 1 and comprise lithium (Li) through francium (Fr). They all have very similar chemical behaviors and characteristics; notably these elements produce a violent and explosive reaction when exposed to water. Although, hydrogen is technically classified as a Group 1 element, it exhibits few characteristics of an alkali metal and is often grouped with the nonmetals and slightly offset to illustrate this.
20 Ca	**Alkaline earth metals**. The alkaline earth metals make up the Group 2 elements of the periodic table, from beryllium (Be) through radon (Ra). The alkaline earth metals have very high melting points and create basic alkaline solutions. They are also somewhat volatile when exposed to water, and can catch fire.
26 Fe	**Transition metals**. The transition metals, which are most commonly called 'metals' are defined by a partially filled 'd' shell and generally comprise Groups 3 through 12 in the periodic table. They also encompass the lanthanides and actinides, in that they are still what we would call metals (although these element groups are put aside into their own categories).
13 Al	**Post transition metals**. These metals reside in the band between those elements that have true metallic properties, and those that are semiconductors. Which elements are included in this region is still the subject of intense debate, so this band of classification may be different depending on whose periodic table you look at.

Table 3.1 (cont.): Classifications of Elements Within the Periodic Table

5 B	**Metalloids** are sometimes called 'semi-metals', or 'poor metals', in that they can conduct electricity under certain circumstances. The metalloids consist of boron (B), silicon (Si), germanium (Ge), arsenic (As), antimony (Sb), tellurium (Te), and sometimes-radioactive polonium (Po). Some of these might look familiar; we commonly classify them as semiconductors.
63 Eu	**Lanthanides**. The lanthanides comprise elements numbered 57 (lanthanum, hence the name of the set) through 71 (lutetium). They are grouped together because they have similar chemical properties, and in fact were thought to be one element for a long time. The lanthanides, along with the actinides, are often called *'the f-elements'* because they are the elements that have valence electrons in their 'f' shell.
92 U	**Actinides**. The actinides comprise elements numbered 89 (actinium) through 103 (lawrencium). They are all radioactive 'f' shell elements. Only thorium (Th) and uranium (U) are naturally occurring actinides abundantly found on Earth.
1 H	**Nonmetals**. The term 'nonmetals' is used to classify the elements that have absolutely no electrical conductivity properties at all and would be wholly classified as electrical insulators. These elements include hydrogen (H), carbon (C), nitrogen (N), oxygen (O), phosphorous (P), sulfur (S), and selenium (Se).
9 F	**Halogens**. The halogen elements comprise Group 17 of the periodic table and range from fluorine (F) through astatine (At). They are very chemically reactive and will react with most other elements.
18 Ar	**Noble gases**. Finally, the noble gases comprise Group 18 of the periodic table. They are generally very stable chemically and exhibit similar properties of being colorless and odorless. They do not react with other elements (hence the name), with some exceptions that we'll mention later.

Now that we know what the various types of elements look like, and how and why they are organized the way they are, let's take a look at each of the elements one-by-one to see what makes them unique universal building blocks. We'll start with the Period 1 elements, and work down the chart from there.

The Guru Says...

The names of chemical elements and their associated compounds are common nouns, rather than proper nouns (i.e., proper names). They are capitalized at the beginning of a sentence, but are used in a lowercase fashion everywhere else. This is a standard writing convention you will see everywhere.

In chemical notation we write them out using their abbreviations only (i.e., table salt = sodium chloride = NaCl).

For each element, the Guru will draw the Bohr atomic model at roughly the same size for visual clarity (although as we have previously shown in Figure 3.1, they differ greatly in size). At the end of this chapter you'll find Table 3.2 showing a comparison of some the important physical properties of each element.

The Period 1 Elements – Hydrogen and Helium

Hydrogen (H) gets its name from the Greek words for water, *hydro*, with the suffix *'gen'* as in, *'to produce'*. Thus the name hydrogen implies a *'water producing'* element. This name was originally coined by Antoine-Laurent de Lavoisier. Hydrogen has an atomic number of one, and an atomic mass of approximately one. This means that it has one proton, one electron, and no neutrons in its most common form found in the universe.

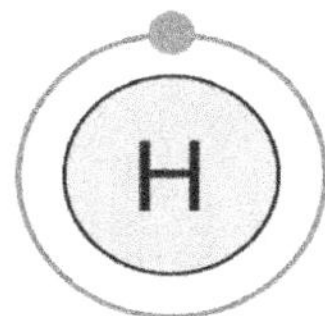

Hydrogen (H)

Cavendish discovered the element hydrogen, a colorless gas, in the year 1766. Hydrogen is classified as a nonmetal with an electron configuration of $1s^1$ meaning that its electron is in the first orbital shell, and that it only has one electron.

Hydrogen is light enough to actually escape from Earth's gravity (helium, which is the next element following hydrogen also has this ability as well). Hydrogen is a *potentially explosive* gas, which does not make it a good candidate for filling hot air balloons or dirigibles – *think Hindenburg disaster*. It does, however make a good fuel for space flight and was used as the primary propellant in the space shuttle's main engine. It is a fairly harmless substance when treated correctly and despite popular myth, does not automatically explode in pure form when ignited, it merely softly burns. Pre-mix hydrogen with oxygen and, well, that's a different story.

Hydrogen is an important staple in both the organic and inorganic laboratories. It is used in the Haber ammonia process to commercially produce fertilizer as well as for a rocket fuel, potential energy storage mechanism for automobiles, and for producing many types of acids. The largest sources of hydrogen on Earth are from volcanic gas emissions, while the largest reserves of it are our oceans. Free-floating hydrogen gas is a relatively rare find on Earth outside a laboratory.

Helium (He) is derived from the Greek word for the Sun – *Helios*, where the element was first discovered using a light-bending method known as *interferometry*. Helium has an atomic number of two, and an atomic mass of approximately four amu, which means that it has two protons, two electrons, and two neutrons in its most common form.

Janssen discovered helium in the year 1868. Like hydrogen, this element is a colorless gas at room temperature. It is classified as a noble gas with an electron configuration of $1s^2$ (i.e., its first orbital 's' shell has two electrons).

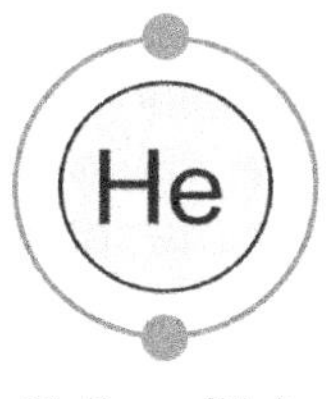

Helium (He)

Helium is a light, inert gas with the lowest melting point of any element (it remains a liquid right down to absolute zero and can only be solidified by applying extreme pressure). As a noble gas, helium can only combine with other elements to form weak-bonded compounds. Although helium is inert, it should not be inhaled since it will quickly displace the oxygen in your lungs, which is not a good thing.

Helium's primary use today (other than filling party balloons) is in the field of cryogenics and superconductivity because of its very low (near absolute zero) boiling point. Helium is also used as a shielding gas for certain types of arc welding and as a cooling 'liquid' for atomic reactors. Deep-sea divers often use a mixture of oxygen and helium in their scuba tanks due to the favorable compressibility properties of helium at high ocean pressures.

Helium is the second most abundant element in the universe since it is the byproduct of the fusion reactions within stars. Sources of helium on Earth occur due to radioactive decay within heavy elements, notably radium and uranium through a process called *alpha decay*.

The Period 2 Elements – Lithium through Neon

3	4		5	6	7	8	9	10
Li	Be		B	C	N	O	F	Ne

Lithium (Li) is also a derivation of a Greek word. In this case the word *lithos*, meaning 'stone'. Lithium has an atomic number of three, and an atomic mass of approximately seven amu, which means that it has three protons, three electrons, and four neutrons in its most common form.

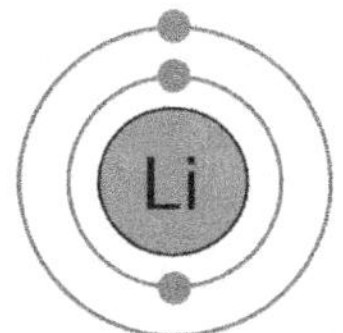

Lithium (Li)

Arfvedson discovered lithium in the year 1817. It is classified as an alkali metal with an electron configuration of [He] $2s^1$. It is also the lightest of all solid metal elements, with a density about half that of water.

Lithium is a silvery-white substance in pure form and, because it really doesn't want the lone electron its valence shell, will react with almost everything it touches, and spontaneously combust when it comes in contact with water. Lithium is a highly corrosive agent, so be nice to that next lithium battery you touch.

Lithium is used for a multitude of applications including automobile grease, medical applications, lightweight batteries, and as an alloying material. Lithium metal does not occur freely in nature due to its high reactivity, although it is found in most igneous rocks in trace amounts. Lithium metal is commercially produced through an electrochemical process involving lithium chloride ($LiCl_2$).

Beryllium (Be) gets its name from the Latin word *beryllos*, meaning roughly 'sapphire mineral' although sapphire is chemically an oxide of aluminum. Beryllium has an atomic number of four, and an atomic mass of approximately nine amu, which means that it has four protons, four electrons, and five neutrons in its most common form.

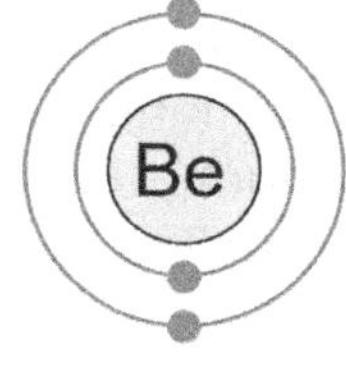

Beryllium (Be)

Vauquelin discovered beryllium in the year 1798. Physically, this element is a steel-gray, solid metal at room temperature. It is classified as an alkali earth metal with an electron configuration of [He] $2s^2$. This is just like helium, but with a fully filled second 's' shell – *starting to make sense?*

This element conducts heat extremely well, is remarkably light, is nonmagnetic, and has an elastic strength that is higher than that of steel. It also really likes X-rays and will absorb them releasing neutrons at a ratio of approximately 30:1 (so if you were curious as to where scientists get these free neutrons to fire at an element to artificially make a bigger element, think high voltage, X-ray tube, and sheet of beryllium metal).

...and oh yes, beryllium metal is sweet to the taste, but is also highly toxic to humans so it would not be a good idea to substitute this stuff for sugar in your coffee.

Boron (B) originates from the Persian word *burah*, meaning roughly 'borax'. *Borax*, the powered soap is actually sodium borate ($Na_2B_4O_7 \cdot 10H_2O$) and is the most common form of the stuff found in nature. Boron has an atomic number of five, and an atomic mass of approximately 11 amu, which means that it has five protons, five electrons, and six neutrons in its most common form.

The Guru Says...

Ok, so what's with this new electron notation?

Well, our notation would get pretty long if we wrote out everything starting at the $1s^1$ shell for each and every element. Chemists derived a shorthand notation by writing the last preceding noble gas up to the point in question, and then continuing from there.

In this case, writing [He] $2s^1$ is the same as saying 'helium plus a $2s^1$ shell', which in long hand would be $1s^2\ 2s^1$ (take a look at the Bohr model picture for this element). This space-saving format becomes more valuable when we get to the larger elements.

Three researchers named Gay-Lussac, Thénard, and Davy discovered boron in the year 1808. Physically, this element is a black or dark gray solid, which can be either crystalline or amorphous. It is classified as a metalloid with an electron configuration of [He] $2s^2\ 2p^1$ (helium with a filled second 's' shell and a new 'p' shell with one electron – remember, the 'p' shell can hold six electrons, so we'll see this shell for a while).

Boron exhibits some natural optical properties with its ability to transmit select frequencies of infrared light. It is also an extremely poor electrical and thermal conductor, and is not generally considered toxic, which gives it a really interesting engineering use. Boron can stop an atomic chain reaction dead in its tracks and is used as a *'last resort'* safety mechanism to perform an emergency shut down of a runaway nuclear reactor. The largest boron deposits on Earth (in the form of *Borax*) are located in California's Mojave Desert.

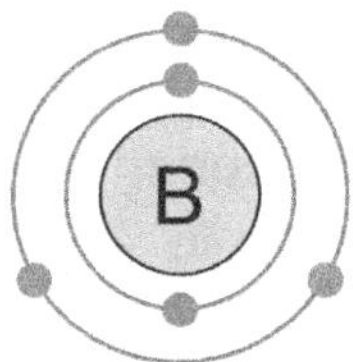

Boron (B)

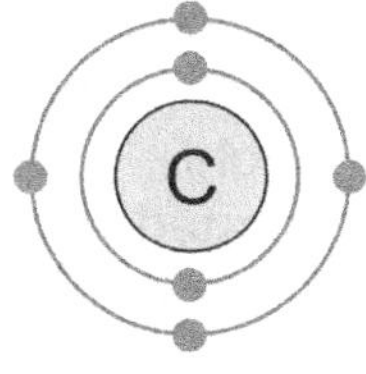

Carbon (C)

Although there are several possible origins of the name of our next element such as *Kohlenstoff* in German, and *carbone* in French, the origin is actually in the Latin word *carbo*, roughly equating to the word for 'charcoal'. **Carbon** (C) has an atomic number of six, and an atomic mass of approximately 12 amu, which means that it commonly has six each of protons, electrons, and neutrons.

Carbon has been around so long that no one knows who or when it was discovered. This element is a black, or colorless, solid element. It is classified as a nonmetal with an electron configuration of [He] $2s^2\ 2p^2$.

Carbon, besides being the official reference mass for the atomic weights shown in the periodic table, is also responsible for anything with the classification as being 'organic'. An entire field within the discipline of chemistry is dedicated just to the study of carbon and its compounds – the field of *organic chemistry*.

Pure carbon is found in nature in three different natural forms, namely that of graphite, lampblack (soot), and diamond (yes, a diamond is nothing more that a chunk of pure crystallized carbon – sorry for the let down).

Carbon is used for every imaginable application including the hardening of steel, making rubber tires black (they're actually white before lampblack is added), making graphite-based greases, medical applications, air filtration, and diamonds. In addition to jewelry, diamonds are used for cutting and grinding, since they are one of the hardest substances known to humans. Chemists refer to carbon as the most important element on the periodic table (and it's a non-metal to boot, a rarity amongst the elements).

Adding one more proton and electron changes a solid into a gas. **Nitrogen** (N) has its origin in the Latin word *nitrum* and the Greek word *nitron*. Nitrogen has an atomic number of seven, and an atomic mass of approximately 14 amu, which means that it has equally seven protons, seven electrons, and seven neutrons in its most common form.

Rutherford discovered nitrogen in the year 1772. Physically, this element is a colorless, odorless, inert gaseous element. It is classified as a nonmetal with an electron configuration of [He] $2s^2$ $2p^3$.

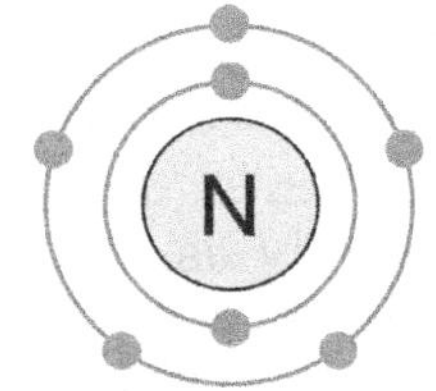

Nitrogen (N)

In medieval times, nitrogen gas was referred to as *'dephlogisticated air'* since it was observed that nothing could live within it. Given that nitrogen is plentiful (it makes up about 78-percent of Earth's atmosphere) and is easily extracted from air, it has a multitude of engineering and scientific uses. For starters, a German scientist that we've mentioned before by the name of Haber discovered a process whereby you can take plentiful nitrogen and combine it with even more plentiful hydrogen to form anhydrous ammonia (a very effective plant fertilizer). His method, known appropriately as the *Haber process*, is a chemical *short-circuiting* of a natural process called the *Nitrogen Cycle*, which normally feeds plants. So, all one needs to do is add water, sunshine, dirt, and some plants, *and there really is no scientific reason why anyone should ever go hungry*.

Besides this important use, nitrogen is also compressed to great pressures (which generates a lot of heat), and allowed to cool down. The resulting product is known as *liquid nitrogen*, which bubbles on Earth at a temperature of roughly −196 °C (or 77 K, less than one hundred degrees above absolute zero). This created another industry called *cryogenics* (not to be confused with *cryonics*, which are those weird guys who freeze dead bodies in the hopes of bringing them back to life at a later date). Liquid nitrogen cryogenics is used to flash-freeze everything from laboratory samples to fresh fish and can make great ice cream in about five minutes. Nitrogen is also used in pretty much every chemical explosive known since the byproduct of the reaction is a rapid release of nitrogen back to its gaseous state, which produces extremely large pressures and thus breaks things with a 'bang'.

Oxygen (O) has its naming origin in the Greek word *oxys*, meaning 'sharp' or 'acidic' with the addition of the suffix 'gen' meaning to produce. Oxygen has an atomic number of eight, and an atomic mass of approximately 16 amu, which means that it also has the same number of protons, electrons, and neutrons, in this case eight.

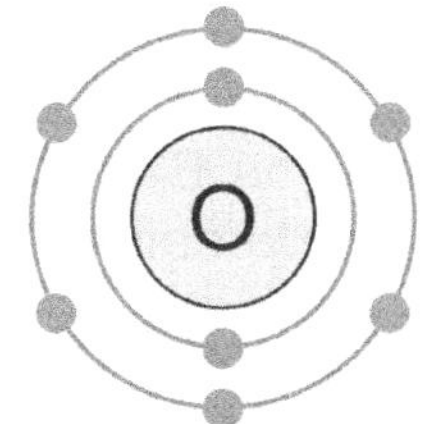

Oxygen (O)

Priestley discovered oxygen in the year 1774. This element is a colorless, odorless gas at room temerature, or pale blue liquid when very cold. It is classified as a nonmetal with an electron configuration of [He] $2s^2$ $2p^4$.

If you like breathing, then oxygen is the element for you, since without it you could not convert the food you eat to energy (not to mention that whole nasty suffocation thing). Oxygen acts as a catalyst for matter to burn and in fact can spontaneously ignite cotton clothing without an ignition source if you were unfortunate enough to spill some of the liquid stuff on you. With liquid nitrogen, the danger is frostbite; with liquid oxygen, the dangers are frostbite and setting your clothes on fire.

About 70-percent of your body is made up of water (H_2O), and water is almost 90-percent oxygen by weight; so you might be called a *carbon based life form*, but most of your mass is due to oxygen. Oxygen is also the third most abundant element on Earth and is used in everything from industrial applications (such as steel production and welding) to medical uses (oxygen tanks for breathing, respirators, etc.). Plants and animals utilize oxygen for respiration; plants convert carbon dioxide (CO_2) back into oxygen during daylight hours through a process called photosynthesis.

By the way, even though oxygen has no smell, three of them *'bolted'* together forms *ozone* (O_3) which does have a sharp or acidic smell – hence the origin of the name. You can generate ozone anywhere there are large electric sparks, such as from a laboratory *Van de Graaff* generator, or in nature from lightning strikes.

Fluorine (F) is derived from the French word *fluere* meaning 'flow' or 'flux'. Fluorine has an atomic number of nine, and an atomic mass of approximately 19 amu, which means that it has nine protons, nine electrons, and 10 neutrons in its most common form.

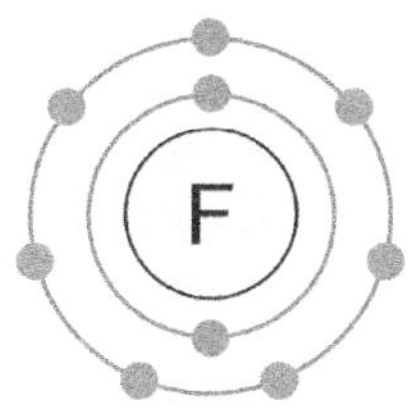

Fluorine (F)

Researchers Frederic and Moissan discovered fluorine in the year 1886. This element is a pale yellow, corrosive gas. It is classified as a halogen with an electron configuration of [He] $2s^2$ $2p^5$.

Fluorine is a violently reactive element, much like the Group 1 elements, but for an entirely different reason. While Group 1 elements like to get rid their outer electron to achieve electronic stability, fluorine is so close to having a full valence shell that it can taste it (a concept known as *electronegativity*); so it will steal an electron from anything it comes into contact with. All Group 17 halogens have this property; fluorine is just the worst offender.

Free elemental fluorine is dangerous at any level and will react with virtually any organic or inorganic substance; so as you can imagine, storing the stuff is an interesting prospect (it will dissolve glass, unless the 'glass' is made out of some mineral that contains fluorine, like Fluorite). It has a pungent odor and can in some circumstances react with some of the 'unreactive' *noble gasses*, giving them a bad reputation for not being as 'noble' as they pretend to be.

Despite the dangers in working with the stuff in its pure form, compounds of fluorine are quite beneficial in the creation of refrigerant gasses, plastics, reducing tooth decay, and yes, refining uranium for both proper and nefarious uses. It is commonly found on Earth as either *Fluorite* (CaF_2, calcium fluoride) or *Cryolite* (Na_3AlF_6, sodium hexafluoroaluminate).

Neon (Ne) is derived from the Greek word *neos* meaning 'new'. Neon has an atomic number of 10, and an atomic mass of approximately 20 amu, which means that it has 10 protons and the same number of electrons and neutrons in its most common form.

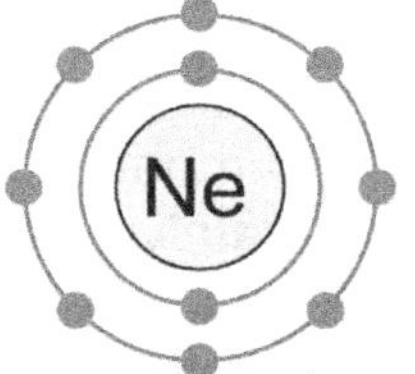

Neon (Ne)

Ramsay and Travers discovered neon in the year 1898. Physically, this element is a colorless, odorless, unreactive gas. It is classified as a noble gas with an electron configuration of [He] $2s^2$ $2p^6$.

Neon is a rare noble gas, which is obtained from the air we breathe. There is about one atom of neon in every 65,000 atoms of 'air'. Its principal use is in illumination applications since when exposed to electricity it produces the brightest glow of any of the noble gasses. In fact, neon is ubiquitous with the signs that bears its name – *neon signs*.

Neon also has applications in gas lasers (such as the helium neon, He-Ne laser, which produces the characteristic red dot of laser light associated with such devices), early glass television and power tubes, lightning arrestors (converting the high voltage electricity into light), lightning and high voltage indicators (because again, it converts all the electricity to light), pilot lamps, and so on. Occasionally liquid neon is used for cryogenic purposes since it has the ability to remove more heat than either liquid helium or nitrogen, but this is only in special applications since liquid nitrogen is much, much less expensive to use.

The Period 3 Elements – Sodium through Argon

11	12	13	14	15	16	17	18
Na	Mg	Al	Si	P	S	Cl	Ar

Sodium (Na) is the first of several elements where the symbol (Na in this case) really does not coincide with the name. This is a clue that this element has been known about for a very long time. Sodium derives its name from the Latin word *natrium*, or 'natural', and was a reference to a naturally occurring stomach remedy, *sodium bicarbonate*. Sodium has an atomic number of 11, and an atomic mass of approximately 23 amu, which means that it has 11 protons, 11 electrons, and 12 neutrons in its most common form.

Davy discovered sodium in the year 1807. Physically, this element is a silvery-white metal solid. It is classified as an alkali metal with an electron configuration of [Ne] $3s^1$.

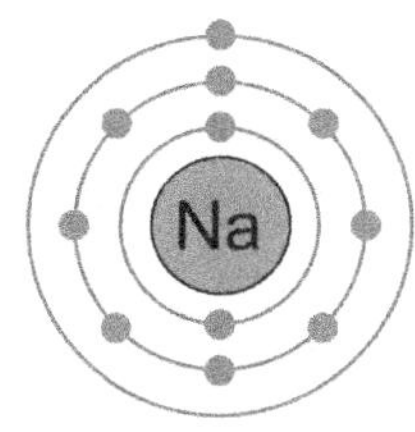

Sodium (Na)

Sodium is an extremely lightweight metal whose specific gravity is less than water; thus it will float. Since it has a high volatility, it will rapidly combine with substances like water to form a hydroxide. This process releases copious quantities of hydrogen (premixed with oxygen due to the bubbling action involved), which will explode. For this reason, pure sodium metal is stored away from air and moisture in containers typically filled with mineral oil.

Sodium has a variety of uses from common table salt (sodium chloride, NaCl) to the production of glass, soap, and paper. Because sodium is a metal, it can conduct quite a bit of heat and is used in a liquid form in some atomic reactors as a cooling fluid in the heat exchangers. Sodium is the sixth most abundant element on Earth making up approximately three-percent of the crust of the planet. It is the most abundant of all the alkali metals and is most commonly found as the salt we all know and love to excess, sodium chloride.

The Guru Says...

...and now with sodium we have a new shorthand of [Ne] $3s^1$, which means we write everything up to neon and pop on a new 's' shell.

Written out completely this would be, $1s^2 2s^2 2p^6 3s^1$.

Magnesium (Mg) simply takes its name from the place where it was originally found, the area known as Magnesia, which is a district in Thessaly, Greece. As we will find out as we look through the elements, many are named for either their discoverer, or their place of discovery. Magnesium has an atomic number of 12, and an atomic mass of approximately 24 amu, which means that it has 12 protons, 12 electrons, and 12 neutrons in its most common form.

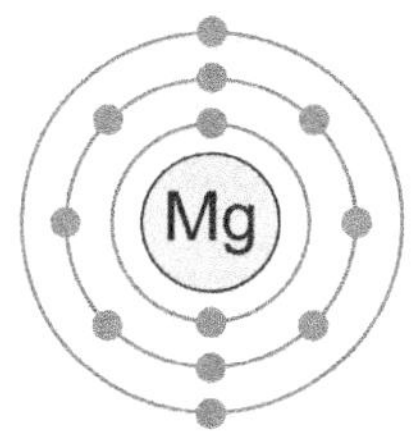

Magnesium (Mg)

Davy also discovered magnesium in the year 1808. Like sodium, this element is also a silvery-white metal solid. However, it is classified as an alkali earth metal with an electron configuration of [Ne] $3s^2$.

Magnesium is a lightweight metal, which is sometimes used all by itself (such as in the cylinder heads of air-cooled Volkswagen engines) or used as an alloy with other metals. It is the eighth most abundant element on Earth and is easily machined and welded making it a very good construction material for aerospace uses. It is often produced in a powdered or granular form for use in fireworks providing the observed brilliant white sparkling effect.

Magnesium has various medical uses including the production of *Milk of Magnesia* (magnesium hydroxide, $Mg(OH)_2$, a laxative) and *Epsom salt* (magnesium sulfate, $MgSO_4$). It is readily obtained from seawater or extracted from the minerals *Magnesite* (magnesium carbonate, $MgCO_3$) and *Dolomite* (calcium magnesium carbonate, $CaMg(CO_3)_2$).

Aluminum (Al) has its name origin from the Latin word *alumen*, which was an ancient astringent and dyeing agent (and strangely a flame retardant and food additive). Aluminum has an atomic number of 13, and an atomic mass of approximately 27 amu, which means that it has 13 protons, 13 electrons, and 14 neutrons in its most common form.

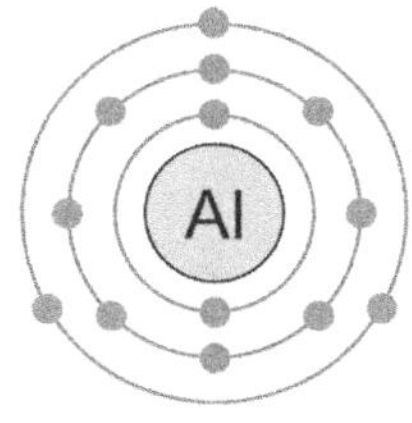

Aluminum (Al)

Oersted discovered aluminum in the year 1825. Physically, this element is a silvery solid. It is classified as a metal with an electron configuration of [Ne] $3s^2$ $3p^1$.

Aluminum is nonmagnetic, lightweight, easy to machine, and forms a rustproof layer instantly that does not degrade due to environmental exposure. *Well almost.* Just don't tell the element mercury we said this. Mercury has the unique ability to corrode aluminum extremely rapidly, and since it cannot 'play nice' with aluminum, these two metals must be kept separate from each other. The principal oxide of aluminum, aluminum oxide (Al_2O_3) has another name in crystalline form; it's called *sapphire*. Aluminum makes up over eight-percent of Earth's crust.

Aluminum's uses are everywhere from the airframes of airplanes, to cookware, to electrical transmission lines. It used to be a rare element, so rare in fact that it was more valuable than gold in mid 1800's. President Grant once received a medal made of extremely valuable 'aluminum' instead of the more common gold. Shortly after that, however, some fellow discovered that aluminum is all over the place in the mineral *Bauxite* ($Al(OH)_3$) and the value of the President's medal dropped to the price of a recycled aluminum can. *Sorry Mr. Grant.*

Silicon (Si) is derived from the Latin word for 'flint', *silicis* or *silex*, since the sedimentary rock that early flints were made from contained silica. Silicon has an atomic number of 14, and an atomic mass of approximately 28 amu, which means that it has 14 protons, 14 electrons, and 14 neutrons in its most common form.

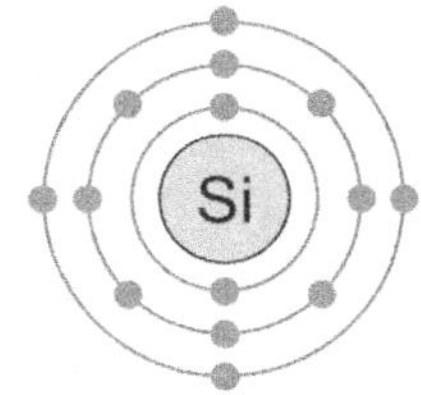

Silicon (Si)

Berzelius discovered silicon in the year 1824. Physically, this element is a silvery dark blue-gray semi-metal. It is classified as a metalloid with an electron configuration of [Ne] $3s^2$ $3p^2$.

Silicon makes up about 26-percent of Earth's crust. The stuff is literally all over the place. Just take a walk on the beach and you're *sloshing* mostly through the element silicon. Silicon has high-tech uses as a semiconductor in electronics and simpler uses as the principal ingredient in clay and glass. Silicon is not found freely in nature, rather it occurs as compounds such as the minerals *Quartz* (silicon dioxide, SiO_2) and *Amethyst* (also silicon dioxide, SiO_2).

Phosphorus (P) is derived from the Greek word *phosphoros*, meaning 'light bearing'. Phosphorus has an atomic number of 15, and an atomic mass of approximately 31 amu, which means that it has 15 protons, 15 electrons, and 16 neutrons in its most common form.

A researcher by the name of Brand discovered phosphorus in the year 1669. This element is a redish-silvery or pale yellow-tinged white solid at room temperature. It is classified as a nonmetal with an electron configuration of [Ne] $3s^2$ $3p^3$.

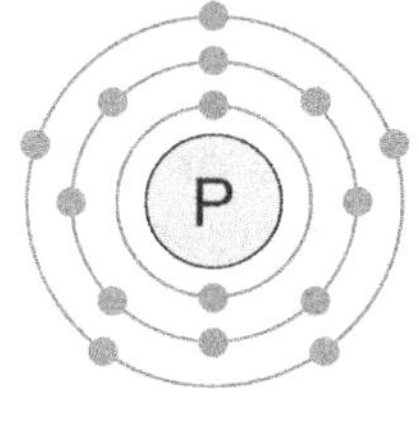

Phosphorus (P)

Phosphorous is a highly poisonous substance, which comes in two flavors: red and white. Red phosphorus is relatively stable and does not spontaneously ignite in air, requiring a higher temperature to combust (a good thing to make the 'red' part of safety matches out of). White phosphorous, on the other hand, burns on exposure to oxygen (air) and keeps on burning. White phosphorus is a very dangerous substance indeed.

Phosphorous is never found by itself and typically comprises about 0.1-percent of any given mineral making it relatively abundant on Earth, although spread out pretty thin. Other than its importance as an element in biochemical processes, its primary commercial use is in detergents (as trisodium phosphate, Na_3PO_4), pesticides, and of course matches.

The next element is also a principal ingredient in safety matches. **Sulfur** (S) is derived from the Latin word for 'brimstone' – *sulphurium*. Sulfur has an atomic number of 16, and an atomic mass of approximately 32 amu, which means that it has 16 protons, 16 electrons, and 16 neutrons in its most common form. *Are you starting to see a pattern between the protons, electrons and neutrons?*

Sulfur is old stuff. So old in fact, that no one knows when it first arrived in the hands of humans. Physically, sulfur is a lemon-yellow solid. It is classified as a nonmetal with an electron configuration of [Ne] $3s^2$ $3p^4$.

Sulfur has uses ranging from the active ingredient to make rubber tires harder (a process known as *vulcanization*) to forming the basis of the useful compound known as gunpowder. Numerous medical uses for sulfur exist, including a line of antibiotics known as the *sulfonamides*.

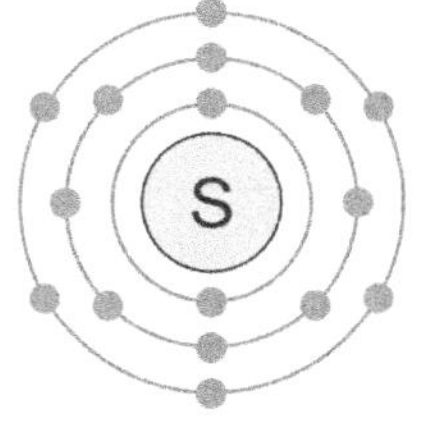

Sulfur (S)

Although sulfur is a useful element, several compounds are deadly poisons, such as hydrogen sulfide (H_2S) and sulfur dioxide (SO_2). Sulfuric acid is made from this element in the form of H_2SO_4. Sulfur also naturally occurs as a byproduct of in crude oil and natural gas refinement.

Chlorine (Cl) is derived from the Greek word *khloros*, meaning a 'greenish-yellow color'. Chlorine has an atomic number of 17, and an atomic mass of approximately 35 amu, which means that it has 17 protons, 17 electrons, and 18 neutrons in its most common form.

Scheele discovered chlorine in the year 1774. Physically, this element is a yellowish-green gas. It is classified as a halogen with an electron configuration of [Ne] $3s^2$ $3p^5$.

Chlorine, as a member of the halogen group, directly combines with almost all of the other elements (like fluorine, it really wants that extra electron). It is a highly toxic gas and not something to be played with. As a result, chlorine is used extensively as a disinfecting agent, since it kills by brute force alone, stripping electrons off of anything it comes in contact with, such as bacteria, viruses, and your fingers if you touch it.

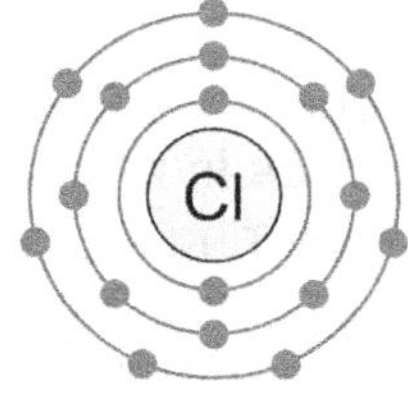

Chlorine (Cl)

Chlorine has a variety of uses in the production of textiles, paper products, dyes, petroleum products, medicines, insecticides, disinfectants, foods, solvents, plastics, and paints. It is primarily found in nature as sodium chloride (NaCl) or common table salt (it's interesting to note that two very dangerous and reactive elements combine to form a benign substance that we put on our French fries).

Argon (Ar) is from the Greek word *argos*, which means 'inactive'. This is a very good descriptor for this element. Argon has an atomic number of 18, and an atomic mass of approximately 39 amu, which means that it has 18 protons, 18 electrons, and 21 neutrons in its most common form.

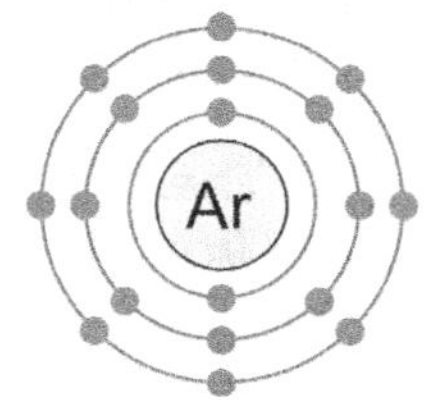

Argon (Ar)

Researchers Ramsay and Rayleigh discovered argon in the year 1894. Physically, this element is a colorless, odorless, tasteless gas. It is classified as a noble gas with an electron configuration of [Ne] $3s^2$ $3p^6$.

Argon makes up about one-percent of Earth's atmosphere and is solely obtained from the same source. Argon, like most other noble gasses, has principally lighting applications associated with fluorescent lights and lasers. Since it is a common enough element, it is also used in arc welding to keep oxygen in the air from interfering with the bonding of the metals being welded (a process known as *shielding*). In this function argon is also used to insulate or isolate reactive elements from the air when storing them.

The Period 4 Elements – Potassium through Krypton

19 K	20 Ca	21 Sc	22 Ti	23 V	24 Cr	25 Mn	26 Fe	27 Co	28 Ni	29 Cu	30 Zn	31 Ga	32 Ge	33 As	34 Se	35 Br	36 Kr

Potassium (K) is a hybrid of the English words 'pot' and 'ash' with the obligatory *'-ium'* added to the end to make it sound *elemental-like.* It is a principal constituent in the ashes of fires and has numerous scientific uses. Potassium has an atomic number of 19, and an atomic mass of approximately 40 amu, which means that it has 19 protons, 19 electrons, and 21 neutrons in its most common form.

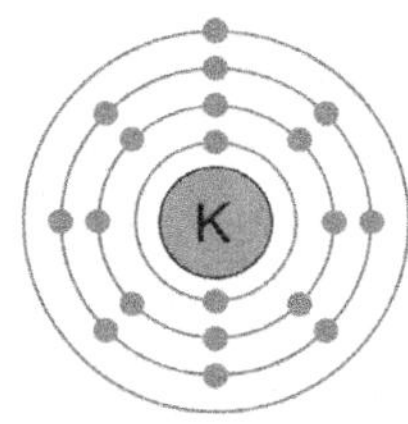

Potassium (K)

Davy also discovered potassium in the year 1807. This element is a silvery-white metal, solid at room temerature, but with a rather low melting point. It is classified as an alkali metal with an electron configuration of [Ar] $4s^1$.

Potassium is another one of those very abundant elements on Earth. Roughly 2.5-percent of this planet's crust is composed of potassium. It is an extremely reactive and soft element, which can be cut with a plastic knife, and must be stored in oil to prevent it from spontaneously igniting by contact with air.

Potassium is used extensively as a soil additive, which is essential for plant growth and is never found as a free element. Its typical sources are *Sylvite* (potassium chloride, KCl) and *Carnallite* (hydrated potassium magnesium chloride, $KMgCl_3 \cdot 6(H_2O)$) and caustic potash (KOH).

Calcium's (Ca) name has its origin in the Latin word *calx* or *calcis* meaning 'lime' or 'limestone'. Calcium has an atomic number of 20, and an atomic mass of approximately 40 amu, which means that it has 20 protons, 20 electrons, and 20 neutrons in its most common form.

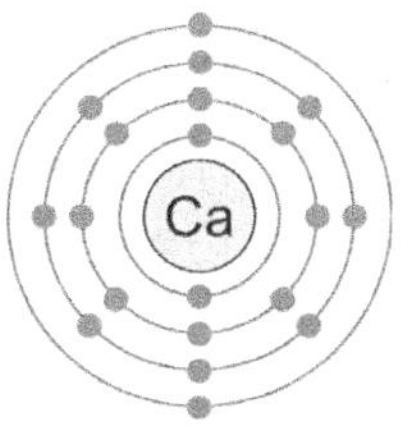

Calcium (Ca)

Davy did it again with the discovery of calcium in the year 1808. In pure form, this element is a silvery-gray metal. It is classified as an alkali earth metal with an electron configuration of [Ar] $4s^2$.

Calcium is essential for the health of all animals and makes up about three-percent of Earth's crust making it the fifth most abundant element on the planet. It is used in the manufacture of quicklime, masonry, Portland cement, glass, paint, paper products, and multiple food applications.

Although never found as an isolated element, natural forms of calcium include *Limestone* (calcium carbonate, $CaCO_3$), *Gypsum* (calcium sulfate dihydrate, $CaSO_4 \cdot 2H_2O$), and the *Fluorite* we've seen before (calcium fluoride, CaF_2).

Scandium (Sc) is named after the country in which it was first discovered: Scandinavia. Scandium has an atomic number of 21, and an atomic mass of approximately 45 amu, which means that it has 21 protons, 21 electrons, and 24 neutrons in its most common form.

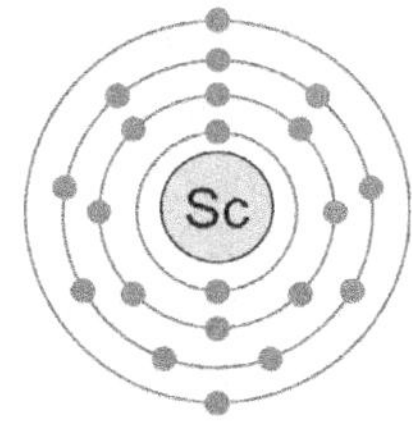

Scandium (Sc)

A chemist by the name of Nilson discovered scandium in the year 1879. Physically, this element is a silvery-white metal. It is classified as a transition metal with an electron configuration of [Ar] $3d^1$ $4s^2$.

Scandium is used as an additive to paints and pigments to produce the blue color 'aquamarine'. Its primary use is in the lighting industry where it is added to incandescent and mercury vapor lamps to make the light spectrum more closely resemble natural daylight. Scandium is a fairly common element found in the minerals *Euxenite* (a radioactive ore also containing calcium, niobium, tantalum, cerium, titanium, yttrium, uranium and thorium), and *Gadolinite* (another mixture of elements containing cerium, lanthanum, neodymium, yttrium, beryllium, and iron).

Titanium (Ti) has its origins in Greek mythology. The Titans were the first sons of Earth and were a primitive race of powerful deities. Titanium has an atomic number of 22, and an atomic mass of approximately 48 amu, which means that it has 22 protons, 22 electrons, and 26 neutrons in its most common form.

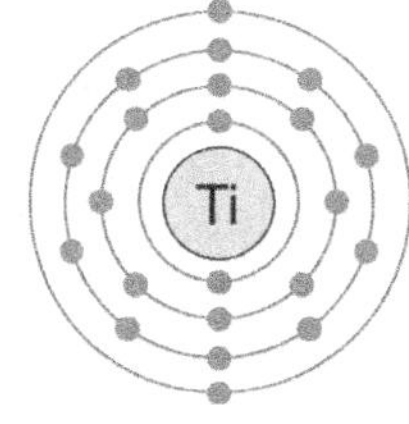

Titanium (Ti)

Gregor discovered titanium in 1791. This element is another silvery-white metal. It is classified as a transition metal with an electron configuration of [Ar] $3d^2$ $4s^2$.

Titanium is a great all-around metal with a low density, high strength and temperature resistance, and high corrosion resistance, making it suitable for aerospace applications. It is the ninth most abundant element in Earth's crust and has the tensile strength of steel, but is 45% lighter. It is machined and fabricated easily and is used as an additive to paints and pigments to produce the color white.

Titanium is also considered *physiologically inert* and thus has numerous applications in the medical field, including replacement joints and surgical bolts, hooks and pins. Titanium occurs naturally in multiple igneous rocks.

The name of **Vanadium** (V) originates from the Scandinavian goddess of beauty, due to the multicolored compounds this element produces. Vanadium has an atomic number of 23, and an atomic mass of approximately 51 amu, which means that it has 23 protons, 23 electrons, and 28 neutrons in its most common form.

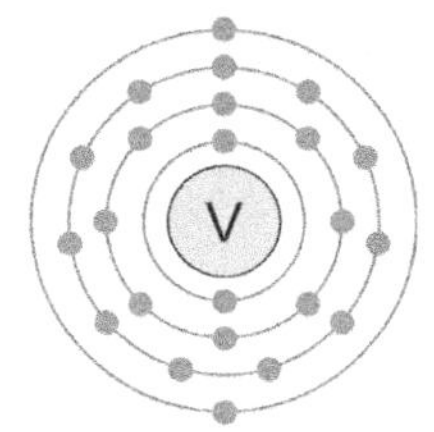

Vanadium (V)

Andrés discovered vanadium in the year 1830. Physically, this element is a grayish-silver, soft, solid metal. It is classified as a transition metal with an electron configuration of [Ar] $3d^3$ $4s^2$.

Vanadium is a toxic substance that makes a great additive to metal alloys giving them durable rust-resistant properties. This is an especially important property for high-speed tool steels like those used to make drill bits. Vanadium is also used in the fabrication of superconducting magnets. Vanadium occurs in approximately 65 minerals, with *Vanadinite* ($Pb_5(VO_4)3Cl$) being the principal source.

Chromium (Cr) comes from the Greek word *chroma* for 'color', again due to the multicolored compounds this element produces especially when heated. Chromium has an atomic number of 24, and an atomic mass of approximately 52 amu, which means that it has 24 protons, 24 electrons, and 28 neutrons in its most common form.

Vauquelin discovered chromium in the year 1797. Physically, this element is a silver-gray, hard metal. It is classified as a transition metal with an electron configuration of [Ar] $3d^5$ $4s^1$.

Chromium is typically alloyed with vanadium in steel to produce high quality tools, and used to make glass green. It is also used as a corrosion resistant metallic plating for other metals. On Earth, chromium is principally found in the mineral *Chromite* (iron chromium oxide, $FeCr_2O_4$). All chromium compounds, as well as the element itself, are toxic to humans.

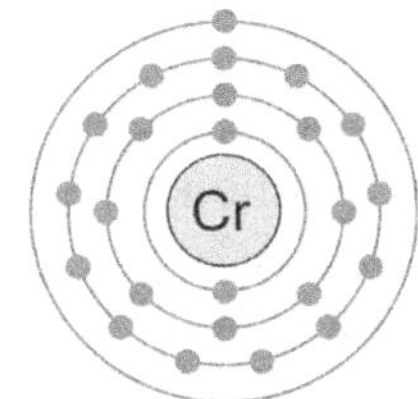

Chromium (Cr)

Manganese (Mn) originates from the Latin word *magnes*, which means 'magnet', since early-observed forms of the element had magnetic properties and were attracted to iron. Manganese has an atomic number of 25, and an atomic mass of approximately 55 amu, which means that it has 25 protons, 25 electrons, and 30 neutrons in its most common form.

Gahn discovered manganese in 1774. This element is a gray-white, brittle, hard metal. It is classified as a transition metal with an electron configuration of [Ar] $3d^5$ $4s^2$.

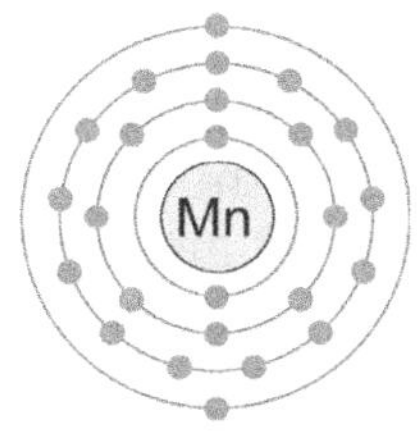

Manganese (Mn)

Manganese is primarily used as an important alloying agent. It is added to improve the strength, toughness, stiffness, and hardness of steels. Manganese is also used in standard dry cell batteries, and as a nutritional mineral supplement (just not in high doses, where it is toxic). *Pyrolusite* (manganese dioxide, MnO_2) and *Rhodochrosite* (manganese carbonate, $MnCO_3$) are the most common forms of this element in nature.

Iron (Fe) as an element has been known about for a very long time, hence the abbreviation that looks nothing like the name. Iron was originally called *ferrum* in Latin but was later changed to iron by the Anglo-Saxons in the middle ages. Iron has an atomic number of 26, and an atomic mass of approximately 56 amu, which means that it has 26 protons, 26 electrons, and 30 neutrons in its most common form.

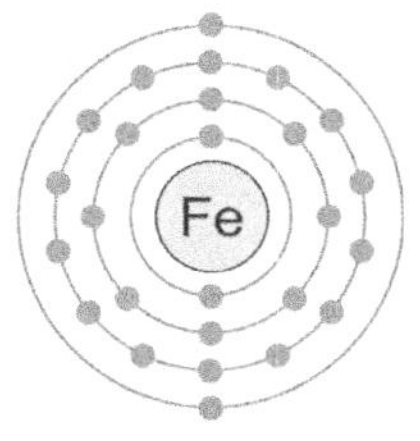

Iron (Fe)

Physically, this element is a silvery-gray, magnetic metal. It is classified as a transition metal with an electron configuration of [Ar] $3d^6$ $4s^2$.

Iron is used everywhere steel is used, since steel is by definition processed iron. It is a cheap, durable, heavy, magnetic, and abundant building material, which is easily alloyed into many different flavors depending on the final use.

The principal iron source used by humans consists of iron ore, also called Hematite (iron oxide, Fe_2O_3). Although the core of Earth itself is composed entirely of molten iron, no one has made a straw long enough to tap that resource yet.

Cobalt (Co) is a name derived from the German word *Kobald* meaning 'evil spirit' or 'goblin' due to the unusual properties this element imparted to alloyed metals. Cobalt has an atomic number of 27, and an atomic mass of approximately 59 amu, which means that it has 27 protons, 27 electrons, and 32 neutrons in its most common form.

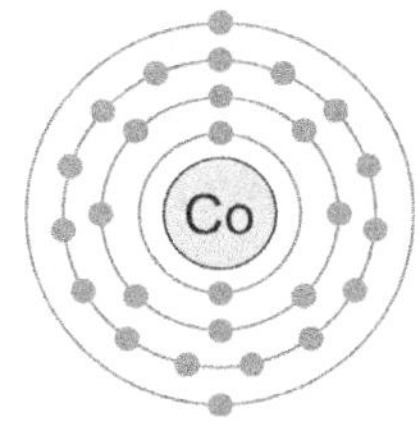

Cobalt (Co)

Brandt discovered Cobalt in the year 1739. This element is a silvery bluish-white, hard, brittle, magnetic metal. It is classified as a transition metal with an electron configuration of [Ar] $3d^7$ $4s^2$.

Cobalt's principal use is as an alloying agent for other metals and to impart magnetic properties to the material. It is also used to improve a material's hardness and resistance to oxidation, as well as to produce a product known commonly as *stainless steel*. Powered cobalt has been used for centuries to provide a blue pigment to materials such as paint, pottery, and glassware.

Cobalt is typically found in the minerals *Erythrite* (hydrated cobalt arsenate, $Co_3(AsO_4)_2 \cdot 8H_2O$) and *Smaltite* (cobalt iron nickel arsenide, $(Co,Fe,Ni)As_2$) and likes to hide out with deposits of iron, nickel, silver, lead, and copper.

Nickel (Ni) is another one of those 'evil' elements that the Germans alloyed with elements like iron to produce materials with unusual properties. It originated from the idiom *kupfernickel*, which literally translates to 'Devil's copper'. Today we call this compound *niccolite* (or nickel arsenide), one of the more prevalent forms of the element. Nickel has an atomic number of 28, and an atomic mass of approximately 59 amu, which means that it has 28 protons, 28 electrons, and 31 neutrons in its most common form.

Cronstedt discovered nickel in 1751. This element is a silvery-white, hard metal. It is classified as a transition metal with an electron configuration of [Ar] $3d^8$ $4s^2$.

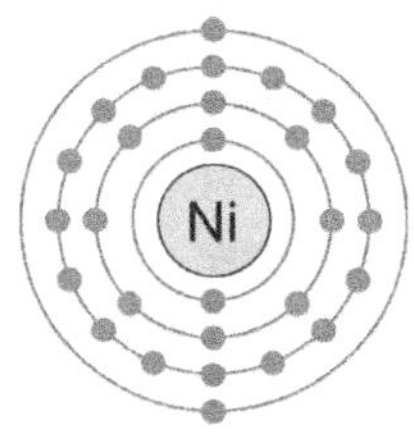

Nickel (Ni)

Although we toss the term 'nickel' around each and every day, some nickel compounds are highly toxic and carcinogenic (like nickel carbonyl and nickel sulfide). Nickel's primary use is in coating other metals (because it can be polished to a mirror finish), acting as an alloying material imparting ductile and magnetic properties, and in dry cell batteries. Nickel is also used to give objects a green pigment. Nickel is primarily found in the mineral *Pentlandite* (iron-nickel sulfide, $(Fe,Ni)_9S_8$).

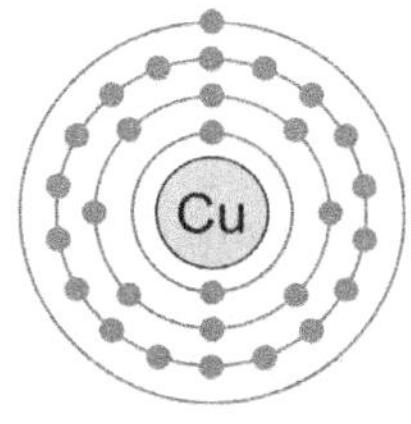

Copper (Cu)

Copper (Cu) is another really old element named after the Latin word *cuprum*. Cuprum is a shortened version of *cyprium* (meaning 'metal of Cyprus'), named after the Roman mines located on this island. Copper has an atomic number of 29, and an atomic mass of approximately 64 amu, which means that it has 29 protons, 29 electrons, and 35 neutrons in its most common form.

Copper is another metal whose existence has been known for several thousand years. Physically, this element is a reddish metal. It is classified as a transition metal with an electron configuration of [Ar] $3d^{10}$ $4s^1$.

Copper is used everywhere electricity flows. It is the primary metal in all electrical conductors. It is also used extensively as an alloying agent with two notable examples, brass (copper and zinc), and bronze (copper and tin). Copper occurs naturally in numerous minerals with the three major ones being *Malachite* (copper carbonate hydroxide, $Cu_2CO_3(OH)_2$), *Cuprite* (copper oxide, Cu_2O), and *Azurite* (a different copper carbonate hydroxide, $Cu_3(CO_3)_2(OH)_2$).

Zinc (Zn) is attributed to the early German word *zinke*, which translates to 'tin' (which is a completely different heavier element located 20 atomic numbers away). Whatever the origin, it is important to know that zinc is not the same as tin even though they look somewhat alike. Zinc has an atomic number of 30, and an atomic mass of approximately 65 amu, which means that it has 30 protons, 30 electrons, and 35 neutrons in its most common form.

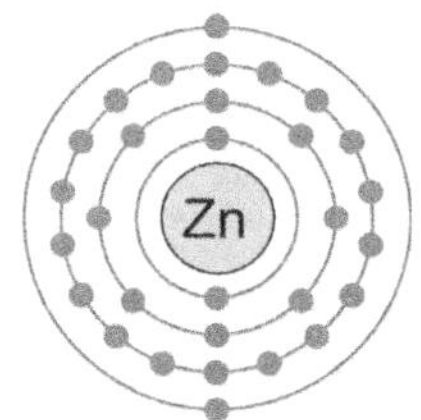

Zinc (Zn)

Zinc is another ancient human discovery. Physically, this element is a bluish-white, brittle metal. It is classified as a transition metal with an electron configuration of [Ar] $3d^{10}$ $4s^2$.

Zinc is also another one of those elements that are used as an alloying agent and is used to *galvanize* (reduce the corrosion susceptibility of) steel pipes. Compounds of zinc, such as zinc oxide, are used for everything from sunscreen to the manufacture of soap and batteries. Zinc sulfide is the coating on the hands of wrist watches that make them glow in the dark after exposure to light.

Although zinc is not classified as a toxic metal to humans (it's actually important in the metabolization of food), breathing in the dust of zinc or its oxides can cause a temporary muscle spasm condition known as *zinc chills* or the *oxide shakes*. Zinc is the 24th most abundant element in Earth's crust and is found everywhere.

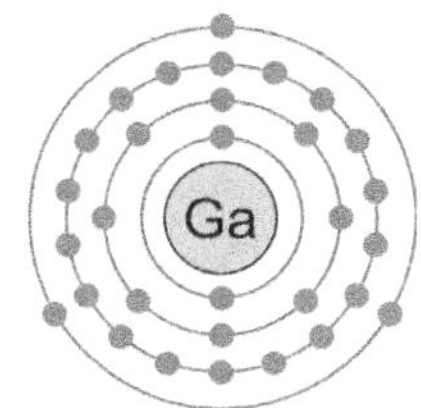

Gallium (Ga)

Gallium (Ga) takes its name from the ancient Latin name for France – *Gallia*. Gallium has an atomic number of 31, and an atomic mass of approximately 70 amu, which means that it has 31 protons, 31 electrons, and 39 neutrons in its most common form.

De Boisbaudran discovered gallium in the year 1875. This element is a bluish silvery-white metal that can melt on a warm day. It is classified as a metal with an electron configuration of [Ar] $3d^{10}$ $4s^2$ $4p^1$.

Gallium has two very important uses. First, it provides that mirror finish on a pane of glass turning it into a mirror (before gallium people used silver backed mirrors); and secondly, it is used to *dope* semiconductors and in the manufacturing of solid-state devices (in the form of gallium arsenide, GaAs, which converts electricity to light in *light emitting diodes*, or LED's). Gallium is reclaimed from coal dust as well as from various minerals. It is not considered to be toxic.

Germanium (Ge) takes its name from the ancient Latin name for Germany – *Germania*. Germanium has an atomic number of 32, and an atomic mass of approximately 73 amu, which means that it has 32 protons, 32 electrons, and 41 neutrons in its most common form.

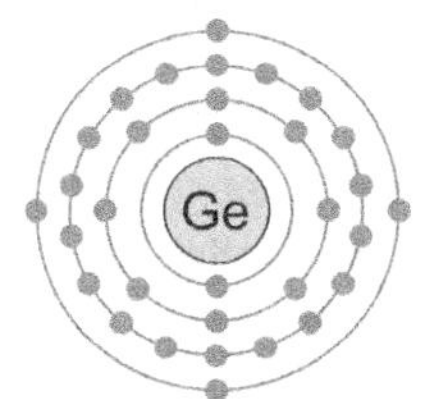

Germanium (Ge)

Winkler discovered germanium in the year 1886. This element is a grayish-white, brittle, crystalline semi-metal. It is classified as a metalloid with an electron configuration of [Ar] $3d^{10}$ $4s^2$ $4p^2$.

Germanium is another one of those good semiconductor elements, which is typically alloyed with arsenic (As). Since germanium oxide crystals have a high index of refraction (i.e., a measure of how well they bend light that passes through them), this compound is used extensively in the manufacture of microscope, telescope, and camera lenses. Additionally, germanium is used as a phosphor in fluorescent lamps.

Germanium compounds are considered non-toxic to humans, but are lethal to various strains of bacteria. Thus, this element has antibiotic properties. Germanium is principally found in the mineral *Germanite* (copper iron germanium sulfide, $Cu_{26}Fe_4Ge_4S_{32}$).

Arsenic (As) is from the Latin word *arsenicum* in reference to the compound in which it was observed (called 'yellow orpiment'). Arsenic has an atomic number of 33, and an atomic mass of approximately 75 amu, which means that it has 33 protons, 33 electrons, and 42 neutrons in its most common form.

Physically, this element is a yellow or gray, brittle semi-metal. It is classified as a metalloid with an electron configuration of [Ar] $3d^{10}$ $4s^2$ $4p^3$.

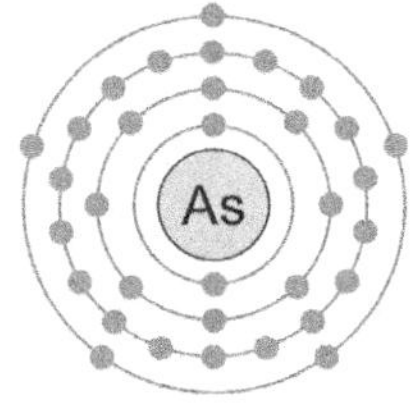

Arsenic (As)

Arsenic is typically used in the production of semiconductors, certain alloys requiring fine formation, in fireworks to produce a reddish color, and in commercial insecticides and poisons. Arsenic fumes smell like garlic; so if you find a mystery chemical that has this property – *get away from it!* Arsenic is naturally found in the mineral *Realgar* (arsenic sulfide, As_4S_4).

Selenium (Se) is derived from the Greek word *Selene* for 'moon', probably due to its general color and appearance. Selenium has an atomic number of 34, and an atomic mass of approximately 79 amu, which means that it has 34 protons, 34 electrons, and 45 neutrons in its most common form.

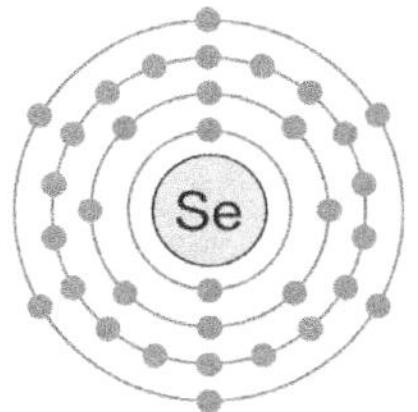

Selenium (Se)

Berzelius discovered selenium in the year 1817. Physically, this element is a red or gray, crystalline or amorphous solid. It is classified as a nonmetal with an electron configuration of [Ar] $3d^{10}$ $4s^2$ $4p^4$.

Selenium is used extensively in photovoltaic devices (i.e., light and motion sensors, digital camera imaging chips, etc.) due to its ability to convert light into an electric current, as well as in rectifiers (such as diodes), since it can also convert alternating current into direct current by allowing electricity to flow in one direction only through its crystals. It is also the *active ingredient* in copy machine toner that makes the whole process work.

Selenium as an element itself is nontoxic, while some of its compounds can be quite dangerous. Locoweed, which every rancher in the American west knew to keep their cattle and horses away from, contains toxic compounds containing selenium that would drive their livestock mad. Selenium occurs in the mineral *Clausthalite* (lead selenide, or PbSe).

Bromine (Br) is from the Greek word *bromos*, which literally means 'stink' or 'stench' because when it was originally separated from seawater, it smelled awful. Bromine has an atomic number of 35, and an atomic mass of approximately 80 amu, which means that it has 35 protons, 35 electrons, and 45 neutrons in its most common form.

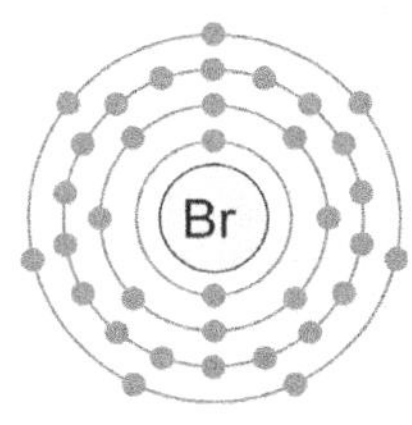

Bromine (Br)

Balard discovered bromine in the year 1826. This element is a reddish-brown liquid at room temperature. It is classified as a halogen with an electron configuration of [Ar] $3d^{10}$ $4s^2$ $4p^5$.

Bromine is used to produce the color purple in several different pigments and as a fire suppression agent in some fire extinguishers. Bromine compounds are used in the medical industry to produce certain classes of sedatives. Bromine is one of the five elements that is a liquid at, or very near, room temperature (most elements are solids and a handful are gasses).

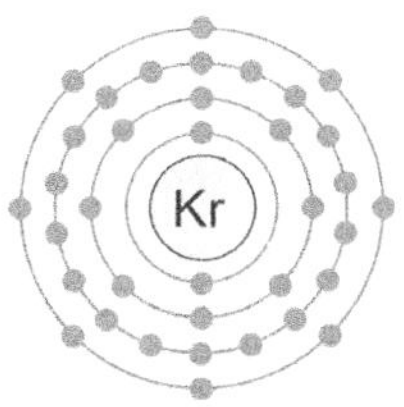

Krypton (Kr)

Krypton (Kr) is from the Greek word *kryptos*, meaning 'hidden', a reference to its obscure presence in the air we breathe. Krypton has an atomic number of 36, and an atomic mass of approximately 84 amu, which means that it has 36 protons, 36 electrons, and 48 neutrons in its most common form.

Ramsay and Travers discovered krypton in 1898. This element is a colorless, odorless gas. It is classified as a noble gas with an electron configuration of [Ar] $3d^{10}$ $4s^2$ $4p^6$.

Krypton is obtained from the atmosphere where it exists in concentrations of approximately one part-per-million (or 1 ppm). Its principal use is as an additive gas to common incandescent light bulbs to give them a brilliant white color turning them into what we call *halogen lamps*.

Krypton is generally inert, but can be corrupted by the over-ambitious element fluorine to form krypton difluoride (KrF_2), thereby ruining its noble status, and making a whole lot of chemists quite irritated over this finding.

The Period 5 Elements – Rubidium through Xenon

37	38	39	40	41	42	43	44	45	46	47	48	49	50	51	52	53	54
Rb	Sr	Y	Zr	Nb	Mo	Tc	Ru	Rh	Pd	Ag	Cd	In	Sn	Sb	Te	I	Xe

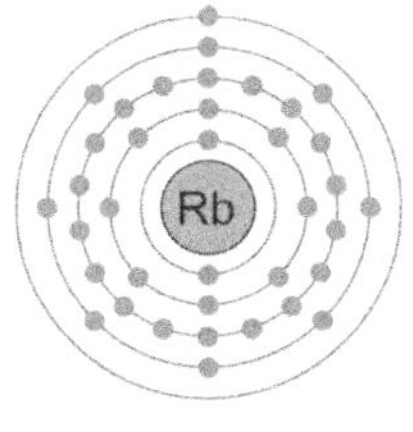

Rubidium (Rb)

The name **Rubidium** (Rb) comes from the Latin word *rubidus*, meaning 'deepest red'. This is in reference to the spectral lines the material emits when heated (that whole interferometer thing again). Rubidium has an atomic number of 37, and an atomic mass of approximately 85 amu, which means that it has 37 protons, 37 electrons, and 48 neutrons in its most common form.

Bunsen and Kirchhoff discovered rubidium in the year 1861. Physically, this element is a silvery-white metal. It is classified as an alkali metal with an electron configuration of [Kr] $5s^1$.

Rubidium is another one of those pesky metals that likes to catch fire or explode when put in contact with water or air. A low melting point element (it can become a liquid at room temperature), its isotope Rb_{87} emits a brilliant and characteristically constant red frequency spectrum when exposed to electricity, which makes an extremely accurate timepiece and is useful in constructing atomic clocks. Rubidium is also used in fireworks to produce red and violet colors. Rubidium is the 23rd most abundant element in Earth's crust and is located in numerous minerals.

The element **Strontium** (Sr) is named after the town in Scotland where it was discovered – *Strontian*. Strontium has an atomic number of 38, and an atomic mass of approximately 88 amu, which means that it has 38 protons, 38 electrons, and 50 neutrons in its most common form.

Crawford discovered strontium in the year 1790. This element is another silvery-white metal. It is classified as an alkali earth metal with an electron configuration of [Kr] $5s^2$.

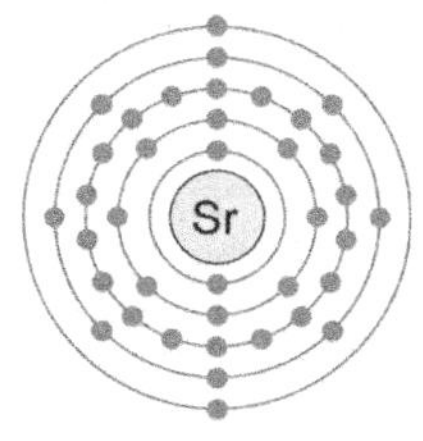

Strontium (Sr)

Strontium has primary uses as an incendiary source in fireworks and flares (adding a reddish crimson color) as well as in the production of some metallic alloys and in the glass of optical instruments. An isotope of strontium Sr_{90} is extremely dangerous as it is, a) radioactive, and, b) can easily replace calcium in the human body having serious carcinogenic effects. This isotope is carefully used to develop portable 'nuclear batteries' called *Systems for Nuclear Auxiliary Power*, or SNAP.

The element **Yttrium** (Y) is named after the town in Sweden where it was discovered – *Ytterby*. Incidentally, Ytterby has the claim to fame in that the quarry located in this town is the site where three other elements were discovered (namely, erbium, terbium, and ytterbium). Yttrium has an atomic number of 39, and an atomic mass of approximately 89 amu, which means that it has 39 protons, 39 electrons, and 50 neutrons in its most common form.

Gadolin discovered yttrium in the year 1789. This element is yet another silvery-white metal. It is classified as a transition metal with an electron configuration of [Kr] $4d^1$ $5s^2$.

Yttrium is primarily applied in the electronics industry where it is used as a microwave filter material as well as being an important alloy for acoustic speaker magnets, low frequency filters, and amplifier circuits. Yttrium is commonly used to alloy chromium, molybdenum, aluminum, zirconium, titanium, and magnesium. Yttrium aluminum oxide compounds ($Y_3Al_5O_{12}$), are used to create faux diamonds with a high degree of visual similarity to the real thing. Yttrium occurs naturally in many different mineral compounds.

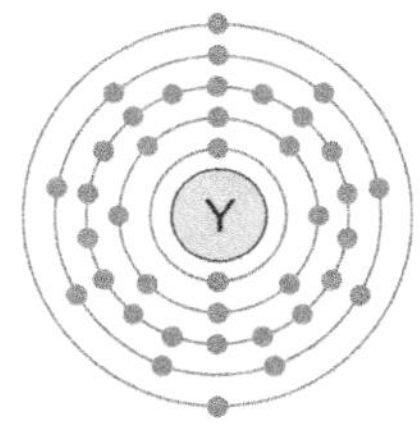

Yttrium (Y)

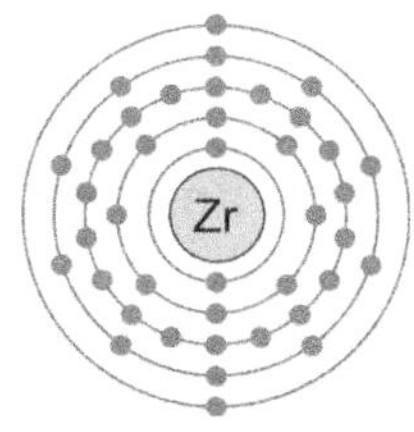

Zirconium (Zr)

Zirconium (Zr) is named after the Persian word *zargun*, which translates to '*gold-like*'. It was first observed in the minerals *Jargon*, *Hyacinth*, *Ligure* and *Zircon*, all of which exhibit a golden color in natural daylight. Zirconium has an atomic number of 40, and an atomic mass of approximately 91 amu, which means that its most common form has 40 protons, 40 electrons, and 51 neutrons.

A researcher by the name of Klaproth discovered zirconium in the year 1789. Physically, this element is a grayish-white metal. It is classified as a transition metal with an electron configuration of [Kr] $4d^2$ $5s^2$.

Zirconium has a high corrosion resistance, making it an excellent alloying material for steels used under extreme environmental conditions. This element is also utilized in a variety of commercial applications including electronics fabrication, explosives, electric light filaments, nuclear reactor containment vessels, and faux gemstones (i.e., crystalline zirconium oxide, ZrO_2 known as *cubic zirconium*). Zirconium is principally obtained from the mineral *Zircon* (zirconium silicate, $ZrSiO_4$).

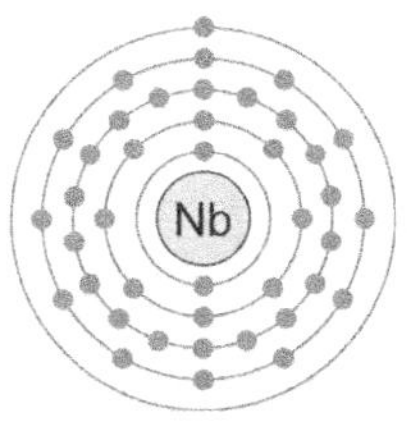

Niobium (Nb)

Niobium (Nb) is named after the goddess *Niobe*, the daughter of Tantalus (since niobium is often found in tantalum deposits – *cute huh?*). This element is still called by many under its older name – *columbium*. Niobium has an atomic number of 41, and an atomic mass of approximately 93 amu, which means that it has 41 protons, 41 electrons, and 52 neutrons in its most common form.

Hatchett discovered niobium (columbium) in the year 1801. Physically, this element is a silvery-white soft metal. It is classified as a transition metal with an electron configuration of [Kr] $4d^4$ $5s^1$.

Crystalline niobium, like the similar element tantalum, can act as a diode and signal rectifier by allowing electric current to flow in one direction only, thereby making it a valuable element for electronics purposes. It is also used to produce a type of superconducting magnet. Niobium is typically found in the mineral *Columbite* (niobite-tantalite (Fe, Mn)Nb_2O_6).

Molybdenum (Mo) arrives at its name from the Greek word *molybdos* meaning *'lead'*; although as we shall see, lead is a much heavier element and doesn't actually behave anything like molybdenum. Molybdenum has an atomic number of 42, and an atomic mass of approximately 96 amu, which means that it has 42 protons, 42 electrons, and 54 neutrons in its most common form.

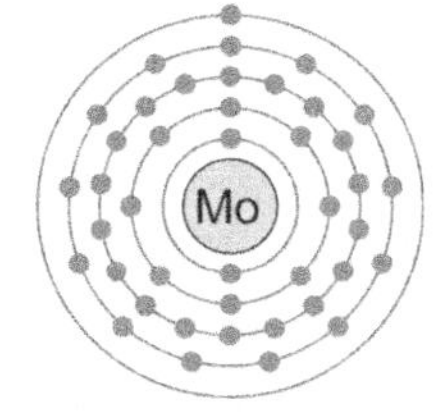

Molybdenum (Mo)

Scheele discovered molybdenum in the year 1778. Physically, this element is a silvery-white metal. It is classified as a transition metal with an electron configuration of [Kr] $4d^5$ $5s^1$.

Molybdenum is used as an alloying agent to produce high-strength steels. It is also alloyed with chromium steels to produce tools of exceptional nature. Its durability is applied in many places, such as the aircraft and aerospace industries, in gun fabrication, in nuclear boilers and blast furnaces, and as a *catalyst* (i.e., something that speeds-up another chemical reaction) in refining crude oil into gasoline. Molybdenum does not occur freely in nature; it is typically extracted from the mineral *Molybdenite* (molybdenum disulfide, MoS_2).

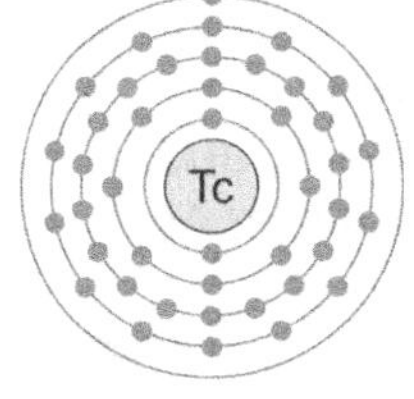

Technetium (Tc)

Technetium (Tc) derives its name from the Greek word *technetos* meaning 'artificial'. Technetium was the first element to be synthesized artificially, and is the one of the few radioactive

elements before bismuth on the periodic table that does not occur naturally in nature. Technetium has an atomic number of 43, and an atomic mass of approximately 98 amu, which means that it has 43 protons, 43 electrons, and 55 neutrons in its most common form.

Researchers Perrier and Segrè discovered/invented technetium in the year 1937, after noticing that there is a great big hole in the periodic table at space number 43. Physically, this element is a silvery-gray metal. It is classified as a transition metal generally having an electron configuration of [Kr] $4d^5$ $5s^2$.

Technetium is one of the elemental *oddballs.* A non-naturally occurring, radioactive element that is chemically similar to rhenium with some great corrosion resistance and hardness alloying properties – *unfortunately, it also makes anything it's alloyed with radioactive.* Technetium is a great low temperature superconducting material – *but again it's radioactive.*

The Guru Says...

Technetium is sort of an oddity in the universe. It is the first element that was predicted to exist long before it was actually 'found' since scientists saw that there was a missing element between molybdenum and ruthenium. It occurs naturally as part of uranium fission, but can be easily manufactured in a lab by bombarding molybdenum with neutrons.

The Guru has actually told you how to make the stuff if you were paying attention.

Technetium has found a home in the medical industry where its isotope Technetium-99 (Tc_{99}) is used as a radioactive tracer in tests. Today technetium is a wholly manmade element.

Ruthenium (Ru) is named from the Latin word for Russia – *Ruthenia.* Ruthenium has an atomic number of 44, and an atomic mass of approximately 101 amu, which means that it has 44 protons, 44 electrons, and 57 neutrons in its most common form.

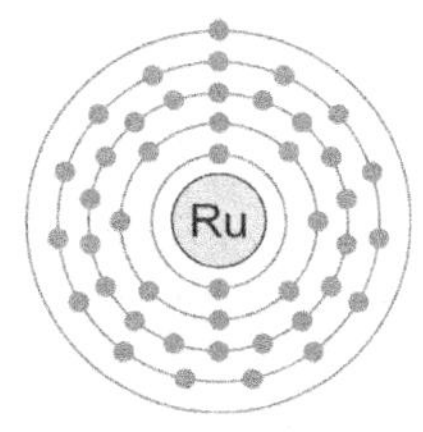

Ruthenium (Ru)

Klaus discovered ruthenium in the year 1844. Another silvery-white, hard metal, this element is classified as a transition metal with an electron configuration of [Kr] $4d^7$ $5s^1$.

Ruthenium through the element indium (In) are a group of *semi-precious metals* that you will find listed on the commodity trading sheets due to their proximity to the element silver. You'll find the group of *precious metals* right around where gold occurs in the periodic table.

Ruthenium's uses are as an alloying agent for precious and semi-precious metals (notably platinum and palladium) as well as an excellent corrosion resistance additive. Ruthenium occurs in trace amounts and is a by-product of nickel, copper, and platinum mining.

Rhodium (Rh) is named after the Greek word *rhodon* for 'rose' since most salts made from Rhodium have a rose-colored appearance. Rhodium has an atomic number of 45, and an atomic mass of approximately 103 amu, which means that it has 45 protons, 45 electrons, and 58 neutrons in its most common form.

Wollaston discovered Rhodium in the year 1803. This element is a silvery-white metal. It is classified as a transition metal with an electron configuration of [Kr] $4d^8$ $5s^1$.

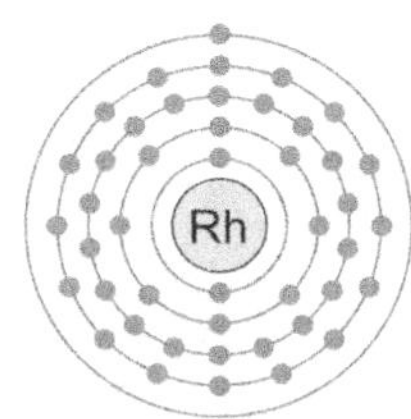

Rhodium (Rh)

Rhodium is used as an alloying agent to harden platinum and palladium. It also has various uses in electronics due to its low electrical resistance and high corrosion resistance. Rhodium is also used in optical instruments due to its ability to be polished to a mirror finish that does not tarnish. Rhodium principally occurs with other platinum ores and is a byproduct of platinum mining.

Palladium (Pd) has an unusual name origin. It is named after the asteroid Pallas that was also discovered in the same year. Pallas was the Greek goddess of wisdom. Palladium has an atomic number of 46, and an atomic mass of approximately 106 amu, which means that it has 46 protons, 46 electrons, and 60 neutrons in its most common form.

Wollaston discovered palladium in the year 1803. This element is a steel-white metal. It is classified as a transition metal with an electron configuration of [Kr] $4d^{10}$.

Palladium is used extensively in the medical industry to manufacture surgical instruments. It is also used to purify hydrogen gas (hydrogen readily passes though palladium metal, while other gasses do not), alloy precious metals, make electrical contacts, and produce dental amalgam. It is also the active ingredient in automotive catalytic converters. Palladium is a trace element found while mining nickel.

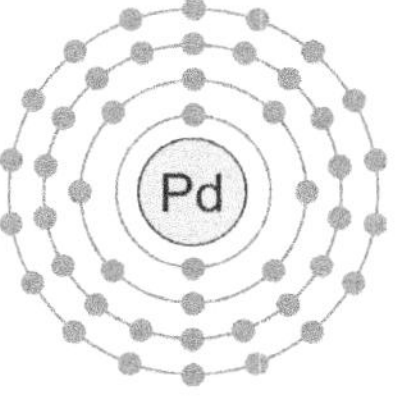

Palladium (Pd)

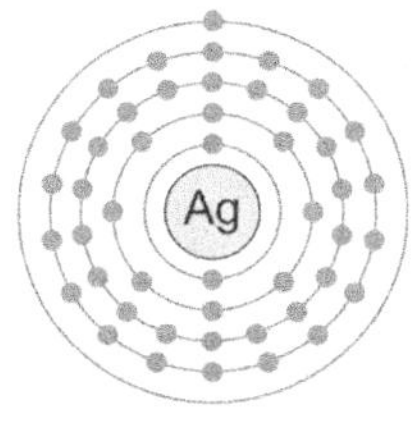

Silver (Ag)

Silver (Ag) has also been around forever, hence its abbreviation, which in Latin stands for *argentums*. This name was changed in the Middle Ages to the Anglo-Saxon version *Seolfor* or *siolfur*, and later into the English version *silver*. No matter how it's spelled, it's still the same stuff. Silver has an atomic number of 47, and an atomic mass of approximately 108 amu, which means that it has 47 protons, 47 electrons, and 61 neutrons in its most common form.

Physically, this element is a pale silvery-white metal, and is classified as a transition metal with an electron configuration of [Kr] $4d^{10}$ $5s^1$.

Silver has the highest electrical and thermal conductivity of any metal in the universe and thus has numerous uses in photography, dentistry, electrical solder and brazing compounds, electrical contacts, batteries, mirrors, and printed circuits.

Silver normally does not tarnish in air, but will develop a gray coating when exposed to ozone (O_3) or compounds of sulfur. Before the industrial revolution (around 1750) tarnished silver was an unknown concept to humans. Although silver is considered non-toxic, compounds of silver are poisonous. Silver occurs naturally in silver ore, called *Argentite* (cubic silver sulfide, or Ag_2S).

Cadmium (Cd) has its name origin in the Latin word *cadmia*, which means 'calamine' (as in the lotion). Calamine (the pharmaceutical product) is actually zinc carbonate – cadmium (the metal) was first located in this mineral. Cadmium has an atomic number of 48, and an atomic mass of approximately 112 amu, which means that it has 48 protons, 48 electrons, and 64 neutrons in its most common form.

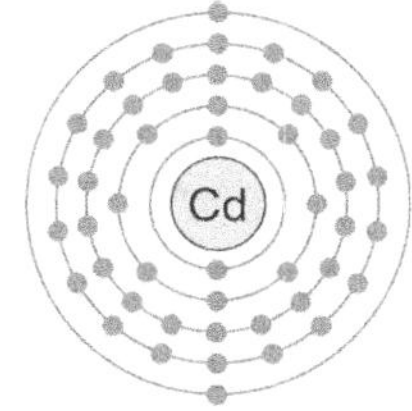

Cadmium (Cd)

Strohmeyer discovered cadmium in 1817. This element is a bluish-white, soft metal. It is classified as a transition metal with an electron configuration of [Kr] $4d^{10}$ $5s^2$.

Cadmium is used to create metal alloys with low melting points. Principally, cadmium is used as a metal plating element, in the construction of rechargeable batteries (nickel cadmium batteries, or NiCd's), and to create the color yellow in pigments. Cadmium and all of its compounds are toxic to humans. Cadmium is found in the mineral *Greenockite* (cadmium sulfide, CdS).

Indium (In) derived its name from the Latin word *indicum*, meaning 'indigo' or 'violet', which is how its spectrum looks when it is heated. Indium has an atomic number of 49, and an atomic mass of approximately 115 amu, which means that it has 49 protons, 49 electrons, and 66 neutrons in its most common form.

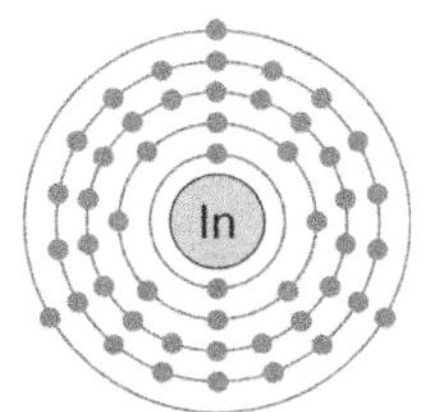

Indium (In)

Researchers Hieronymus and Richter discovered indium in the year 1863. This element is a silvery-white, soft metal. It is classified as a metal with an electron configuration of [Kr] $4d^{10}$ $5s^2$ $5p^1$.

Indium is the last of our semi-precious metals and is also used in low melting point alloys. It has the weird property that it will emit an audible high-pitched 'scream' when bent (you can actually hear the indium crystals breaking). Indium is principally used in the electronics industry in the fabrication of transistors, thermistors, photoconductors, and rectifiers. Indium is typically found as a trace element in zinc, iron, lead, and copper ores.

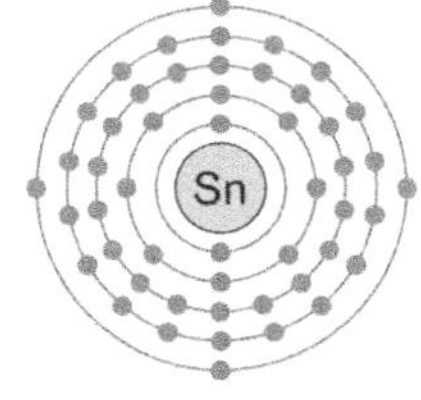

Tin (Sn)

Tin (Sn) is another oldie. The Romans called it *stannum*, hence its abbreviation. Its current name is patterned after the Etruscan god, *Tinia*. Tin has an atomic number of 50, and an atomic mass of approximately 119 amu, which means that it has 50 protons, 50 electrons, and 69 neutrons in its most common form.

Physically, this element is a silvery-white metal. It is classified as a metal with an electron configuration of [Kr] $4d^{10}$ $5s^2$ $5p^2$.

Tin also has the unusual property that it makes noise when it is bent (try it sometime). The sound of the crystalline structure of the piece of tin breaking produces a phenomenon known as 'crying tin'. Tin is typically used in applications where corrosion (salt) resistance is desired and thus is used as a plating substance. It is also used as a 'glue' to hold harder elements to a rotating surface to reduce wear (a technique known as *babbitting*). The primary source of tin is the mineral *Cassiterite* (tin oxide, SnO_2).

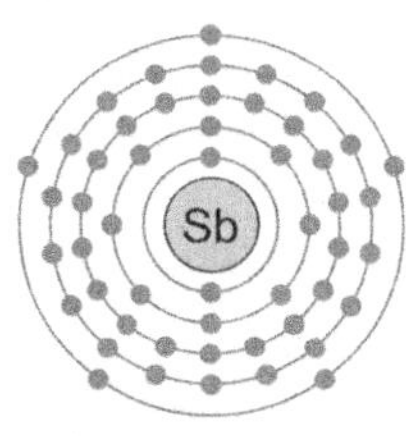

Antimony (Sb)

Antimony (Sb) has a name that is kind-of a Greek word puzzle. In ancient Greek it combines the words *anti* (not) with *monos* (to be alone) implying that it is a metal not occurring by itself. Its symbol, well that's another story, it's based upon the primary mineral where antimony is found, *Stibnite*. Antimony has an atomic number of 51, and an atomic mass of approximately 122, which means that it has 51 protons, 51 electrons, and 71 neutrons in its most common form.

Antimony, known to mankind since before the time of the Romans, is a bluish silvery-white, brittle metalic element. It is classified as a metalloid, with an electron configuration of [Kr] $4d^{10}$ $5s^2$ $5p^3$.

Antimony is widely used in alloying to increase hardness and mechanical strength of many different metals. It is also used in the production of semiconductors, batteries, bullets, fire retardant compounds, glass, ceramics, paints, and pottery. Antimony and most of its compounds are toxic to humans. Antimony is fairly common and it usually found in the mineral *Stibnite* (sometimes called *Antimonite*, or antimony sulfide Sb_2S_3).

Tellurium derives its name from the Latin word for Earth – *Tellus*. Tellurium (Te) has an atomic number of 52, and an atomic mass of approximately 127 amu, which means that it has 52 protons, 52 electrons, and 75 neutrons in its most common form.

RADIOACTIVE

German researcher Von Reichenstein discovered tellurium in the year 1782. Physically, this element is a silvery-white semi-metal. It is classified as a metalloid, with an electron configuration of [Kr] $4d^{10}$ $5s^2$ $5p^4$.

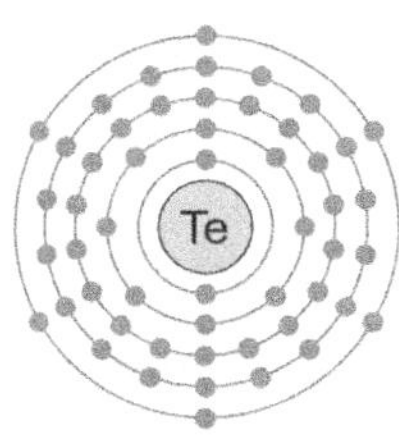

Tellurium (Te)

Tellurium is used in making semiconductors and solar panels, as well as having applications in ceramics, tinting glass, and explosives. Tellurium is toxic to humans and is moderately radioactive. Tellurium is a byproduct of copper and lead production.

Iodine (I) derives its name from the Greek word *iodes*, which means 'violet'. The violet color is most prevalent in the vapors released from solid iodine, which are a deep purple (and dangerous to breathe). Iodine has an atomic number of 53, and an atomic mass of approximately 128 amu, which means that it has 53 protons, 53 electrons, and 75 neutrons in its most common form.

Courtois discovered iodine in the year 1811. Physically, this element is a bluish-black solid, with a blue-violet vapor when it sublimates. It is classified as a halogen with an electron configuration of [Kr] $4d^{10}$ $5s^2$ $5p^5$.

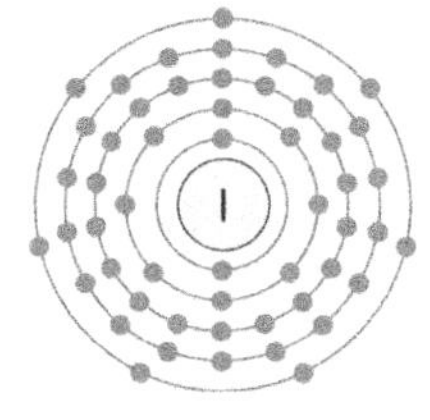

Iodine (I)

Iodine forms compounds with many elements, but is not as reactive as the other halogens. Interestingly enough, iodine has some of the properties of metals (such as ductility and conductivity), but that's where the similarity ends.

Iodine is slightly soluble in water and fully dissolvable in other solutions creating a purple tint. Iodine is required by humans for proper thyroid function and is typically added to table salt for this purpose. It is also used in the medical industry as a disinfectant. Iodine is principally extracted from seawater and seaweed.

Xenon's (Xe) name is derived from the Greek word for 'stranger' – *xenos* due to some of the strange chemical behaviors it exhibits. Xenon has an atomic number of 54, and an atomic mass of approximately 131 amu, which means that it has 54 protons, 54 electrons, and 77 neutrons in its most common form.

Ramsay and Travers also discovered xenon in the year 1898. It is a colorless gas, classified as a noble gas, with an electron configuration of [Kr] $4d^{10}$ $5s^2$ $5p^6$.

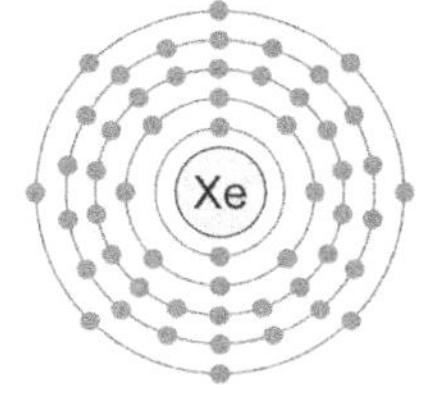

Xenon (Xe)

Xenon gas is used in vacuum tubes, lighted 'neon' signs to produce the color blue, bactericidal lamps, strobe lamps, and anywhere a high molecular weight shielding gas is needed. Although xenon is a non-toxic noble gas, it has been known to form toxic compounds when exposed to the influence of the halogen group. Xenon is found in the atmosphere at levels of approximately one part in twenty million and is obtained as a by-product of liquid nitrogen production.

The Period 6 Elements – Cesium through Radon and the Lanthanides

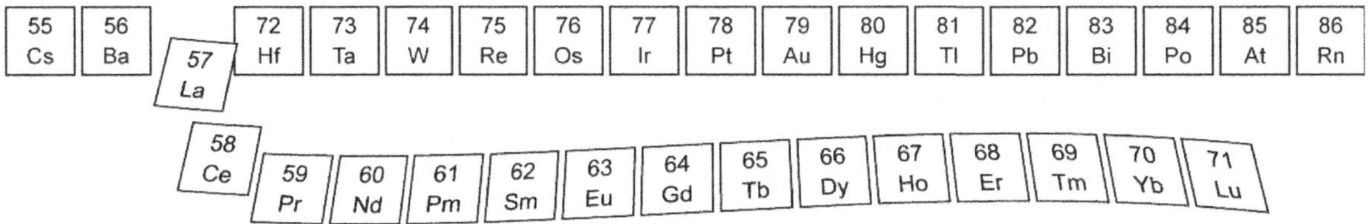

Cesium (Cs) is from the Latin word *coesius*, which means 'sky blue' again due to its spectral content when heated. Cesium has an atomic number of 55, and an atomic mass of approximately 133 amu, which means that it has 55 protons, 55 electrons, and 78 neutrons in its most common form.

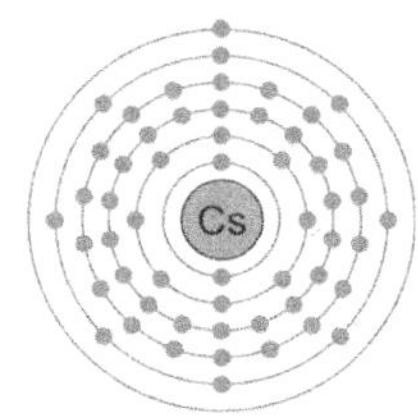

Cesium (Cs)

Bunsen and Kirchhoff discovered cesium in the year 1860. Physically, this element is a silvery-gold, soft metal, which will melt on a warm day. It is classified as an alkali metal with an electron configuration of [Xe] $6s^1$.

Most people, who have seen too many 1950's 'B' movies with the word 'atomic' somewhere in the title, believe that cesium is radioactive. As you can see from the lack of a radiation symbol, naturally occurring cesium (non isotopic) is not radioactive and is a strongly reactive metal, predominately used to react with other gaseous elements in a vacuum environment (like a light bulb) to prevent them from interfering with the electrical operation.

By and large, cesium's most important use is in the isotope cesium-137 (Cs_{137}), which has electrons that vibrate so precisely when excited by electricity that they are actually used by the SI to define the unit of the *second.*

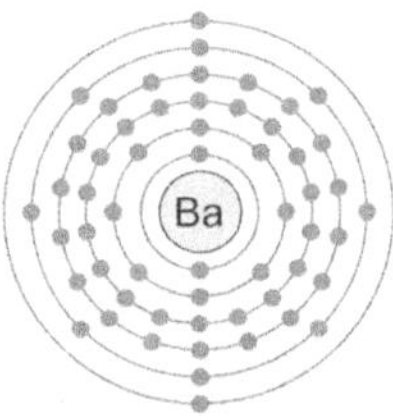

Barium (Ba)

Barium derives its name from the Greek word *barys* meaning 'heavy' or 'dense'. It's just that, compounds containing barium have some heft to them. Barium (Ba) has an atomic number of 56, and an atomic mass of approximately 137 amu, which means that it has 56 protons, 56 electrons, and 81 neutrons in its most common form.

Davy was still busy looking for new elements when he discovered barium in 1808. This element is a silvery-white, soft metal. It is classified as an alkali earth metal with an electron configuration of [Xe] $6s^2$.

Barium is used as a pigment in paints and glassmaking as well as in the manufacture of rubber products, X-ray machines, and to produce the green color in fireworks. Isotopes of barium are used as radioactive tracers in the medical industry. It is typically found in the mineral *Barite* (barium sulfate, $BaSO_4$).

The **Lanthanide (or Lanthanoid) series** is a collection of 15 metallic elements ranging from lanthanum through lutetium that are crammed into the periodic table between barium and hafnium. These elements, along with scandium and yttrium are typically called the *rare Earth elements*.

These elements essentially take a side trip from the normal Period 6 electron shell filling scheme to fill electrons in their inner 'f' orbital shell instead (refer back to Figure 3.2 to refresh your memory on the order of the electron shells in the Bohr model).

Figure 3.4: The *'Chemically Boring'* and Identically Looking Lanthanide Series

Since this 'f' shell is inside the already filled '4d' and '5p' shells, these elements do not have any net changes in their valence electron shells, so they all look chemically alike, which is why they are all relegated to a single Group 3, Period 6 spot in the periodic table.

The lanthanides, shown in Figure 3.4, consist of the following elements: **Lanthanum** (which is from the Greek word – *lanthaneis, or* 'to lie hidden'*)*; **Cerium** (named after the asteroid Ceres); **Praseodymium** (in Greek, *prasios* (green) and *didymos* (twin), or 'green-twins'); **Neodymium** (in Greek, *Neos* (new) and *didymos* (twin), or 'new twins'); **Promethium** (dedicated to one of the Greek Titans, *Prometheus);* **Samarium** (named after the mineral in which this element is found, *Samarskite*); and **Europium** (named after the continent of Europe).

But wait, there's still more. The lanthanides also contain the elements **Gadolimium** (named after the Finnish chemist Johan Gadolin, the discoverer of Yttrium); **Terbium** (named in honor of the town in Sweden where several new elements were discovered – *Ytterby)*; **Dysprosium** (from the Greek word *dysprositos*, which literally means 'hard to get at'*);* **Holmium** (named after the Latin version of the name for Stockholm, Sweden – *Holmia);* **Erbium** (also named after the town of Ytterby, Sweden); **Thulium** (named after the ancient name of Scandinavia – *Thule);* **Ytterbium** (again named after the town of Ytterby, Sweden); and finally **Lutetium** (named after the ancient name of Paris, France – *Lutecia).*

The lanthanides have a wide variety of applications in today's modern world, particularly in the fields of communications, physics, and electronics. Exotic devices from superconductors to cell phones, lasers to hybrid car batteries, and microwave and radar equipment all owe a debt of gratitude to these rare Earth elements, all hiding in the same location within the periodic table.

Most lanthanide elements are found in ore deposits containing yttrium. All non-radioactive lanthanides are of low toxicity to humans. All of the lanthanide elements are non radioactive with the exception of promethium.

Now, picking up right after the copycat lanthanide series, we continue across the periodic table with some of the most expensive and sought-after metals in the universe. We'll start with hafnium at atomic number 72.

Hafnium (Hf) is named after the ancient name of Copenhagen, Denmark – *Hafnia*. Hafnium has an atomic number of 72, and an atomic mass of approximately 178 amu, which means that it has 72 protons, 72 electrons, and 106 neutrons in its most common form.

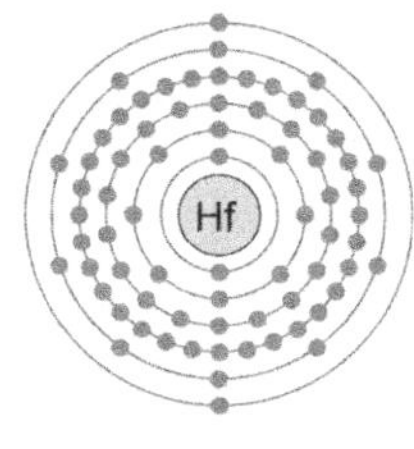

Hafnium (Hf)

Coster and deHevesy discovered hafnium in 1923. Physically, this element is a silver-gray metal. It is classified as a transition metal with an electron configuration of [Xe] $4f^{14}$ $5d^{2}$ $6s^{2}$.

Hafnium is a great absorber of neutrons and is therefore used in atomic power plants. Chemically similar to zirconium, it is also used to provide corrosion resistance to alloyed materials. Hafnium is considered non-toxic to humans. Hafnium is not found free in nature and is found in most minerals containing zirconium.

Tantalum (Ta) is based on the Greek mythological king *Tantalos* (who was also the father of *Niobe* where the name niobium comes from). Tantalum has an atomic number of 73, and an atomic mass of approximately 181 amu, which means that it has 73 protons, 73 electrons, and 108 neutrons in its most common form.

Ekenberg discovered tantalum in 1802. This element is a steel-gray, heavy, hard metal. It is classified as a transition metal with an electron configuration of [Xe] $4f^{14}$ $5d^{3}$ $6s^{2}$.

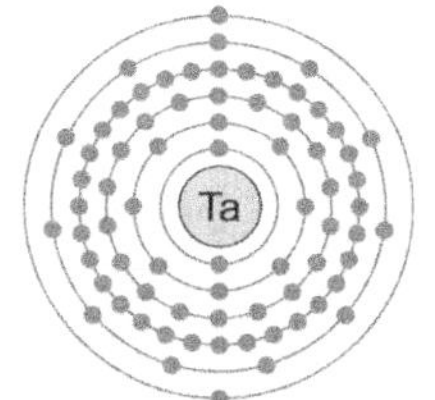

Tantalum (Ta)

Tantalum is used in chemical process equipment, furnaces, electronics, nuclear reactors, and aircraft components. Wire made from this element is used as a filament for evaporating other metals in a process called *deposition*. Tantalum is used in a variety of alloys, providing high melting points, strength, and corrosion resistance (i.e., tantalum carbide is one of the hardest materials ever created by man). Additionally, tantalum oxide films produce electrical capacitors with high levels of tolerance.

Tantalum is also considered physiologically inert and therefore is used in a multitude of medical applications. Tantalum is found primarily in the mineral *Tantalite* (*Ferrotantalite* and *Manganotantalite*, $(Fe,Mn)Ta_2O_6$).

Tungsten (W) is a combination of two Swedish words, *Tung* (heavy) and *sten* (stone), or *heavy stone,* after the physical property of rock specimens containing this element. Its symbol is based upon the mineral in which tungsten is found, *Wolframite.* Tungsten has an atomic number of 74, and an atomic mass of approximately 184 amu, which means that it has 74 protons, 74 electrons, and 110 neutrons in its most common form.

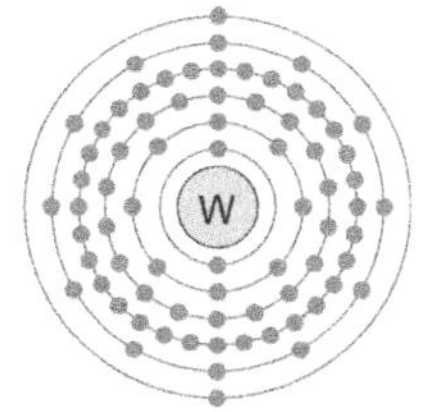

Tungsten (W)

De Elhuyar discovered tungsten in the year 1783. Physically, this element is a heavy, steel-gray to tin-white metal. It is classified as a transition metal with an electron configuration of [Xe] $4f^{14}$ $5d^4$ $6s^2$.

Being one of the densest elements around, the thermal expansion of tungsten is similar to that of borosilicate glass, so this element is used for glass/metal seals for applications were high temperatures are present. Tungsten and its alloys are used to make filaments for electric lamps, electrical contacts, and high temperature heating elements. Tungsten carbide (WC) is an extremely durable metal alloy used extensively in the mining industry as it can drill through most hard rocks with very little difficulty. Tungsten predominately occurs in *Wolframite* (iron manganese tungstate, $(Fe,Mn)WO_4$).

Tungsten is harmless to humans – unless you were to drop a large chunk of the stuff on your toe; then that's another matter.

Rheniuim (Re) is derived from the Latin name for the Rhine River – *Rhenus*. Rhenium has an atomic number of 75, and an atomic mass of approximately 186 amu, which means that it has 75 protons, 75 electrons, and 111 neutrons in its most common form.

Physicists Tacke-Noddack and Berg discovered rhenium in the year 1925. This element is another silvery-white metal. It is classified as a transition metal with an electron configuration of [Xe] $4f^{14}$ $5d^5$ $6s^2$.

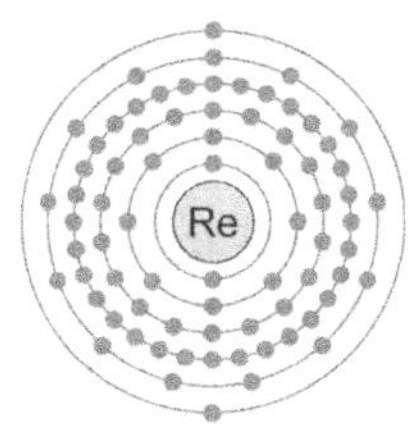

Rhenium (Re)

Rhenium is used in photographic flash bulbs and strobes, but is most frequently used as an alloying agent in tungsten and molybdenum. Rhenium is obtained in trace amounts in platinum ores.

Osmium (Os) derives its name from the Greek word *osme*, meaning a 'smell' or 'odor'. It is another one of those stinky elements whose name is based on a specific compound, in this case osmium tetroxide, a highly toxic oxidizer with a terrible smell (if you live long enough to be able to comment on the experience). Osmium has an atomic number of 76, and an atomic mass of approximately 190 amu, which means that it has 76 protons, 76 electrons, and 114 neutrons in its most common form.

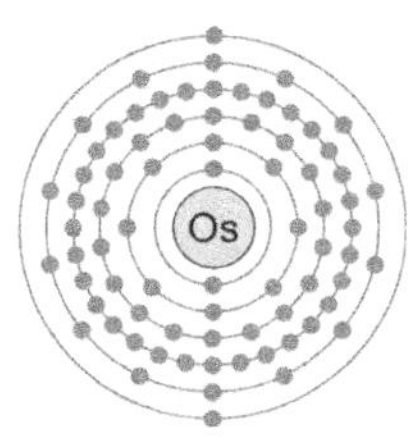

Osmium (Os)

Tennant discovered osmium in the year 1803. Physically, this element is a bluish-white, hard, brittle metal. It is classified as a transition metal with an electron configuration of [Xe] $4f^{14}$ $5d^6$ $6s^2$.

Osmium is 'technically' classified as the densest element on the periodic table. It is used to add hardness to metal alloys, particularly silver and steel. It is also used for instrument pivots, bearings in watches, and electrical contacts due to its durability.

Osmium tetroxide (OsO_4) is used sparingly in forensic work to detect the presence of fingerprints. Osmium powder is extremely toxic to humans. Osmium is principally located in the mineral *Osmiridium* (a natural alloy of osmium and iridium, Os-Ir).

Starting with iridium below, we now enter the world of the *precious metals*, which ends with everybody's favorite – *gold*.

Iridium (Ir) is derived from the Latin word *iris*, which means 'rainbow'. This is a reference to the salts of iridium, which are highly colored. Iridium has an atomic number of 77, and an atomic mass of approximately 192 amu, which means that it has 77 protons, 77 electrons, and 115 neutrons in its most common form.

Tennant discovered iridium in the year 1803. Physically, this element is a silvery-white, hard, brittle metal. It is classified as a transition metal with an electron configuration of [Xe] $4f^{14}$ $5d^7$ $6s^2$.

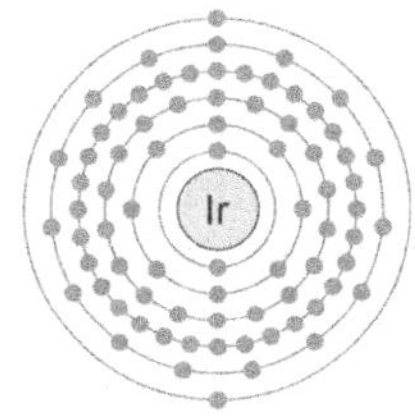

Iridium (Ir)

There is currently a scientific debate as to whether or not iridium is the densest of all the elements (a title currently held by osmium). What we do know is that iridium is impervious to all known acids, but can be attacked by simple table salt (NaCl). Iridium is primarily used for hardening platinum. It is also used in crucibles and other high temperature applications.

Iridium occurs in nature as the mineral *Osmiridium*, which is typically found in platinum ores. It is also commercially recovered as a byproduct from the nickel mining industry.

Platinum (Pt) is from the Spanish word *platina*, meaning 'little silver'. Platinum has an atomic number of 78, and an atomic mass of approximately 195 amu, which means that it has 78 protons, 78 electrons, and 117 neutrons in its most common form.

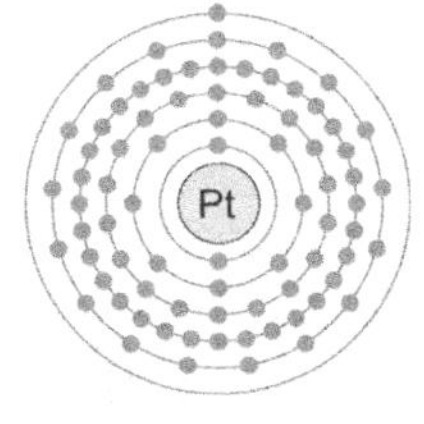

Platinum (Pt)

De Ulloa discovered platinum in the year 1735. Physically, this element is a silvery-white soft metal. It is classified as a transition metal with an electron configuration of [Xe] $4f^{14}$ $5d^9$ $6s^1$.

Platinum is used in jewelry, wire, electronic and laboratory equipment, thermocouples, and for plating items that must be exposed to high temperatures or corrosive environments for long periods of time. Platinum-cobalt alloys have interesting magnetic properties. Also, since platinum can absorb large amounts of hydrogen gas, the metal is often used as a catalyst to burn gasses without a source of external ignition or heat for that matter (which is a peculiar sight to behold). A gaseous mixture of hydrogen and oxygen will explode without a flame at room temperature in the presence of platinum – *so never mix these elements together.*

Platinum occurs as the elemental metal and is usually found with small amounts of other metals belonging to the same region of the periodic table, such as osmium, iridium, ruthenium, palladium, and rhodium. Even though platinum typically costs more than gold, there is substantially more of it on Earth.

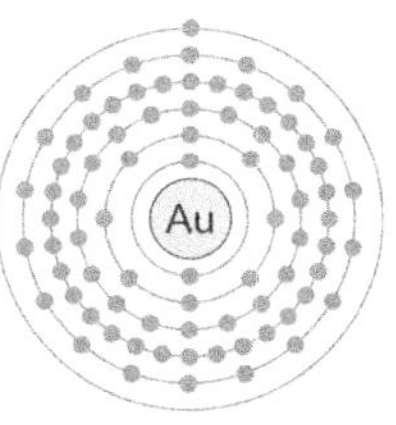

Gold (Au)

Gold (Au). If there is one element that everyone knows about, it's this one. Gold was, during the times of the Romans, called *Aurum*, hence its abbreviation. Gold has an atomic number of 79, and an atomic mass of approximately 197 amu, which means that it has 79 protons, 79 electrons, and 118 neutrons in its most common form.

Physically, this element is a golden-yellow, soft metal. It is classified as a transition metal with an electron configuration of [Xe] $4f^{14}$ $5d^{10}$ $6s^1$.

Gold is used in coinage and is the monetary standard for several countries. It is also used for jewelry, dental work, and plating applications. It is a good conductor of electricity and heat, and is typically alloyed with copper to improve its strength, since pure gold is relatively soft and easily bendable. Elemental gold metal is considered non-toxic and occasionally used as a decorative food additive.

Gold is found as the elemental metal in ore deposits and is typically associated with the minerals *Quartz* (silicon dioxide, SiO_2) and *Pyrite* (iron sulfide, FeS_2). Gold can be synthesized from the element lead fulfilling the ancient alchemists' dream; unfortunately it requires having an atomic reactor to do this and is not very cost efficient.

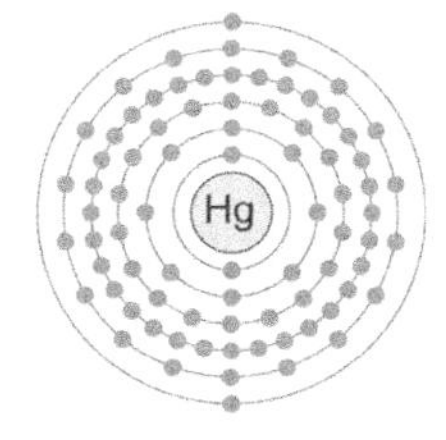

Mercury (Hg)

Mercury (Hg) is named after the planet Mercury (at least it is now). In early times the Romans knew it as *Hydrargyrum*, which literally means 'liquid silver'. Mercury has an atomic number of 80, and an atomic mass of approximately 201 amu, which means that it has 80 protons, 80 electrons, and 121 neutrons in its most common form.

Mercury has been known since pre-Roman times. Physically, this element is a silvery-white liquid metal. It is classified as a transition metal with an electron configuration of [Xe] $4f^{14}$ $5d^{10}$ $6s^2$.

Mercury is a cumulative neurological poison, which over a long enough time can cause severe dementia. In 18th and 19th century England the mercury compound mercuric nitrate was used to process felt hats, resulting in more than one hat worker going mad due to inhaling the fumes, hence the term – '*mad as a hatter*'.

Mercury has many uses associated with the physical fact that it is a liquid metal at room temperature. Mercury is used to make thermometers, barometers, mercury vapor lamps, mercury switches, batteries, dental preparations (believe it or not), paints, pigments, and catalysts. Mercury easily combines with other elemental metals to form alloys, called *amalgams*. The principal application of this feature of mercury is in the recovery of gold dust from crushed ores. Mercury primarily occurs in the mineral *Cinnabar* (red mercury sulfide, HgS).

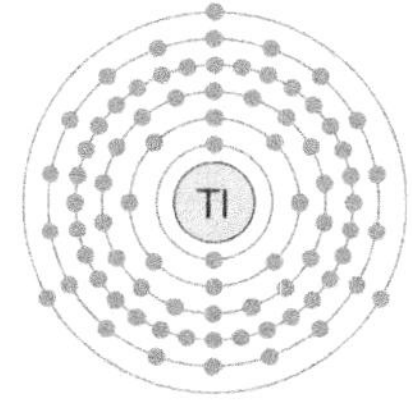

Thallium (Tl)

Thallium (Tl) is named after the Greek word *thallos,* meaning 'green twig', based upon its bright green light spectrum when heated. Thallium has an atomic number of 81, and an atomic mass of approximately 204 amu, which means that it has 81 protons, 81 electrons, and 123 neutrons in its most common form.

The physicist Crookes (who did a lot of work with electrons and electricity) discovered thallium in the year 1861. Physically, this element is a silvery-white to bluish-gray, soft metal. It is classified as a metal with an electron configuration of [Xe] $4f^{14}$ $5d^{10}$ $6s^2$ $6p^1$.

Thallium has very few practical uses that cannot be duplicated with other elements. All salts of thallium are highly toxic and were for many years used as a rat poison, until many countries eventually banned them, since these tasteless salts also killed anything else they came into contact with. Referred to as the *"the poisoner's poison"*, thallium has a very dark, murderous history.

Thallium does have a couple of practical uses other than homicide; notably, thallium iodide crystals are used to construct infrared optical systems and thallium sulfide is used to fabricate photoresistors, owing to the element's sensitivity to infrared light. Thallium is produced in trace amounts through the processing of copper, lead, and zinc ores.

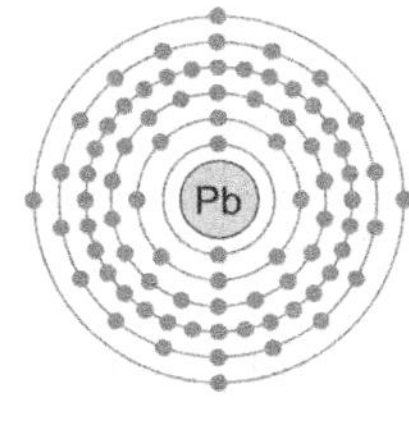

Lead (Pb)

Lead (Pb) goes as far back as recorded history. It was a soft workable metal, easily extracted from surrounding minerals, and fun to play with. The Romans called the stuff *plumbum* (hence the abbreviation). Lead has an atomic number of 82, and an atomic mass of approximately 207 amu, which means that it has 82 protons, 82 electrons, and 125 neutrons in its most common form.

Physically, this element is a heavy, soft, bluish-gray metal. It is classified as a metal with an electron configuration of [Xe] $4f^{14}$ $5d^{10}$ $6s^2$ $6p^2$.

The Romans thought that the sweet taste of lead improved wine, so it was added as a flavoring. We now know that lead is also a cumulative neurological poison, which over time will short your brain out (literally) and cause you to go foaming-at-the-mouth crazy. Is it any wonder when you read some of the historical accounts of Roman behavior; it appears to be a little bit unusual. Now you know why.

Today, lead is used for everything, except the manufacture of 'lead pencils', which never had lead in them to start with; it was graphite (carbon, C). As a soft, low melting point metal, lead is used in the electronics industry for solder, anywhere weight or ballast is needed, as a radiation shielding material in nuclear reactors and x-ray rooms, and in pretty much every wet cell battery used on the planet. Lead is usually found in the ores of zinc, silver and copper, but also is found in plentiful quantities in the mineral *Galena* (lead sulfide, PbS).

Bismuth (Bi) is named after the German word *bisemutum*, meaning 'white mass', although a superficial oxide layer can create a multi-colored appearance. Bismuth has an atomic number of 83, and an atomic mass of approximately 209 amu, which means that it has 83 protons, 83 electrons, and 126 neutrons in its most common form.

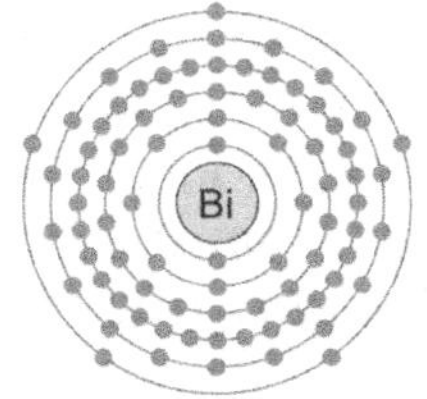

Bismuth (Bi)

Geoffroy discovered bismuth in the year 1753. Physically, this element is a silvery-white, crystalline, brittle metal. It is classified as a metal with an electron configuration of [Xe] $4f^{14}$ $5d^{10}$ $6s^2$ $6p^3$.

Bismuth is the last of the naturally occurring elements, which is not considered radioactive, due to its stable nuclear configuration. It is also edible; it is the active ingredient in *Pepto Bismol*. It is interesting that you can eat this element without any ill effects, considering that the element just before this is lead (a toxic poison), and the one after this is polonium (a radioactive poison). See the difference one electron and proton can make.

Bismuth is used in cosmetics and pigments, and is becoming a replacement for lead, since it is non-toxic (a rarity for a heavy metal actually). It is a fairly common element (twice as abundant as gold) and is principally extracted from the minerals *Bismuthinite* (bismuth sulfide, Bi_2S_3) and *Bismite* (bismuth trioxide, Bi_2O_3).

Polonium (Po) is named after the Country of Poland. Polonium has an atomic number of 84, and an atomic mass of approximately 209 amu, which means that it has 84 protons, 84 electrons, and 125 neutrons in its most common form. Polonium is highly radioactive and a favorite poison of Russian spies.

The Guru Says...

Starting with polonium, every element from here on out is radioactive. Thus, we're going to stop drawing the radioactivity symbol to save on paper and ink.

By the way, bismuth is also technically radioactive – it just has a half life longer than the age of the universe, so for all practical purposes no one cares.

A husband and wife team with the last name of Curie discovered polonium in the year 1898. This element is a silvery-gray metal. It is classified as either a metalloid or a metal with an electron configuration of [Xe] $4f^{14}\ 5d^{10}\ 6s^2\ 6p^4$.

Polonium is used in trace quantities in industrial processes where the removal of static electricity is of high importance (such as in photo and xerographic devices, etc.). Polonium is also used to produce thermoelectric nuclear batteries for space applications, and is a good stand-alone neutron generator all by itself.

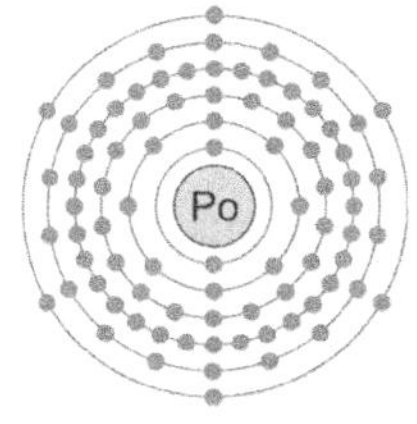

Polonium (Po)

There are no naturally occurring compounds of polonium, only manmade ones. Polonium is found in the presence of uranium (U) ores.

Astatine (At) is derived from the Greek word *astatos*, which means 'unstable'. Astatine has an atomic number of 85, and an atomic mass of approximately 210 amu, which means that it has 85 protons, 85 electrons, and 125 neutrons in its most common form.

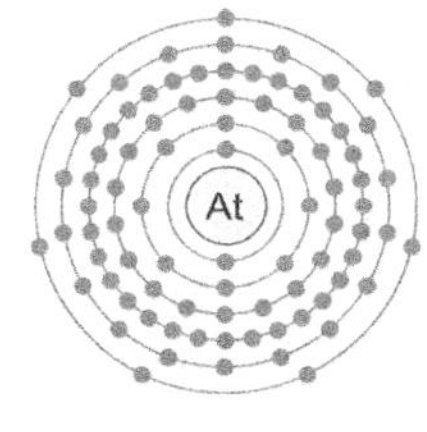

Astatine (At)

Researchers Carson, MacKenzie, and Segrè discovered astatine in the year 1940. This element is classified as a halogen with an electron configuration of [Xe] $4f^{14}\ 5d^{10}\ 6s^2\ 6p^5$.

Astatine is a radioactive halogen (imagine what a frightful thought that must be). It behaves similarly to iodine, except with more metallic properties. Astatine is sometimes used as a radioactive tracer element in medical work and has been used in some forms of cancer treatment. It is toxic to humans in a cumulative sense.

Astatine is a pretty scarce substance that occurs from thorium or uranium decay. At any given moment, there are about 25 grams of the stuff naturally on Earth since its most-stable half-life is a mere 8.1 hours; thus it is typically produced by bombarding bismuth-209 (Bi_{209}) with radioactive helium atoms.

Radon (Rn) has its origin in the word Radium (another element we'll see shortly). It was originally called *niton* after the Latin word *nitens*, which means 'shining', since radon gas has a propensity to glow in the dark. Radon has an atomic number of 86, and an atomic mass of approximately 213 amu, which means that it has 86 protons, 86 electrons, and 127 neutrons in its most common form.

The Guru Says...

In the radiation world, the term half-life is often used. It is a measure (in minutes, days, or years) of how long a particular sample of an element will take to reach a point where only half of the original sample is left.

The longer the half-life, the less radioactive the substance.

A physicist by the name of Dorn discovered radon in the year 1900. Physically, this element is a colorless gas. It is classified as a noble gas with an electron configuration of [Xe] $4f^{14}$ $5d^{10}$ $6s^2$ $6p^6$.

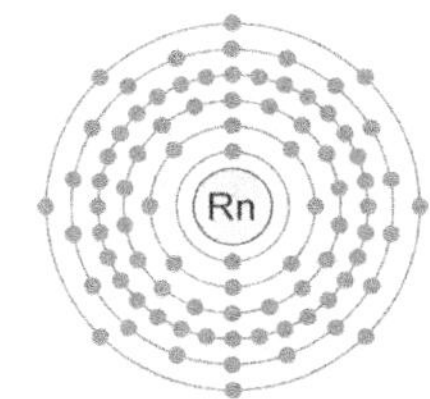

Radon (Rn)

Radon gas, although a novel element to experiment with in the laboratory due to its *'glow in the dark'* properties, especially when super cooled, is more of a carcinogenic hazard than anything else. It is typically present anywhere there are granitic rocks, and because it is much heavier than air, accumulates in underground portions of buildings such as basements.

Since it is also produced by the radioactive decay of uranium, it is a significant health hazard in uranium mines. As radon decays, it transforms itself into different radioactive elements and toxic dust (*great stuff – huh?*). Radon has a half-life of 3.8 days.

The Period 7 Elements – Francium through Ununoctium and the Actinides

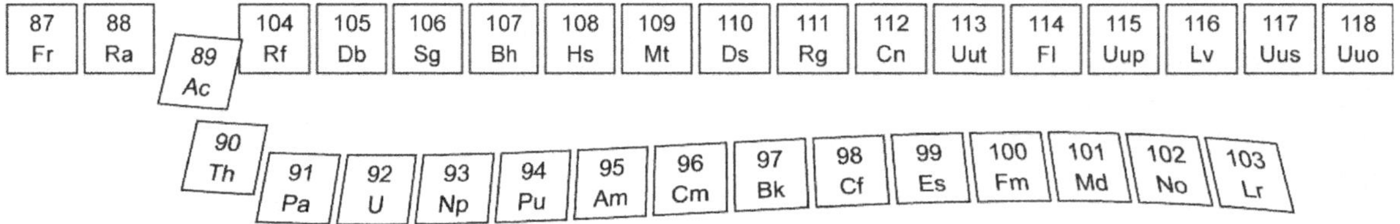

As we start our progress across the last period of the periodic table, we start finding more and more elements that, for all intents and purposes, have no practical value (with a couple of notable exceptions). All of the good stuff in the universe, chemically speaking, is contained within the earlier periods of the elements and this is where you'll find the most variety in chemical compounds. Let's briefly cover these remaining heavier, and mostly unstable, elements of the periodic table starting with the worst radioactive offender of all.

Francium (Fr), while being named after the modern name of the Country of France, is otherwise a bad customer. Francium has an atomic number of 87, and an atomic mass of approximately 222 amu, which means that it has 87 protons, 87 electrons, and 135 neutrons in its most common form.

Francium (Fr)

Perey discovered francium in the year 1939. This element is classified as an alkali metal with an electron configuration of [Rn] $7s^1$.

Francium occurs naturally as uranium decays. It has no known uses, and is so tenuous that there is probably less than a few grams on the planet at any given time. Thus, relatively little is known about this element.

The Guru Says...

Francium is one mean element. It is so radioactive that half of any given sample will disintegrate into pure radiation (energy) in a mere 22 minutes. It is the heaviest known member of the alkali metals series, and is the *most unstable* of any of the naturally occurring elements.

Given this, it's really hard to know exactly what a lump of the stuff would look like anyway, although chemists speculate that it should be a golden metallic color based upon the periodic trends seen in earlier occurring elements, such as cesium. Francium can be produced artificially by bombarding thorium atoms with protons.

Radium (Ra) was named by its discoverer, Marie Skłodowska-Curie, after the Latin word for *'ray'*. She also coined the word *'radioactive'* (the cool-looking triangular radiation symbol that everyone has seen was invented later in 1946 at the University of California at Berkley). Radium has an atomic number of 88, and an atomic mass of approximately 223 amu, which means that it has 88 protons, 88 electrons, and 135 neutrons in its most common form.

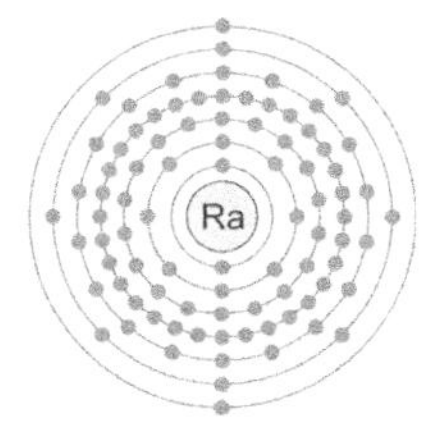

Radium (Ra)

Radium was discovered in 1898. This element is a white metal, classified as an alkali earth metal with an electron configuration of [Rn] $7s^2$.

Radium is another one of those elements that has a somewhat dubious past. It was originally believed that the so-called health benefits from mineral springs were due to the presence of radon gas, which in turn was generated by the element radium decaying; thus was born the business of early 20th century medical quackery, wherein it was believed that being exposed to radium was actually a healthful thing. Needless to say, radium has been responsible for many deaths due to shear ignorance. Even its discoverer, Madame Curie was a victim of exposure to this element.

Despite the bad karma associated with early exposure to this element, radium is used in various applications to produce reliable neutron sources, luminous paints, and medical radioisotopes and compounds to treat cancer. Radium occurs naturally as part of the uranium ore *Uraninite* (uranium oxide, UO_2).

The **Actinide (or Actinoid) series** is a collection of 15 metallic elements ranging from actinium through lawrencium. These elements, along with scandium and yttrium are also called *rare Earth elements*. Like the lanthanide series, these elements take a side trip from their Period 7 electron-filling scheme to fill electrons in the closest 'f' orbital shell (in this case the '5f' electron shell). In this regard, they again appear chemically similar to each other (with the added curiosity of being moderately to highly radioactive).

Elements up to uranium are generally considered to be naturally formed (although small quantities of neptunium and plutonium have been detected in nature, and are always located in trace amounts in uranium ore called *Pitchblende*). Elements above uranium are fabricated through manmade nuclear reactions. If you're looking for the stuff that makes atomic weapons work, or is produced as a byproduct of atomic power, then this is the collection of elements for you.

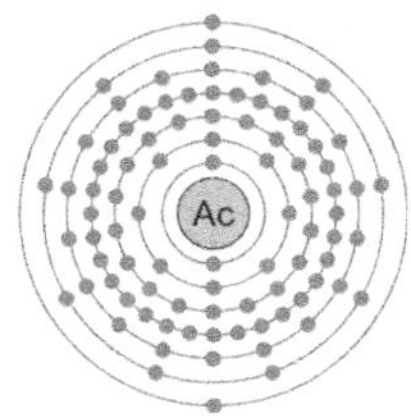

Actinium (Ac)

Actinium (Ac) has a name that is a copycat of Radium, just only in a different language. In Greek, *aktinos* also means 'ray'. Actinium has an atomic number of 89, and an atomic mass of approximately 226 amu, which means that it has 89 protons, 89 electrons, and 137 neutrons in its most common form. Actinium really doesn't have any current uses.

Debierne discovered Actinium in 1899. Physically, this element is a silvery metal. It is classified as a radioactive rare earth with an electron configuration of [Rn] $6d^1\ 7s^2$.

Thorium (Th) is named after the Nordic god of war – *Thor*. Thorium has an atomic number of 90, and an atomic mass of approximately 227 amu, which means that it has 90 protons, 90 electrons, and 137 neutrons in its most common form. Thorium has been used in the past as an alloying material, as a coating for spark plugs (due to its incredible hardness and corrosion resistance), and in the mantles for gas lanterns. This practice has been discontinued because thorium is – *radioactive*. Thorium is currently used today as a nuclear power source.

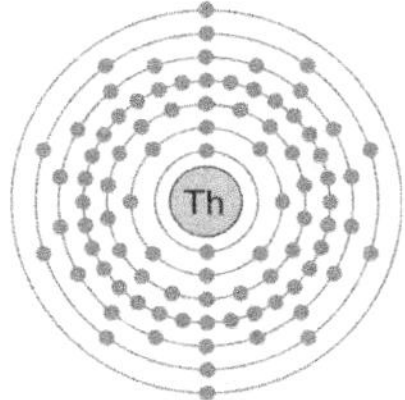

Thorium (Th)

Berzelius discovered Thorium in 1828. Physically, this element is a silvery-white metal. It is classified as a radioactive rare earth with an electron configuration of [Rn] $6d^2\ 7s^2$. Thorium is found principally in the radioactive ores thorium nitrate ($Th(NO_3)_4 \cdot 4H_2O$) and thorium fluoride ($ThF_4 \cdot 4H_2O$).

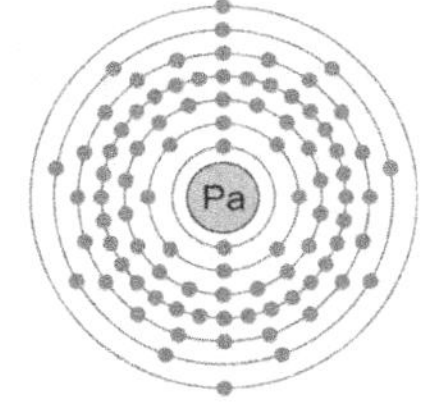

Protactinium (Pa)

Protactinium (Pa) has an atomic number of 91, and an atomic mass of approximately 232 amu, which means that it has 91 protons, 91 electrons, and 141 neutrons in its most common form. It was originally named *brevium* because its dominant form Pa_{234} had a short half-life. It was renamed with the Greek prefix protos (meaning 'first') because this element always decays into actinium, thus was born protoactinium. This became quite a mouthful to say, so in 1949 the name was shortened to protactinium.

Göhring discovered protactinium in the year 1913. Physically, this element is a silvery-white metal. It is classified as a radioactive rare earth with an electron configuration of [Rn] $5f^2$ $6d^1$ $7s^2$.

Protactinium has no practical uses due to its high radioactivity and chemical toxicity, although it is the rarest and most expensive of the naturally occurring elements. Protactinium is found in trace amounts with other radioactive ores, notably *Pitchblende*.

With the element **Uranium** (U), named after the planet Uranus, we have arrived at our last naturally occurring element – *generally speaking*. Anything above atomic number 92 has to be synthesized in an atomic reactor or high-speed neutron-accelerating device such as a *cyclotron* or *linear accelerator*.

Uranium (U)

Uranium has an atomic number of 92, and an atomic mass of approximately 237 amu, which means that it has 92 protons, 92 electrons, and 145 neutrons in its most common form.

Klaproth discovered uranium in 1789. Physically, this element is a metallic gray metal. It is classified as a radioactive rare earth with an electron configuration of [Rn] $5f^3$ $6d^1$ $7s^2$.

Uranium is used extensively as a nuclear fuel to power everything from spacecraft to aircraft carriers. Much of the internal heat radiating from Earth is thought to be due to the presence of uranium and thorium within the mantle, so there's actually quite a bit of the stuff here. Uranuim-238 (U_{238}) has a half-life of 4.5 billion years (roughly the current age of the sun), so it's also safe to say that uranium is a fairly stable element.

Uranium is a toxic radioactive metal, which is extremely ductile and slightly magnetic. Although requiring special handling, it is often used to harden and strengthen steel. Since uranium is approximately 70-percent denser than lead, it is used in military applications where compact weight is desired, such as armored plating for tanks, missile ballast, etc. Uranium salts have been used for producing colored glazes in glass and ceramic dishware (called Vaseline glass). The result is beautiful tableware that you can find at night during a power outage using only a Geiger counter.

The Guru Says…

Uranium used for purposes other than making a nuclear weapon is typically referred to as *'depleted uranium'*. Depleted uranium (U_{238}) is what's left over after this element is processed to remove the fissionable isotope (U_{235}).

Despite having the name 'depleted', depleted uranium is still very much a radioactive element.

Uranium occurs in various minerals, primarily including *Pitchblende* (a uranium-rich amalgam principally containing of *Uraninite*, uranium dioxide, UO_2), *Carnotite* (potassium uranium vanadate, $K_2(UO_2)_2(VO_4)_2 \cdot 3H_2O$), *Autunite* (hydrated calcium uranyl phosphate, $Ca(UO_2)_2(PO_4)_2 \cdot 12H_2O$), and *Uranophane* (calcium uranium silicate hydrate, $Ca(UO_2)_2(SiO_3OH)_2 \cdot 5H_2O$).

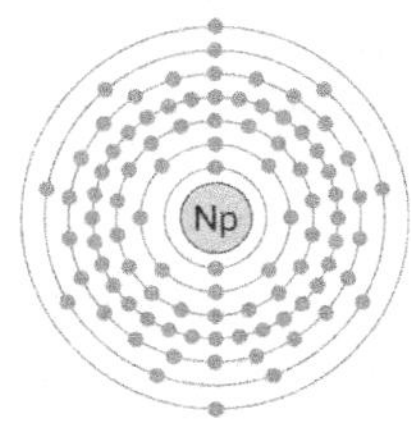

Neptunium (Np)

Neptunium (Np) was named after the planet Neptune, and was one of the first man-made (or transuranium) elements to be synthesized in a laboratory. In this case, bombarding uranium with neutrons from a cyclotron produced this new element. The resulting product was fairly stable with a half-life of 2.3 days. Neptunium has an atomic number of 93, and an atomic mass of approximately 238 amu, which means that it has 93 protons, 93 electrons, and 145 neutrons in its most common form.

McMillian and Abelson discovered/invented neptunium in the year 1940. Physically, this element is a silvery-gray metal. It is classified as a radioactive rare earth with an electron configuration of [Rn] $5f^4$ $6d^1$ $7s^2$.

Small amounts of neptunium have been discovered in naturally occurring uranium ore, hence the Guru's *'generally speaking'* caveat under the discussion of uranium. Other than being a byproduct of atomic power production, and maybe a fissionable material for weapons, it is primarily used in devices, which detect high-energy neutrons.

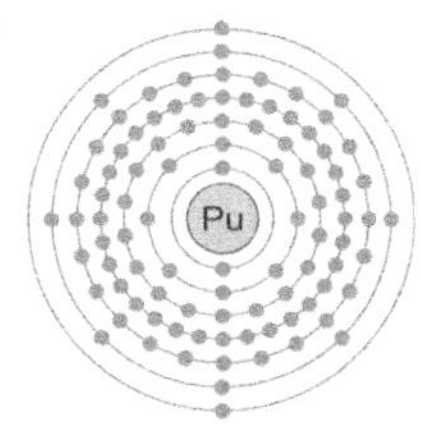

Plutonium (Pu)

Plutonium (Pu), named after the planet Pluto, was, in the 1970's, incorrectly declared by the mainstream media to be *'the deadliest of all the elements'.* Plutonium has an atomic number of 94, and an atomic mass of approximately 243 amu, which means that it has 94 protons, 94 electrons, and 149 neutrons in its most common form.

Seaborg, Kennedy, McMillan, and Wohl discovered plutonium in the year 1941 while exploring the various transuranium elements. Physically, this element is a silvery-white metal. It is classified as a radioactive rare earth with an electron configuration of [Rn] $5f^6$ $7s^2$.

While it is true that Plutonium is a radiological poison, the chemical toxicity of this particular element is rather low – no more than that of mercury, lead or cadmium. Rattlesnake venom is far more chemically toxic than plutonium could ever be.

The Guru Says...

Plutonium is the metal of choice if you're adament on constructing a doomsday weapon, however that would be a great waste of this remarkable material.

A large chunk of about five kilograms of plutonium will actually get hot enough to boil water and will do so, all by itself for 24,100 years before half the chunk is gone. In fact, plutonium is so energetic that one mere kilogram of the material can produce 22 million kilowatt-hours of heating energy, so clearly plutonium is an important power source.

The trick is just keeping the genie in the bottle, since too big of a piece of plutonium can spontaneously detonate into an atomic explosion. *This adds a little bit of an engineering challenge, huh..*

A fraction of a microgram of plutonium can kill you, but it's from the neutron radiation released by that small amount – which is quite considerable if all absorbed by internal organs (a highly unlikely scenario). The element polonium is actually far more of a radiotoxic substance; radium is about 200 times worse than that, and it was once used to paint the hands of glow-in-the-dark watches. Treat plutonium the same way you would treat fissionable uranium and you won't get into trouble.

Finally, **Americium** (Am) is named after America and is the key ingredient in common household smoke alarms. Americium has an atomic number of 95, and an atomic mass of approximately 244 amu, which means that it has 95 protons, 95 electrons, and 149 neutrons in its most common form. Other than being used as an ion source in smoke detectors and as a portable gamma radiation source for oil exploration and small x-ray machines, americium is pretty useless stuff.

Researchers Seaborg, James, Morgan, and Ghiorso discovered Americium in the year 1944. This element is a silvery-white metal. It is classified as a radioactive rare earth with an electron configuration of [Rn] $5f^7$ $7s^2$.

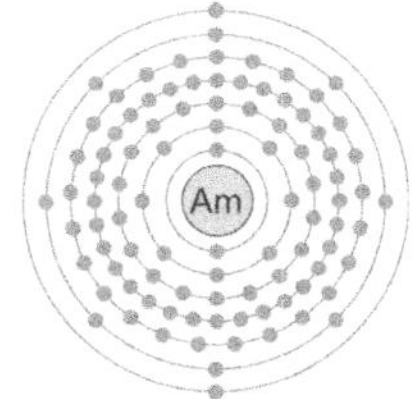

Americium (Am)

It is curious to note that americium naturally decays into, *ready, wait for it – plutonium*. You're either really scared of your smoke detectors at this point, or just had the epiphany that what you hear on television regarding radioactive material is over-sensationalized hype. The Guru is not impressed with hype; it's bad for the digestion.

The elements above americium start to become pretty much research projects with no practical application other than honoring a famous chemist or physicist by naming it after them. These *transuranium elements* are big and bulky, all manmade, and in many cases there have been only a couple of atoms of the stuff ever observed, so figuring out their physical properties is tenuous at best. These elements include *copernicium, ununtrium, flerovium, ununpentium, livermorium, ununseptium* (which is theoretically possible, but has never been synthesized), and *ununoctium* (number 118 discovered in 2006).

Now that you've made it through the crash course on cosmology, matter, chemistry, and some quantum physics, you now know more about what makes up the universe than roughly 99 percent of the population of the planet. It's now time to take a look at some of the basics of how things move about in the universe and ultimately come up with the same findings that Isaac Newton did in 1687. We'll only be about 300 years away from the modern world at that point.

Table 3.2: Various Physical Properties of the Elements

Name & Abbreviation		Atomic Number	Atomic Mass	Atomic Radius	Density (g/cm^3)	Melting Point (°C)	Boiling Point (°C)	Class Code
Hydrogen	H	1	1.00794	0.79 Å	0.09	-259	-253	NM
Helium	He	2	4.002602	0.49 Å	0.18	-272	-269	NG
Lithium	Li	3	6.941	2.05 Å	0.53	180	1,347	AM
Beryllium	Be	4	9.012182	1.4 Å	1.85	1,278	2,970	AE
Boron	B	5	10.811	1.17 Å	2.34	2,300	2,550	MD (NM)
Carbon	C	6	12.011	0.91 Å	2.26	3,500	4,827	NM
Nitrogen	N	7	14.00674	0.75 Å	1.25	-210	-196	NM
Oxygen	O	8	15.9994	0.65 Å	1.43	-218	-183	NM
Fluorine	F	9	18.9984	0.57 Å	1.7	-220	-188	H
Neon	Ne	10	20.1797	0.51 Å	0.9	-249	-246	NG
Sodium	Na	11	22.98977	2.23 Å	0.97	98	883	AM
Magnesium	Mg	12	24.305	1.72 Å	1.74	639	1,090	AE
Aluminum	Al	13	26.98154	1.82 Å	2.7	660	2,467	M
Silicon	Si	14	28.0855	1.46 Å	2.33	1,410	2,355	MD (NM)
Phosphorus	P	15	30.97376	1.23 Å	1.82	44	280	NM
Sulfur	S	16	32.066	1.09 Å	2.07	113	445	NM
Chlorine	Cl	17	35.4527	0.97 Å	3.21	-101	-35	H
Argon	Ar	18	39.0983	0.88 Å	0.86	-189	-186	NG
Potassium	K	19	39.948	2.77 Å	1.78	64	760	AM
Calcium	Ca	20	40.078	2.23 Å	1.55	839	1,484	AE
Scandium	Sc	21	44.95591	2.09 Å	3	1,539	2,832	TM
Titanium	Ti	22	47.88	2.0 Å	4.54	1,660	3,287	TM
Vanadium	V	23	50.9415	1.92 Å	6.11	1,890	3,380	TM
Chromium	Cr	24	51.9961	1.85 Å	7.19	1,857	2,672	TM
Manganese	Mn	25	54.93805	1.79 Å	7.43	1,245	1,962	TM
Iron	Fe	26	55.847	1.72 Å	7.87	1,535	2,750	TM
Cobalt	Co	27	58.6934	1.67 Å	8.9	1,495	2,870	TM
Nickel	Ni	28	58.9332	1.62 Å	8.9	1,453	2,732	TM
Copper	Cu	29	63.546	1.57 Å	8.96	1,083	2,567	TM
Zinc	Zn	30	65.39	1.53 Å	7.13	420	907	TM
Gallium	Ga	31	69.723	1.81 Å	5.91	30	2,403	M
Germanium	Ge	32	72.61	1.52 Å	5.32	937	2,830	MD (M)
Arsenic	As	33	74.92159	1.33 Å	5.72	81	613	MD (NM)
Selenium	Se	34	78.96	1.22 Å	4.79	217	685	NM
Bromine	Br	35	79.904	1.12 Å	3.12	-7	59	H
Krypton	Kr	36	83.8	1.03 Å	3.75	-157	-153	NG
Rubidium	Rb	37	85.4678	2.98 Å	1.63	39	688	AM

Table 3.2 (cont.): Various Physical Properties of the Elements

Name & Abbreviation		Atomic Number	Atomic Mass	Atomic Radius	Density (g/cm^3)	Melting Point (°C)	Boiling Point (°C)	Class Code
Strontium	Sr	38	87.62	2.45 Å	2.54	769	1,384	AE
Yttrium	Y	39	88.90585	2.27 Å	4.47	1,523	3,337	TM
Zirconium	Zr	40	91.224	2.16 Å	6.51	1,852	4,377	TM
Niobium	Nb	41	92.90638	2.08 Å	8.57	2,468	4,927	TM
Molybdenum	Mo	42	95.94	2.01 Å	10.22	2,617	4,612	TM
Technetium	Tc	43	98	1.95 Å	11.5	2,200	4,877	TM
Ruthenium	Ru	44	101.07	1.89 Å	12.37	2,250	3,900	TM
Rhodium	Rh	45	102.9055	1.83 Å	12.41	1,966	3,727	TM
Palladium	Pd	46	106.42	1.79 Å	12.02	1,552	2,927	TM
Silver	Ag	47	107.8682	1.75 Å	10.5	962	2,212	TM
Cadmium	Cd	48	112.411	1.71 Å	8.65	321	765	TM
Indium	In	49	114.818	2.0 Å	7.31	157	2,000	M
Tin	Sn	50	118.71	1.72 Å	7.31	232	2,270	M
Antimony	Sb	51	121.757	1.53 Å	6.68	630	1,750	MD (M)
Tellurium	Te	52	126.9045	1.42 Å	4.93	449	990	MD (NM)
Iodine	I	53	127.6	1.32 Å	6.24	114	184	H
Xenon	Xe	54	131.29	1.24 Å	5.9	-112	-108	NG
Cesium	Cs	55	132.9054	3.34 Å	1.87	29	678	AM
Barium	Ba	56	137.327	2.78 Å	3.59	725	1,140	AE
Lanthanum	La	57	138.9055	2.74 Å	6.15	920	3,469	LS
Cerium	Ce	58	140.115	2.7 Å	6.77	795	3,257	LS
Praseodymium	Pr	59	140.9077	2.67 Å	6.77	935	3,127	LS
Neodymium	Nd	60	144.24	2.64 Å	7.01	1,010	3,127	LS
Promethium	Pm	61	145	2.62 Å	7.3	1,100	3,000	LS
Samarium	Sm	62	150.36	2.59 Å	7.52	1,072	1,900	LS
Europium	Eu	63	151.965	2.56 Å	5.24	822	1,597	LS
Gadolinium	Gd	64	157.25	2.54 Å	7.9	1,311	3,233	LS
Terbium	Tb	65	158.9253	2.51 Å	8.23	1,360	3,041	LS
Dysprosium	Dy	66	162.5	2.49 Å	8.55	1,412	2,562	LS
Holmium	Ho	67	164.9303	2.47 Å	8.8	1,470	2,720	LS
Erbium	Er	68	167.26	2.45 Å	9.07	1,522	2,510	LS
Thulium	Tm	69	168.9342	2.42 Å	9.32	1,545	1,727	LS
Ytterbium	Yb	70	173.04	2.4 Å	6.9	824	1,466	LS
Lutetium	Lu	71	174.967	2.25 Å	9.84	1,656	3,315	LS
Hafnium	Hf	72	178.49	2.16 Å	13.31	2,150	5,400	TM
Tantalum	Ta	73	180.9479	2.09 Å	16.65	2,996	5,425	TM

Table 3.2 (cont.): Various Physical Properties of the Elements

Name & Abbreviation		Atomic Number	Atomic Mass	Atomic Radius	Density (g/cm^3)	Melting Point (°C)	Boiling Point (°C)	Class Code
Tungsten	W	74	183.85	2.02 Å	19.35	3,410	5,660	TM
Rhenium	Re	75	186.207	1.97 Å	21.04	3,180	5,627	TM
Osmium	Os	76	190.2	1.92 Å	22.6	3,045	5,027	TM
Iridium	Ir	77	192.22	1.87 Å	22.4	2,410	4,527	TM
Platinum	Pt	78	195.08	1.83 Å	21.45	1,772	3,827	TM
Gold	Au	79	196.9665	1.79 Å	19.32	1,064	2,807	TM
Mercury	Hg	80	200.59	1.76 Å	13.55	-39	357	TM
Thallium	Tl	81	204.3833	2.08 Å	11.85	303	1,457	M
Lead	Pb	82	207.2	1.81 Å	11.35	327	1,740	M
Bismuth	Bi	83	208.9804	1.63 Å	9.75	271	1,560	M
Polonium	Po	84	208.9824	1.53 Å	9.3	254	962	MD (M)
Astatine	At	85	209.9871	1.43 Å	?	302	337	H
Radon	Rn	86	213.0359	1.34 Å	9.73	-71	-62	NG
Francium	Fr	87	222	?	?	27	677	AM
Radium	Ra	88	223	?	5.5	700	1,737	AE
Actinium	Ac	89	226.0254	?	10.07	1,050	3,200	AS
Thorium	Th	90	227.0728	1.88 Å	15.4	1,750	4,790	AS
Protactinium	Pa	91	232.0381	?	11.72	1,568	?	AS
Uranium	U	92	237.0482	?	20.2	1,132	3,818	AS
Neptunium	Np	93	238.0289	?	18.95	640	3,902	AS
Plutonium	Pu	94	243.0614	?	13.67	640	3,235	AS
Americium	Am	95	244.0642	?	19.84	994	2,607	AS
Curium	Cm	96	247	?	13.5	1,340	?	AS
Berkelium	Bk	97	247	?	14.78	986	?	AS
Californium	Cf	98	251	?	15.1	900	?	AS
Einsteinium	Es	99	252	?	?	860	?	AS
Fermium	Fm	100	257	?	?	1,527	?	AS
Mendelevium	Md	101	258	?	?	?	?	AS
Nobelium	No	102	259	?	?	827	?	AS
Lawrencium	Lr	103	260	?	?	1,627	?	AS
Rutherfordium	Rf	104	261	?	?	?	?	TA
Dubnium	Db	105	262	?	?	?	?	TA
Seaborgium	Sg	106	262	?	?	?	?	TA
Bohrium	Bh	107	263	?	?	?	?	TA
Hassium	Hs	108	265	?	?	?	?	TA
Meitnerium	Mt	109	266	?	?	?	?	TA

Table 3.2 (cont.): Various Physical Properties of the Elements

Name & Abbreviation		Atomic Number	Atomic Mass	Atomic Radius	Density (g/cm^3)	Melting Point (°C)	Boiling Point (°C)	Class Code
Darmstadtium	Ds	110	271	?	?	?	?	TA
Roentgenium	Rg	111	272	?	?	?	?	TM
Copernicium	Cn	112	?	?	?	?	?	TM
Ununtrium	Uut	113	?	?	?	?	?	M
Flerovium	Fl	114	?	?	?	?	?	M
Ununpentium	Uup	115	?	?	?	?	?	M
Livermorium	Lv	116	?	?	?	?	?	M
Ununseptium	Uus	117	?	?	?	?	?	M
Ununoctium	Uuo	118	?	?	?	?	?	M

Element Classification Codes:

- NM = Nonmetal
- M = Metal
- MD = Metalloid
- NG = Noble Gas
- AM = Alkali Metal
- AE = Alkali Earth Metal
- H = Halogen
- TM = Transition Metal
- LS = Lanthanide Series Rare Earth
- TA = Transactinide Element
- AS = Actinide Series Radioactive Rare Earth

Section II – The Nuts and Bolts of Physics

Chapter 4: Motion, Momentum, and Friction

How things move about and why they stop

The universe is not a static place. Things are moving, and most of the time, moving at a pretty good speed. Consider this: in the time it takes you to read this sentence, a photon of light can easily travel from Earth to the moon and back – *twice!!!* Earth moves along its orbital path through space at a speed of 107,300 km/hr; so in 2.8 minutes you have traveled the equivalent distance through space as a trip across the continental United States (about 5,000 km).

Humans walk at speeds approaching 6.5 km/hr, the moon moves through the sky at over 3,600 km/hr (1.01 km/s), most large rivers move at about 4.5 km/hr, a skydiver can reach speeds of roughly 300 km/hr in a free-fall, and sound waves travel through air at 1,220 km/hr at sea level. Indeed, everything in the universe is moving, in all directions, at the same time. The study of this motion is known as *kinematics*.

Kinematics – The Study of Motion

Kinematics is the study of the motion of objects, and systems of objects (called bodies), without consideration of what causes the motion (i.e., we do not consider any forces). Kinematics is referred to as the *geometry of motion*, since we are only interested in the path that motion takes, how far it travels, how fast it moves, and how it moves under constant or variable accelerations (*Sssshhhhhh – don't let Isaac Newton know about this acceleration part, since he showed that acceleration and force are effectively the same thing*).

We engage in kinematics everyday without thinking much about it. Go ride a bike, hit a baseball, jump over a puddle of water in the street, dive into a swimming pool, or watch a plane fly overhead, and you are subconsciously invoking or analyzing kinematic principles without knowing it. The science behind kinematics, and the associated mathematics are as simple as pie (well, pie sprinkled with a little differential calculus – *but hey, a little hot coffee will help that go right down*).

A buddy of the Guru, who is in the U.S. Army, shoots objects out of cannons as his day job. He calls himself a *mortarman*; the Guru says that he's a *kinematics specialist*, since all he is doing is figuring out how far something will travel without really caring about the actual forces involved. The army has fancy charts and tables to show how to do this; the Guru will demonstrate this by using physics.

We start by invoking the ghost of René Descartes and looking at his coordinate system again, this time in two dimensions. We are only going to concern ourselves with motion in a plane, knowing as clever physicists that we can easily expand what we are doing to three dimensions with relative ease and the addition of one more term on our vector. Recycling our coordinate system from Chapter 2, and drawing an arbitrary path on it, we can see the general problem statement in Figure 4.1.

Suppose we have a rubber ball that we toss into the air and it follows the path shown by the dashed line. We can define the position of the ball at any time by the vector $\overline{R(t)}$, called *'vector R of t'* or the vector 'R' as a function of time.

As time 't' changes, so does the position vector 'R'. The actual value for $\overline{R(t)}$ isn't really all that important since we're going to derive some independent expressions for it shortly.

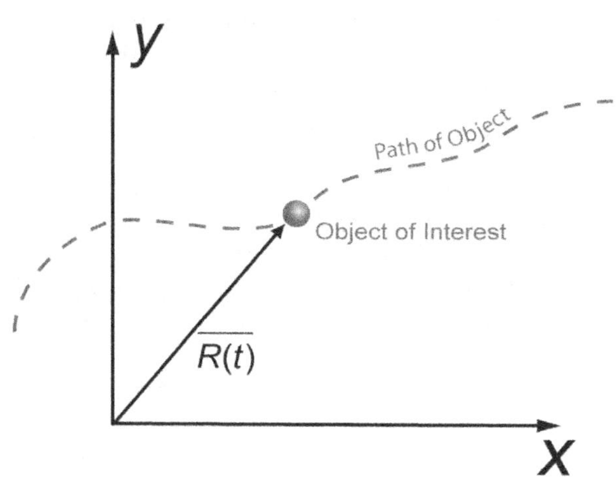

Figure 4.1: The Two-Dimensional Motion of an Object in Space

Notice that $\overline{R(t)}$ defines motion both in the 'x' as well as the 'y' direction (and the 'z' direction if we cared), and that these two different motions are independent of each other. So technically, the path of the rubber ball is actually a function of 'x', 'y' and 't'.

Since it becomes cumbersome to start working with multi-variable expressions, we will split the 'x' and 'y' motions apart and treat them separately with the combined solution forming the two different parts of our vector $\overline{R(t)}$, namely $\overline{R(t)}$ is going to have the form $x\hat{i}+y\hat{j}$ or <*x*, *y*> where 'x' and 'y' are the scalar components of the vector (treated respectively as the motion in the 'x' and 'y' directions).

Even though the 'x' and 'y' motions of the ball appear to be independent of each other, remember, they are still tied together. What synchronizes the two motions is the *time 't'*. The 'x' motion is in terms of time 't' and so is the 'y' motion. Thus, we would expect that our solution for these two dependent parameters 'x' and 'y' would solely depend on the independent parameter 't'. Written another way, our vector should look like $x(t)\hat{i}+y(t)\hat{j}$ or <*x(t)*, *y(t)*>. Now the problem becomes one of finding out what these two expressions look like as a function of time. To do this, we'll start with what we know is physically going on, and work backwards from there.

Kinematic Motion in the Vertical Direction

If an object is thrown into the air, it will eventually fall back down according to how gravity behaves by pulling it towards the center of Earth. The object might be subject to some air drag, or sideways wind gusts, but in all cases, it will fall back down in roughly a straight line.

A simple physical proof of this can be seen in high-altitude parachute jumps, where no matter how close to the edge of space the parachutist jumps from his or her balloon, he or she always lands pretty much near the point of departure. Take the following event as an example.

Problem Solving Tip...

The solution to all kinematics problems involves *solving for the time 't'*, known as the independent variable. Once you know how long it takes for the event to occur, you can then solve for the dependent variables 'x' and 'y'.

On December 10, 1963, Colonel Chuck Yeager took an experimental Lockheed Starfighter NF-104 rocket-assisted airplane to an unheard of {at the time} altitude 37 km. This basically took him straight up to the edge of outer space. Then, his engines *conked-out*. Through all the spinning, and twisting, and upside-down side-sliding that Colonel Yeager experienced under the action of gravity, he wound up bailing out about two kilometers above the ground just a couple of kilometers downrange from where he took off (and yes, the plane crash was spectacular). Gravity acting in a fixed reference frame has a tendency to bring things back roughly to where they started.

So, to model the gravity 'g' of our ball, we note that the ball must fall down – *this is the observable physics of the problem.* Since 'down' is opposite of our 'y' coordinate direction shown in Figure 4.1, we must denote this gravitational action with a minus sign. Therefore, we can write the acceleration due to gravity as,

$$a_{gravity} = -g$$

This is the statement of the physics of the problem, namely, *gravity acts downward towards Earth*. That's it. Pretty simple, huh. The rest of the work will now be to apply mathematics to figure out the relationships for the velocity of the falling ball, and ultimately its position, both as a function of time.

We know that if we accelerate something, its tendency is to move faster and faster (i.e., it increases its speed, or in the vector sense, its velocity increases). The accepted acceleration on Earth due to gravity in SI units is 9.81 meters-per-second-squared (m/s^2). So we know, just from looking at the units that if we multiply acceleration (in m/s^2) times some arbitrary increment of time Δt (in seconds) we'll wind up with units of velocity, namely,

$$Velocity_{y-direction} = a_{gravity} \cdot \Delta t = -g \cdot \Delta t$$

If we assume our time interval begins at time 't=0', we can then simplify this expression by noting that $\Delta t = t - 0 = t$, giving us,

$$V_y = -g \cdot \Delta t = -gt$$

But, what if the object had some arbitrary initially positive starting velocity, say V_{y0}, what then? We would then need to modify our expression to include this additional term.

Thus,

$$V_y = -gt + V_{y0} = V_{y0} - gt$$

This is equivalent to saying that the ball has some initial upward (i.e., thrown upward) velocity of V_{y0} and is slowed down and pulled back to Earth, by the gravitational term $-g \cdot t$. The above equation is our kinematic motion expression for the vertical velocity of a freefalling object acting under the influence of gravity. If we were to throw the ball down from, say an airplane, our initial velocity would be in the same {negative} direction as gravity and we would simply place a minus sign in front of V_{y0}. Let's look at an example.

Example 4.1: The rubber ball shown in Figure 4.1 is thrown into the air with an initial upwards velocity of 5 m/s. At what time does the ball reach its highest point?

We can write our expression for the motion of the ball as follows,

$$\begin{aligned} V_y &= V_{y0} - gt \\ &= 5\text{ m/s} - 9.81\text{ m/s}^{\cancel{2}} \cdot t\ \cancel{\text{s}} \quad \text{Remember, 't' has units of seconds.} \\ &= 5\text{ m/s} - 9.81 \cdot t\text{ m/s} \end{aligned}$$

When does the ball reach its highest point? It reaches the highest point in its path when its velocity goes to zero (and consequently starts to fall back down in the other direction). Thus, we will set V_y equal to zero and solve for 't' giving,

$$0 = 5\text{ m/s} - (9.81\text{ m/s}) \cdot t$$

$$(9.81\ \cancel{\text{m/s}}) \cdot t = 5\ \cancel{\text{m/s}}$$

$$9.81 \cdot t = 5$$

$$\mathbf{t = 0.5097\ s}$$

We know that if we toss a ball into the air with an initial velocity of 5 m/s, it would take slightly over one-half second for it to come to a stop and start falling back to Earth. This is an important concept to know since quite a few *time-of-flight* problems require the solution for the time 't' at a specific point to determine some attribute of the path.

Continuing with our derivation process, we are now desirous of knowing how far an object will fall under the action of gravity. Looking back at our previous expression for the velocity of the object we have,

$$V_y = V_{y0} - gt$$

Again we note that if we multiply velocity (in m/s) times time (in seconds) we will wind up with units of distance. Thus, we can write our expression for the 'y' distance that the ball would travel as,

$$y \approx (V_{y0} - gt) \cdot \Delta t = V_{y0} \cdot \Delta t - gt \cdot \Delta t$$

Fooling around with the units will only get you so far, and we've written our solution in the approximate form shown above.

Our units are correct (i.e., all of them are in units of distance), but we are still missing a term to account for the first time 't' in our original expression. Take a look at Figure 4.2 where we show the ball moving from two different points in the path t_0 to t_e while the height is changing from y_0 to y_e.

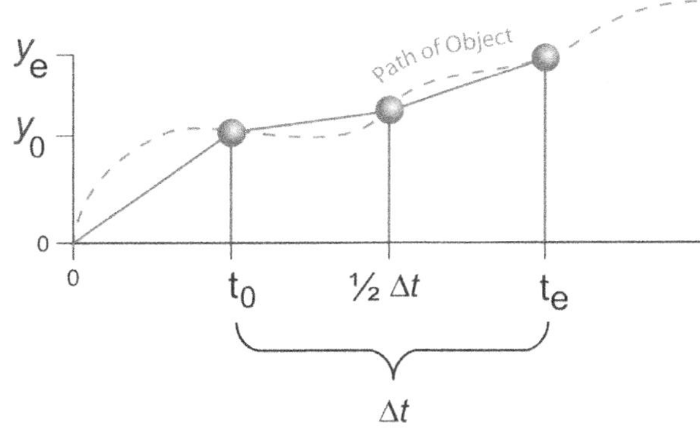

Figure 4.2: The Concept of a Finite Time Difference Between Two Points in Space

The question becomes – *at what time do we evaluate 't'?* Do we evaluate our expression at t_0 or t_e ?

Well, the full explanation to this would drop us into a long-winded explanation of integral Calculus, which the Guru is trying to avoid for the purposes of any derivation within this book. So, we'll split the difference and take the midpoint between the times at t_0 and t_e, or $1/2 \cdot \Delta t$. Our choice of this point is not arbitrary, it is actually the correct choice that you would discover using a Calculus based approach, since we are assuming that the path between t_0 and t_e can be approximated by the midpoint value of 't' (which if the interval was infinitesimally small would indeed be the exact answer).

Substituting this value for 't' into our previous expression, the 'y' motion of the ball is given as,

$$y = V_{y0} \cdot \Delta t - g \cdot \frac{1}{2} \Delta t \cdot \Delta t = V_{y0} \cdot \Delta t - \frac{1}{2} g \Delta t^2$$

$$= V_{y0} t - \frac{1}{2} g t^2$$

As before, we need to make some sort of provision that the ball may have some nonzero starting height, say y_0. We would then need to modify our expression to include this term by stating the final answer as,

$$y = y_0 + V_{y0} t - \frac{1}{2} g t^2$$

This expression is our kinematic motion equation for the vertical position of an object acting under the influence of gravity.

Example 4.2: We now wish to know how high up into the air the rubber ball of Example 4.1 will travel. Assume that the person throwing the ball in the air releases it from his hand at a height of 1.5 meters.

From our previous example we know that the ball stops its upward travel in 0.5097 seconds, and starts falling downward. At this point where the ball stops in mid air, it has reached its maximum height above the ground. So, substituting the values from the example gives,

$$y = 1.5\text{ m} + (5\text{ m/s})\cdot t - \frac{1}{2}(9.81\text{ m/s}^2)\cdot t^2$$

$$= 1.5\text{ m} + (5\text{ m/s})\cdot(0.5097\text{ s}) - \frac{1}{2}(9.81\text{ m/s}^2)\cdot(0.5097\text{ s})^2$$

$$= 1.5\text{ m} + 2.5485\text{ m} - 1.274\text{ m}$$

$$\mathbf{= 2.78\text{ m}}$$

So the ball would travel roughly 2.8 meters into the air before coming to a stop, reverse direction, and start to fall back to Earth. As you can see, a lot of different motion goes on through the simple act of tossing a ball into the air. Let's look at another example.

The Guru Says...

Since the variable time appears in each of the expressions above, we can perform some additional algebraic substitution and completely eliminate the time variable to obtain an alternate relationship linking 'x', 'v', and the gravitational acceleration 'g' together. This works out to be $v^2 = v_0^2 + 2\cdot g\cdot x$, and for any general acceleration, denoted by 'a', as $v^2 = v_0^2 + 2\cdot a\cdot x$.

Example 4.3: The Guru is walking on a suspension bridge across a canyon and is curious to know how high up he is. He didn't bring a tape measure, but found a small rock and dropped it over the edge (first making sure no one was directly below). The rock fell for 2.6 seconds. Approximately how high up is he?

Neglecting air friction, we can easily solve for the vertical falling height using our kinematic expression. We must note in this case, that the distance traveled is going to be entirely due to gravitational acceleration with no initial velocity, and a zero elevation point set to the height of the bridge, thus making everything positive as it falls (go back to Figure 4.1 and flip the coordinates upside down to prove this to yourself).

Given this, our expression reduces to,

$$y = \frac{1}{2}\left(9.81\,\text{m/s}^2\right)\cdot t^2$$
$$= \frac{1}{2}\left(9.81\,\text{m/s}^2\right)\cdot\left(2.6\text{ s}\right)^2$$
$$\mathbf{= 33.2\ m}$$

See, you can use this stuff for everyday practical applications. Who needs a tape measure when you have a rock and a stopwatch handy?

Kinematic Motion in the Horizontal Direction

Deriving the expressions for the kinematic motion in the horizontal, or 'x' direction, is now a snap since they have exactly the same form as the 'y' direction expressions with a few simple changes in terms. Let's start with the expression for the velocity in the 'x' direction. We'll note that gravity does not act in the horizontal direction, (it only acts vertically) and therefore we need to write any constants for acceleration in a more generic way, like with the letter *'a'* as,

$$V_x = V_{x0} + a \cdot t$$

Where we are now assuming that the acceleration is in the positive (to the right) sense as shown in Figure 4.1. This is similar to what would be produced by a rocket under powered flight.

Likewise, there may be initial horizontal velocity and starting position terms that we should probably include just to be safe.

The Guru Says...

Humans dating back to the Romans have been interested in how to lob a projectile under various circumstances and accurately hit someone or something. It is the subject of a category of analysis called *ballistics*.

In fact, the first electronic computer ever invented, named ENIAC for **E**lectronic **N**umerical **I**ntegrator **a**nd **C**alculator, was dedicated to the single task of calculating those artillery firing tables we mentioned earlier.

Thus, we can write our kinematic equation for the horizontal motion of an object as,

$$x = x_0 + V_{x0}t + \frac{1}{2}at^2$$

This equation would describe the independent horizontal motion of an object as a function of time. Each equation can be solved independently (if you know the time of interest) with very little difficulty.

The above process becomes a little more complicated when we combine the 'x' and 'y' motions together into a problem and start asking for the overall motion in terms of each of these terms. Take a look at the following example to see what we're talking about.

Example 4.4: A rocket is observed by radar to be traveling horizontally (downrange) with an initial velocity of 250 m/s, and is accelerating at 5 m/s^2. Its vertical (climb) velocity is initially 100 m/s with a rate of acceleration of 15 m/s^2.

- At what time is the rocket over the next radar station 100 km away?
- What is its change in height at this point?
- What does the motion of the rocket look like?

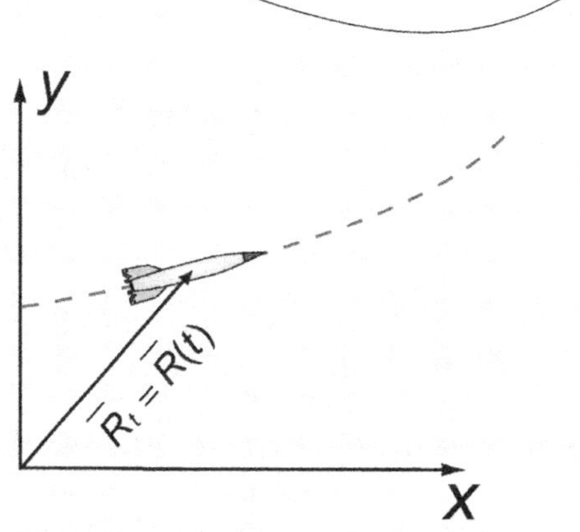

You should already have a feel for how to set up this problem based upon what we've done so far. We need to get a little clever though and realize that our rocket is also accelerating upwardly, so we need to add an additional term 'a_y' to our vertical kinematics equation, to compensate for this motion.

Let's start by setting up the equations for our horizontal and vertical components of motion and plugging in what we know.

Horizontal Motion

$$x = x_0 + V_{x0}t + \frac{1}{2}a_x t^2$$
$$= 0 + (250)t + \frac{1}{2}(5)t^2$$
$$= 250t + 2.5t^2$$

Vertical Motion

$$y = y_0 + V_{y0}t + \frac{1}{2}(a_y - g)t^2$$
$$= 0 + (100)t + \frac{1}{2}(15 - 9.81)t^2$$
$$= 100t + \frac{5.19}{2}t^2$$

We have two semi-independent expressions describing the motion of the rocket. They are linked together by the variable 't' for time.

Our horizontal motion is a function of the initial velocity V_{x0} and any horizontal acceleration due to the rocket motor a_x. Similarly, the vertical motion is a function of the initial vertical velocity V_{y0} and the tug-of-war between the rocket thrust a_y and gravity trying the crash it into the ground. Thus, our first task is going to be to figure out at what time does the rocket reach its downrange destination of 100 km.

Since the predicted motion is downrange, we will use the horizontal kinematic equation, above, and substitute a value of 100 km (100,000 m) for 'x'.

$$100\ \text{km} \cdot \frac{1000\ \text{m}}{1\ \text{km}} = 250 \cdot t + 5 \cdot t^2$$

$$100 \cdot 1000 = 250 \cdot t + 5 \cdot t^2$$

$$100000 = 250 \cdot t + 5 \cdot t^2$$

We discover from this that our solution is in terms of 't' and 't^2', so we need to invoke some previous knowledge from algebra to figure out a solution. Rewriting the above expression into standard polynomial form we have,

$$5t^2 + 250t - 100000 = 0$$

Since our polynomial is of order two (i.e., the highest exponent value of 't' is 't^2'), we know that there are two possible solutions for when the rocket is 100 km away. Our polynomial looks like $at^2 + bt + c = 0$ with a = 5, b = 250, and c = -100000. Recalling the quadratic formula, we can write the solutions for 't' in one step as,

$$t = \frac{-b \pm \sqrt{b^2 - 4 \cdot a \cdot c}}{2 \cdot a}$$

Problem Solving Tip...

The quadratic equation will let us solve for the roots of any polynomial of order two. The trick is to write the polynomial into the standard form shown above and order the variables from highest to lowest, then you can assign the coefficients a, b, and c and solve.

Of course, the clever human will, after doing this a couple of times by hand, run out and get a calculator that has a root solver.

Upon substitution of the problem variables this becomes,

$$t = \frac{-250 \pm \sqrt{(250)^2 - 4 \cdot 5 \cdot (-100000)}}{2 \cdot 5}$$

$$= \frac{-250 \pm \sqrt{62500 + 2000000}}{10}$$

$$= \frac{-250 + 1436.14}{10} \quad or \quad \frac{-250 - 1436.14}{10}$$

$$= \mathbf{118.6\ s} \quad or -168.6\ \text{s}$$

We get two possible answers for the time of flight of the rocket, namely, 118.6 seconds, and minus 168.6 seconds. Both are valid solutions. The first one (+118.6) is the answer we are looking for. The second solution (-168.6) is the other solution for x = 100 km, assuming that the rocket was going the other way and had to counteract the initial positive velocity. As a general rule, when solving time of flight problems, *solutions yielding negative time increments are the ones you toss away.*

Now, we are also asked how high the rocket would be when it is 100 km downrange (i.e., 118.6 seconds away). Let's look at that.

Substituting our time of flight into the vertical equation of motion, we get,

$$y = 100 \cdot t + \frac{5.19}{2} \cdot t^2$$

$$= 100 \cdot (118.6) + \frac{5.19}{2} \cdot (118.6)^2 = 11860 + 36501.17 = 48361.17$$

≈ 48.4 km

Finally, we should note that the geometric solution of $at^2 + bt + c = 0$ is a **parabola**. *All kinematic problems under the action of gravity will trace out a parabolic path in space.*

Our solution is an ideal case neglecting aerodynamic frictional forces on the object, such as 'drag', and changes in the rocket mass as it burns fuel. For our purposes though, this solution will get us 95+ percent of the way there without too much mathematical complexity.

Vector Representation of Kinematic Motion

We can play the same trick in applying vectors to our kinematic expressions, as we did with scalars, and have all the capabilities of vector notation at our disposal. Recall that we could express a position vector as a function of time by $x(t)\hat{\imath} + y(t)\hat{\jmath}$, so that the general form our vector would look like is,

$$\overline{R_t} = \overline{R(t)} = \left(x_0 + V_{x0}t + \frac{1}{2}a_x t^2 \right)\hat{\imath} + \left(y_0 + V_{y0}t - \frac{1}{2}(a_y - g)t^2 \right)\hat{\jmath}$$

For our previous rocket example, at a time corresponding to 118.6 seconds, we would have a position vector of $R_{t=118.6} = 100\hat{\imath} + 48.4\hat{\jmath}$ km. If we were to calculate the magnitude of our vector $\overline{R_t}$ we would have,

$$|R_t| = \sqrt{100^2 + 48.4^2} = 111.1 \text{ km}$$

This distance is called the *slant range distance* of the rocket. Similarly, we could write an expression describing the velocity vector V_t as a function of time as,

$$V_t = (V_{x0} + a_x t)\hat{\imath} + (V_{y0} + a_y t - gt)\hat{\jmath}$$

Substituting what we know from the previous example, the velocity of the rocket at any given time is,

$$V_t = (V_{x0} + a_x t)\hat{\imath} + (V_{y0} + a_y t - gt)\hat{\jmath} \text{ m/s}$$

$$= (250 + 5 \cdot t)\hat{\imath} + (100 + 15 \cdot t - 9.81 \cdot t)\hat{\jmath} \text{ m/s} = (250 + 5 \cdot t)\hat{\imath} + (100 + 5.19 \cdot t)\hat{\jmath} \text{ m/s}$$

Which at our 100 km downrange time of t = 118.6 seconds, would yield the velocity vector of,

$$V_{t=118.6} = (250 + 5 \cdot 118.6)\hat{i} + (100 + 5.19 \cdot 118.6)\hat{j} \text{ m/s}$$
$$= 843.0\hat{i} + 716.8\hat{j} \text{ m/s}$$

This vector has a magnitude of $V_t = \sqrt{843^2 + 716.8^2} = 1{,}106.5$ m/s, which the reader is encouraged to verify. A visual illustration of the complete vector solution to the rocket problem is shown below in Figure 4.3.

Thus, we've taken some simple kinematic notation and developed a powerful vector representation for the motion of an object in space under the influence of gravity. Open up any book on astrodynamics, orbital mechanics, or ballistics and you'll see the same type of notation used – *which you should now have an understanding of.*

It's now time to move on to another topic related to the motion of particles in space – *momentum*.

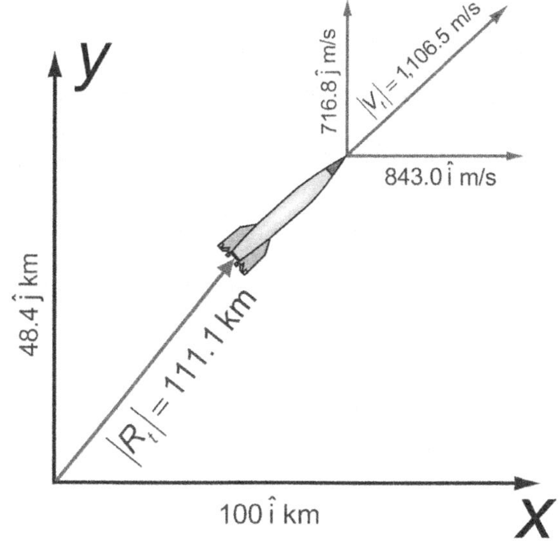

Figure 4.3: Visual Solution to the Rocket Problem from Example 4.4

Momentum and Its Conservation

So far all we've looked at is how an object moves through space in a kinematic sense with very little regard to anything else. In physics, this type of motion is called *particle motion*.

A particle is nothing more than an infinitesimally small point, a *massless* thing actually. In our rocket example we took some liberty by giving the point a tangible form, but you can actually shrink any object mathematically down to a particle and describe its motion with no loss in generality. We will see later on that there is actually a special point in space where we can convert any object, whether it be a rocket or a battleship or a galaxy, to a single mathematical 'dot', and analyze the motion according to the general kinematic principles we've learned in this chapter. That point is called the *center of mass*, or *center of gravity*, and it is an important concept when we get to things that spin about an axis.

As you can see, we can do a lot of tinkering in the universe by only looking at how things move about. Ultimately though, we want to get to the real *nuts-and-bolts* of the universe and be able to manipulate the very forces of nature that actually do the moving. To do this, we need to introduce a new buzzword called *momentum*.

Momentum is the '*quantity of motion of a moving body*'. It is loosely defined as *'mass in motion'*, since the thought process is that once something constructed of matter starts to move in space, important things start to happen. Momentum, or more specifically *linear momentum*, is defined as the product of mass times velocity. Since mass is a scalar and velocity is a vector, we know that momentum is inherently a vector quantity. In the SI system, momentum would be observable with the units of kilogram-meter-per-second (kg-m/s).

Momentum also has another important property in the universe – *it's conserved.* This means that in a *closed-system* the sum of all the individual momentums of the objects in the system is always the same value, regardless of how the objects interact.

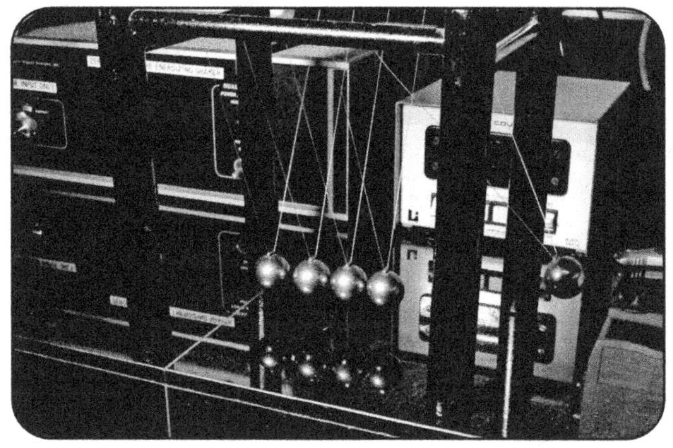

Figure 4.4: Photo of the Guru's Newton's Cradle

Momentum is observed in physics as the <u>effect</u> of force acting on a body, and not the actual cause of the motion itself. It is a physical representation of how the motion of something is perceived. If the momentum of a system is zero, the system is not moving. If it is non-zero, then something has to be moving – *it's as simple as that.*

There's a physics device called a *Newton's Cradle* (Figure 4.4), which consists of five identical steel balls suspended by flexible wires from a wooden frame. When you pull back on one of the steel balls and then release it, it will swing to hit the ball next to it – *but will not move it.* Instead its momentum will be transferred to the third ball, then the fourth, and finally the fifth ball.

Since the there is no sixth ball (and thus no mass to transfer this momentum into), the fifth ball does the next best thing and moves away from the rest of the balls with a velocity equal to the velocity that the first ball possessed when it initially struck the system.

In an ideal world this process would occur forever, but since there is friction in the system, and air drag, and the balls make noise when they 'clack' against each other, energy is lost from the closed-system, and things slow down until they eventually come to a stop.

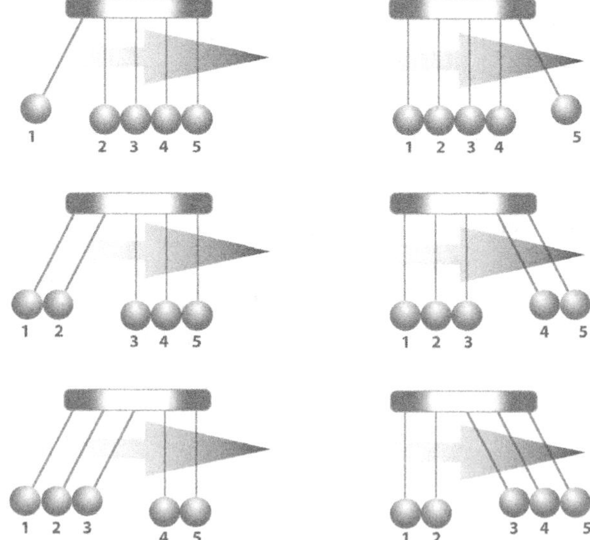

Figure 4.5: Newton's Cradle Demonstration for One, Two, and Three Masses being Released

Take a look at Figure 4.5, where we show some examples of what happens when we release one, two or three of the identical steel balls at the same time. As we can see, releasing one ball (ball #1) makes the last ball (ball #5) fly off the end since momentum is conserved (i.e., the momentum of this system is a constant value, equal to the mass of ball #1 times the speed it has just before it hits ball #2). If the mass of balls #1 and #5 are the same and the product of $m \cdot v$ is conserved, then the speed of ball #5 must be equal to the original 'v'.

This concept is not limited to just one ball. The second inset diagram of Figure 4.5 shows what happens when we release balls #1 and #2 simultaneously. In this example we have '2m' and 'v', thus we can expect balls #4 and #5 to swing off the end. Notice that momentum definitely keeps track of the values of 'm' and 'v'. We would never see them interchanged where we would have the '2m' and 'v' swap places to yield 'm' and '2v' on the other side of the cradle. Similarly, if we drop three balls, we would notice that ball #3 would keep moving non-stop back and forth in order to conserve the mass of the system. The other two balls (#1 and #2, or #4 and #5) would alternate motion, as the velocity vector changes directions.

The Guru Says...

A closed-system is anything of interest that you can put a mental or mathematical boundary around. For example, a pool table is a closed system and the momentum of all the billiard balls on the table is conserved. The same applies for the electrons in an atom, the stars in a galaxy, and so on. Go ahead and think up some closed systems – it's fun!

Finally, we should note that since momentum is defined as the linear product of mass and velocity, $m \cdot v$, this means that for any fixed quantity of momentum, there are theoretically an infinite number of combinations of mass (m) and velocity (v), which mathematically represent the same thing. A fast moving, low mass object can in some cases carry just as much momentum as a slow moving, high mass object (and have the ability to do the same amount of damage in a collision). Let's look at this concept in the following example to get a feel for the scale of the numbers.

Problem Solving Tip...

If we were to replace balls #1 and #2 in this example with a single ball having a mass of '2m', what do you think would happen? Would balls #4 and #5 again fly out, or would something quite different occur?

Well, this problem is actually a little more complex than you would think and requires us to know something about the kinetic energy of the system in order to solve for a solution.

For the time being lets say that balls #4 and #5 would fly out again, but not at the same velocities (#5 would move twice as far as #4). It is very much a different problem.

Example 4.5: A Nolan Ryan fastball has been clocked at up to 170 km/hr. If the baseball weighs 142 grams, how fast must a 20 kg anvil be falling to have the same momentum?

This problem involves a couple of unit conversions and some simple multiplication. Thus, we start by converting the baseball weight into kilograms.

$$142\,\not{g} \cdot \frac{1\,\text{kg}}{1000\,\not{g}} = 0.142\,\text{kg}$$

So, the momentum of a Nolan Ryan fastball across home plate would be the product of the mass of the ball and its velocity.

$$M_{fastball} = 0.142\,\text{kg} \cdot 170\,\frac{\text{km}}{\text{hr}} = 24.14\,\frac{\text{kg} \cdot \text{km}}{\text{hr}}$$

Now we just need to equate the momentum of the baseball to the 20 kg anvil and solve for the equivalent velocity.

$$M_{fastball} = M_{anvil}$$

$$24.14\,\frac{\text{kg} \cdot \text{km}}{\text{hr}} = 20\,\text{kg} \cdot v_{anvil}\,\frac{\text{km}}{\text{hr}}$$

$$\frac{24.14\,\frac{\cancel{\text{kg}} \cdot \text{km}}{\text{hr}}}{20\,\cancel{\text{kg}}} = v_{anvil}\,\frac{\text{km}}{\text{hr}}$$

$$v_{anvil} = \frac{24.14}{20} = 1.21\,\frac{\text{km}}{\text{hr}} = \mathbf{0.33\,\frac{m}{s}}$$

So getting hit with a super fast baseball is equivalent to having a large anvil fall on you at one-third of a meter per second (and we all know how that would hurt). Momentum is serious stuff, and we'll be revisiting it again several times as we piece together how the universe works.

Friction and Frictional Forces

Finally, let's start into our discussion of forces by introducing a simple force that only does one thing – *slow stuff down*. That force is known as *friction*. Frictional forces arise from the fact that, 1) surfaces interact with each other, 2) no surface is perfectly smooth, and 3) electrons act like a form of glue that makes things stick together.

Friction is a good and bad thing. Good from the perspective that without it, pretty much every known activity that occurs on Earth could not happen without it (try driving a car or bicycle down a frictionless road, or hold onto a frictionless cup of coffee). It's bad, because it wears things out (from the gears and clutch in a car, to the wind drag on an airplane, to eroding the mountains themselves – it's all friction). Because of its good and bad nature, humans have figured out a way to exploit the properties of friction to minimize and maximize it, as the need requires.

Frictional forces always remove energy from a system and make things more chaotic. There is no form of friction that actually aids in the motion of anything in the universe – *frictional forces always oppose motion*. The only result of frictional interaction between two objects is to produce *heat* (which on an atomic level is electrons moving faster). Try rubbing your hands together rapidly and see what happens – *they get hot.*

In order to understand what is going on, we need to look at how two objects slide across one another. To do this, take a look at Figure 4.6 where we have a block of stone sliding across a concrete surface due to a force 'P' pushing on it.

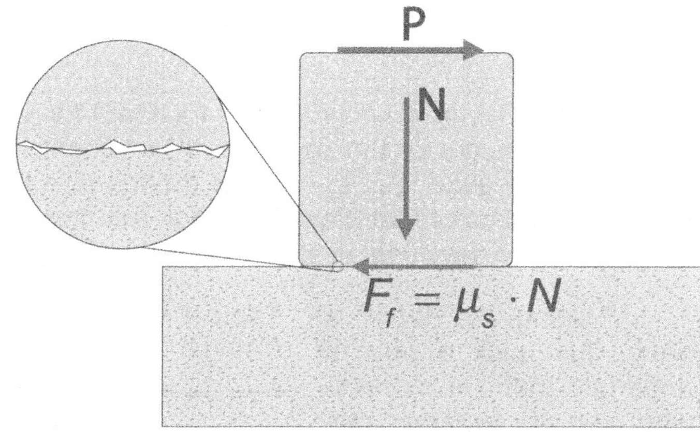

Figure 4.6: Example of Static Friction Between Two Contacting Surfaces

The stone block also has another force, a frictional force, which we'll call 'F_f' opposing the pushing force 'P' of the block. This force always acts in the opposite direction of the applied pushing (or pulling) force and is always located along the interface between the two objects that are in contact with each other. This frictional force, as we can see in the figure, is due to the always-imperfect surfaces between the two materials, which momentarily 'stick' together. The frictional force is quantified as being equal to the weight of the object that is moving (formally called the *normal force*, or 'N'), times a coefficient of static (i.e., non-moving) friction (which is denoted as μ_s).

Stating it another way, *the frictional force is equal to the weight of the object times the coefficient of static friction.* Once the force pushing on the object 'P' is equal to the frictional force 'F_f', the object will start to move (we'll deal with moving friction in a second, right now let's stick to things that aren't moving). Our equation for static friction, therefore, simply becomes,

$$\bar{F}_f = \mu_s \cdot \bar{N}$$

The Guru Says...

Figuring out frictional coefficients is not a terribly difficult task. You simply take an object of material 'A' and a flat board having a surface of material 'B', and place object 'A' atop of board 'B'.

You raise one edge of the board to the point where material 'A' just starts to move and measure this angle with respect to the horizontal. *The inverse-tangent of this angle is the coefficient of static friction.* The inverse-tangent of the minimum angle to keep the object moving is the coefficient of kinetic friction (μ_k).

To be mathematical purists, we'll put the bars above the frictional and normal forces since they are indeed vectors. Most of the time, though, we can get a little relaxed with the notation and treat the above equation as a scalar expression, since it's implied which directions the normal and frictional forces act. Looking at this expression, *what's intuitively missing from it?* How about the area of contact between the two surfaces?

Well, we really don't need to know how much of the block is in contact with the concrete since the frictional force between them is not actually dependent on this physical parameter. So, when you get down to it, the only thing that keeps the tires on a car from breaking traction is the weight of the car on each tire, and the road surface it sits on. Tell that to the next person who says that his/her super-wide tires gives them better traction – *hah*.

Since we now know that all frictional forces are a function of only two variables, namely the weight of the object and the coefficient of friction between them, it would probably be a good idea to run into the lab and figure out various coefficients of friction for different pairs of materials. Fear not, the Guru has tabulated some of the more useful coefficients for daily use as shown in Table 4.1.

Using the table is easy. Just pick a pair of materials that are close to what you are working with, and if the object is not initially sliding, then use the static friction coefficient. If the object is sliding with respect to the contact surface, or starts to move because the applied force 'P' is greater than the static friction force, then use the kinetic friction coefficient in the place of the static friction coefficient in the previous equation.

Table 4.1: Some Typical Coefficients of Friction for Various Material Combinations

TYPE OF SURFACE	COEFFICIENT OF STATIC FRICTION μ_s	COEFFICIENT OF KINETIC FRICTION μ_k
Metal on Metal (Non-lubricated)	0.74	0.57
Metal on Metal (Oil-lubricated)	0.15	0.06
Metal on Wood Surfaces	0.60	0.20
Metal on Stone Surfaces	0.70	0.30
Wood on Wood Surfaces	0.50	0.25
Stone on Stone Surfaces	0.70	0.40
Earth (dirt/sand) on Earth	1.00	0.20
Glass on Glass Surfaces	0.94	0.40
Teflon on Teflon Surfaces	0.04	0.04
Ice on Ice Covered Surfaces	0.10	0.03
Tire on Dry Asphalt/Concrete	1.00	0.80
Tire on Wet Asphalt/Concrete	0.60	0.40
Tire on Snow Covered Road	0.30	0.20

One thing should be instantly obvious from looking at the table. The coefficient of static friction is always greater than the coefficient of kinetic friction. Once an object starts to slide, its tendency is to keep on sliding.

Also, no coefficient of friction can be greater than 1.0 by definition; in other words, the frictional forces can never be greater than the normal force. It now should also be clear why any machine that has moving parts would need some form of lubrication to protect its internal workings, and why driving on a wet road the same way you do when it's dry is probably not a very good idea.

Example 4.6: Most mechanical wear on a car engine occurs during starting, and in the first minute before the lubrication system moves oil from the pan to the top of the engine. Why?

Engine starting implies that parts are not initially moving with respect to each other; thus, we would use the coefficient of static friction for a metal on metal surface. Since some oil always sticks to the parts even when there is no oil pressure, we can split the difference on the coefficients of static and kinetic friction.

$$\mu_{s \text{ (somewhat lubricated)}} = \frac{\mu_{s \text{ (dry metal)}} + \mu_{s \text{ (wet metal)}}}{2} = \frac{0.74 + 0.15}{2} = 0.445$$

We can also do the same thing for the kinetic friction, which would be representative of the engine right after it has started, but hasn't been running for long.

$$\mu_{k \text{ (somewhat lubricated)}} = \frac{\mu_{k \text{ (dry metal)}} + \mu_{k \text{ (wet metal)}}}{2} = \frac{0.57 + 0.06}{2} = 0.315$$

The percent change of something is defined as the starting amount minus the final amount, divided by the initial starting amount, namely, $\%_{change}$ = (starting value – final value) / starting value). The percent drop in friction force that occurs when starting an engine can be found by looking at the changes between the coefficients.

$$\%_{\text{change in friction}} = \frac{0.445 - 0.315}{0.445} = 0.292 = 29.2\%$$

So using our simple approach here, we can estimate that there is roughly a 30% increase in frictional wear to mechanical parts due to starting an engine. Indeed, excessive engine starting wears an engine out faster than anything, which is why there are so many additives put into oil to counteract this problem.

Now, let's examine the issue of 'full' versus 'partial' lubrication, especially on parts far away from the oil pan like the valves and hydraulic lifters. We know that our semi-lubricated engine parts have a coefficient of kinetic friction of 0.315 from above. Fully lubricated parts have a lower value of 0.06. Thus, the change in engine wear is,

$$\%_{\text{change in friction}} = \frac{0.315 - 0.06}{0.315} = 0.809 = \mathbf{80.9\%}$$

So what do you think? It's probably a pretty good idea to let the engine warm up for two or three minutes before taking off, rather than just jumping in a cold car and barreling down the highway at full throttle.

Example 4.7: Power is equal to force 'F' times velocity 'v' and is a scalar. If a car sends 50 kW of power at 40 km/h to the wheels, what is the required vehicle contact force (normal force) on dry pavement? How about on wet pavement? How much should we reduce the engine power if we start to skid on wet pavement?

This problem involves some mixed units, so we need to review Table 1.6 and make note of what our base unit conversions are. We have the unit of power, which has meters, kilograms, and seconds in it, so we will have to make sure that our speed is also in terms of meters and seconds, so that everything will work out correctly.

Converting kilometers-per-hour (km/h) to meters-per-second (m/s) gives,

$$\frac{40\ \cancel{km}}{\cancel{hr}} \cdot \frac{1000\ m}{1\ \cancel{km}} \cdot \frac{1\ \cancel{hr}}{3600\ s} = \frac{40000\ m}{3600\ s} = 11.11\ \frac{m}{s}$$

Since power equals force times the dot product of velocity, power is a scalar, and we have the fully scalar expression,

$$P = F \cdot v \quad \text{or} \quad F = \frac{P}{v}$$

We know that the required vehicle contact force (i.e., the normal force 'N', or how much weight is required to be in contact with the road for a given power setting) is only as good as the frictional force between the tires and the road. Therefore,

$$F_{\text{(road condition)}} = \mu_{s\,\text{(road condition)}} \cdot N_{\text{required}}$$

Substituting this into our previous expression gives us the *kinetic equation* for the car's performance under differing road conditions (*kinetics*, unlike *kinematics*, is the study of forces acting on an object).

The Guru Says...

This is not to say in this example that the vehicle loses or gains mass. Rather, the kinetic expression is a measure of the required contact force between the tires and the road for a given power and speed setting.

When your tires break traction something has to adjust, and from the physics standpoint of this problem, the car 'appears' to instantaneously become too light for the road (as you scream *'wha-hooooo-eeee'* and skid off the highway).

This is the reason that vehicle stability control systems in modern cars immediately reduce engine power (the other independent variable) to the affected wheels to regain traction and get the required contact force of the car to within the range of what the vehicle actually weighs at a standstill.

$$F_{\text{(road condition)}} = \frac{P}{v}; \quad \mu_{s\,\text{(road condition)}} \cdot N_{\text{required}} = \frac{P}{v}; \quad N_{\text{required}} = \frac{P}{v \cdot \mu_{s\,\text{(road condition)}}}$$

We are told that the power and speed settings of the car are 50 kW and 40 km/h (11.11 m/s) respectively. Thus our specific expression representing the required contact force on the road by the automobile can be found by substituting these values into the previous kinetic expression.

$$N_{\text{required}} = \frac{50000\ W}{11.11\frac{m}{s} \cdot \mu_{s\,\text{(road condition)}}} = \frac{50000\ \frac{m^{\cancel{2}} \cdot kg}{s^{\cancel{3}2}}}{11.11\frac{\cancel{m}}{\cancel{s}} \cdot \mu_{s\,\text{(road condition)}}} = \frac{50000}{11.11 \cdot \mu_{s\,\text{(road condition)}}}\ \frac{m \cdot kg}{s^2}$$

$$= \frac{4500.5}{\mu_{s\,\text{(road condition)}}}\ \text{Newtons}$$

Here, we have written out the unit of "Newton's" rather than using its SI abbreviation 'N' to avoid confusion with the required contact force (for which we are also using the letter 'N' to denote).

Substituting for dry road conditions ($\mu_s = 1.0$) we have,

$$N_{\text{required}} = \frac{4500.5}{1.0} = \mathbf{4500.5\ Newtons} \quad \text{or, } \mathbf{1125.13\ Newtons\ per\ tire}$$

So under the assumed engine power and speed, the required contact force of the car is 4500 Newton's (roughly a gross effective mass of 460 kg on all tires). Notice that we used the coefficient of static friction between rubber tires and the road surface – even though the tires are moving. In the physics of friction, rolling is a different activity than sliding, and one is covered by static friction, the other by kinetic friction. Now, what about when the road is wet? Our effective contact force would be,

$$N_{\text{effective}} = \frac{4500.5}{0.6} = \mathbf{7500.8\ Newtons} \quad \text{or, } \mathbf{1875.2\ Newtons\ per\ tire}$$

Thus our car would have to weigh 7500 Newton's (or have an effective mass of 764 kg) in order to not start to slide under the given power settings. Most passenger automobiles have a gross mass in the range of 1,000 to 1,500 kg, so there is a bit of a margin of safety to prevent you from spinning out when it starts to rain (although we do see some bad drivers do this in spite of the governing physics of the problem). Finally, what about the case where we are on a wet road and start to skid out? Going back to our kinetic equation one more time we find out that,

$$N_{\text{effective}} = \frac{4500.5}{0.4} = 11{,}251.25 \text{ Newtons} \quad \text{or, } 2{,}812.8 \text{ Newtons per tire}$$

This would imply that the car would have to momentarily decrease the power to the spinning tire by the following percentage in order to absolutely maintain traction.

$$\%_{\text{change in required power}} = \frac{11251 - 4500}{11251} = 0.60 = \mathbf{60.0\%}$$

Unless we can magically effect a sudden substantial change in the vehicle's mass as we start to skid, it is clearly a simpler engineering solution for the car's computer to reduce the power to the slipping wheel.

Chapter 5: Newton's Laws of Motion

Getting hit on the head with an apple was a pretty good thing

What can we say about Sir Isaac Newton, the mathematician, chemist, physicist, astronomer, and theologian? Newton dabbled in many things during his life, helping to invent the mathematical concepts of differential and integral calculus, developing a method to find the roots of a polynomial, and perfecting the first workable reflecting telescope. He studied light and optics and was the first to discover that white light consists of a combination of different colors. He was the first person to analyze Kepler's *Laws of Planetary Motion*, using kinetic methods, and develop a generalized universal theory of gravitation still used today to describe the most advanced astronomical movements. Yes indeed, Newton was quite the Renaissance man.

Newton also developed the foundation of classical physics (appropriately termed *Newtonian Mechanics*) upon which all future work was to be built.

In 1687 Newton published his most famous work entitled, *Philosophiæ Naturalis Principia Mathematica* (shortened to the *Principia*), which translates from Latin to mean *Mathematical Principles of Natural Philosophy*. In this work, he proposed three fundamental laws of motion (Newton's Laws) that all objects containing matter must follow. Newton's three laws of motion form what physicists term the fundamental *starting principles* in physics, in that when in doubt, you can always find the solution of a problem by starting with Newton's Laws and working forward from there. These three laws detail under what circumstances something will move in the universe, what forces are required to make the movement occur, and what happens to the object that is applying the force. In order to apply Newton's Laws correctly, we need to establish a simple ground rule for their application, which we'll find, translates right to the very core of 20th century quantum mechanics.

The Picture Frame That Doesn't Move

You're probably sitting down in a room somewhere, quietly and with little motion or physical effort, reading this book. But are you really actually sitting still? Let's think about this for a moment. Clearly you're sitting down, in what appears to be a motionless room – at least you don't feel like you are moving. Putting aside the fact that Earth whizzes through space at over 100,000 km/hr, Earth is also spinning on its axis. We complete one full rotation roughly every 24 hours.

The circumference of Earth is about 40,000 km at the equator, so this would imply that a point on the equator is moving at roughly 1,670 km/hr. If you're sitting at the same latitude as the Guru, about 32-degrees north of the equator, this would mean that you are moving at a constant rotational speed of approximately,

$$Speed_{Guru} = 1670 \cdot cos\left(32^{\circ}\right) \approx 1400 \text{ km/hr}$$

Which is faster than the speed of sound in air at sea level (that value is only 1,239 km/hr). So what is going on here? How come we don't feel this motion? How come we don't hear the deafening sound of the air moving by?

We don't experience any ill effects from our moving through space in this manner because in our quiet room reading this book we are not moving with respect to the room. Sure, the room is moving at 1,400 km/hr with respect to the 'outer space' around us, and Earth itself is moving at over 100,000 km/hr with respect to the sun, and our solar system is moving at 70,000 km/hr with respect to the rest of the Milky Way galaxy, and our Galaxy is spinning at a rate of 792,000 km/hr, and on and on it goes, but relative to our room, we are not moving.

Welcome to the world of the inertial reference frame…

An inertial reference frame is, strictly speaking, *a coordinate system that does not accelerate*. Constant velocity is not a problem, just no acceleration. We can even relax this requirement a little bit further, and state without any loss of generality (courtesy of Professor Einstein) that an inertial coordinate system cannot accelerate – with respect to your personal observational position.

Thus, as long as what you are looking at and/or measuring is not accelerating, you're *good to go* and Newton's Laws of motion can be directly applied and translated into any other inertial system.

The Guru Says…

Unlike the legend of George Washington chopping down the cherry tree, the story of Isaac Newton seeing a falling apple and speculating about how gravity behaves is a true story.

Newton himself often recalled that he was inspired to formulate his theory of gravitation by watching apples fall from a tree, noting that the action of the gravitational force did not end when the apple hit the ground, but appeared to extend for a considerable distance into Earth.

If this force works on apples, why not the moon, and every other thing in the universe Newton recogned.

Despite many cartoons and drawings to the contrary, Newton getting hit on the head with an apple was not his inspiration for his gravitational theory.

For example, suppose the Guru is measuring the speed of a baseball using a radar gun. The speed that he measures at any given instant in time is going to be *relative* to the point on Earth where he is standing. Anyone standing next to him, and not moving relative to his position, would also be expected to measure the same thing with his or her radar gun (as would a guy down the street, and so on).

Now, let's say that the person next to the Guru decides to jump into a car and go speeding away at a constant velocity. Would his radar gun measure the same thing? No, of course not, since his inertial coordinate system is moving relative to the one that the Guru is in. Although the measurements of the baseball would not be the same, they are related to each other by virtue of each being relative to their respective inertial reference frames. This brings us to the concept of motion relative to a frame in translation.

If we have a stand-alone coordinate system 'A' that is 'fixed' on Earth, as shown in Figure 5.1 (and remember that physically the concept of being 'fixed' in space is meaningless in a general sense), then its velocity would be zero. A second coordinate system bolted to a passing object (let's call it 'B') would also have zero velocity – with respect to the object. It would have a velocity of V_B as measured by the observer in coordinate system 'A'.

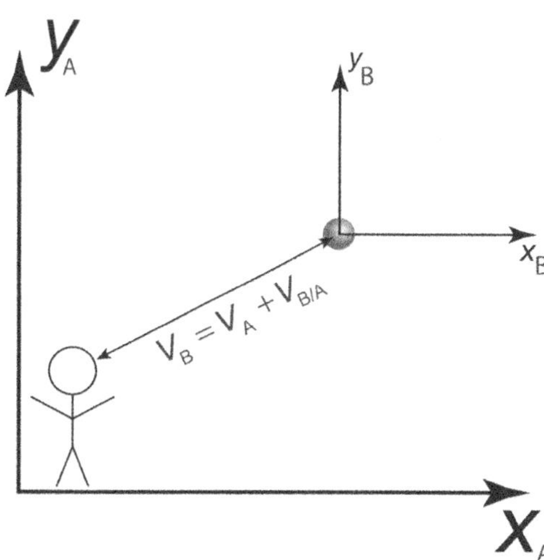

Figure 5.1: Example of Relative Inertial Reference Frames 'A' and 'B'

Consequently, if the moving object in coordinate system 'B' is assumed to be fixed, then the observer in coordinate system 'A' is observed to move with velocity V_A in the opposite direction – so you see, it's all very much relative. The concept of which inertial frame you measure a physical phenomenon in has many more far reaching implications than what we see here, and is the subject of the *Theory of Relativity*, which we'll discuss in Chapter 14.

For the time being though, we can express the motion relative to either system shown in Figure 5.1 by using the following simple geometric relationship.

$$V_B = V_A + V_{B/A}$$

We would read this as, *"the velocity of frame 'B' is equal to the velocity of frame 'A' plus the velocity of frame 'B' relative to 'A'."* Using this concept, we convert from one inertial coordinate system to another, and back again, with ease. This idea also works for displacements as well with,

$$X_B = X_A + X_{B/A}$$

If you think about it, this is nothing more than adding distance vectors together. We've left the expression describing accelerations between the systems out at this time, so as not to confuse things, but we could just as easily write an expression for the summation of constant accelerations between our two inertial frames in the same manner as above. Finally, since we are assuming that these transformations between inertial coordinate systems are linear, we can convert through any number of coordinate systems by hopping from one to the next. To go from hypothetical coordinate system 'A' to system 'E' we could write,

$$V_E = V_A + V_{B/A} + V_{C/B} + V_{D/C} + V_{E/D}$$

The important thing to realize here is that when we apply Newton's laws, we need to make sure that we are doing so in an inertial coordinate system. Once we figure out what things look like in that system, we are then free to translate them to any system of our choosing.

Now, knowing that we are always working with an inertial reference frame, let's see what Newton's laws tell us about the motion inside of one.

Newton's First Law – Getting Lazy Objects to Move

Newton formally stated his first law of motion in the Principia as follows, *"Every body continues in its state of rest, or of uniform motion in a right line, unless it is compelled to change that state by forces impressed upon it."*

We can state this simply as, "An object in motion tends to stay in motion (or an object at rest tends to stay at rest), unless acted upon by an external force". In short – *matter is lazy and it will only do things if you force it to.*

Notice the Guru stresses the point that an object will only change what it is doing if an external force is applied. Internal forces such as those produced by heating or cooling an object, or radioactive decay do not produce any change in motion of the object. It has to be something external (like a baseball bat hitting a ball), which produces the motion.

Also, pay close attention to Newton's own statement about, *"...uniform motion in a right line..."*. Thus, the natural tendency of all motion is to always travel uniformly in a straight line in the absence of external forces. Throw a baseball on Earth and the gravitational force bends the path of the ball back towards Earth. Throw the same ball in outer space far from any planets and it will want to travel in a perfectly straight line forever.

Newton's first law is also called the 'Law of Inertia' since it describes another fundamental property of matter – appropriately called *inertia*. Inertia is defined as the resistance of any physical object to a change in its motion or resting state. The word 'inertia' is derived from the Latin word, *iners*, which literally means 'lazy'.

Inertia is the resistance you feel trying to push or pull on a heavy object to get it to move. Inertia is also the force you feel when the heavy object you were pushing gets away from you and you frantically pull (or push) on it to slow it down. Inertia is a property of matter. If something has mass, then it has inertia. Mix in a little velocity and we have a physical property we've already looked at – momentum.

We should also note that Newton's first law of inertia really doesn't make any distinction between whether or not the object has momentum. Take another look at Figure 5.1. From the perspective of coordinate system 'A', the ball is moving. Viewing the ball from coordinate system 'B', it is at rest (since it is not moving with respect to this coordinate system, the definition of being at rest).

Thus in both coordinate systems (one observing no motion and one observing only constant velocity) there is no acceleration. You can conclude from the first law that no forces must be acting on the system, and it is indeed an inertial coordinate frame.

Newton's Second Law – Why Something Moves at All

Once Newton figured out what mechanical properties matter has, and what tended to keep objects moving or standing still, he still had to wrestle with the problem of why something even moved at all. Enter the second law, which he formally stated as, *"The change of motion is proportional to the motive force impressed; and is made in the direction of the right line in which that force is impressed."*

Newton first postulated that the force that acts to move an object through space was proportional to its velocity. Thus, you could think of the *'first draft'* of the second law as something like,

$$F_{EXT} \approx V$$

This kind-of makes sense, since as a first shot at quantifying the force on an object, we would first observe the momentum of the object ($m \cdot v$) and notice that it takes some sort of force to change this momentum. As good as this looks, though, it still is not the right answer, since after looking at the whole momentum thing for a while, I'm sure he noticed that heavier objects moving at the same speed required more force to stop them. Thus, a good second draft of the second law would be,

$$F_{EXT} \approx m \cdot V$$

This is better, but also violates Newton's first law, since we know that objects having a constant velocity really won't have any external forces acting on them (i.e., $F_{EXT} = 0$). *Hmmmmm, what now...*

What if we said that the force was not equal to the momentum, but rather it was equal to the time rate of change of the momentum of the object, $\Delta(m \cdot V)/\Delta t$? We could then write,

$$F_{EXT} = \frac{\Delta(m \cdot V)}{\Delta t} = \frac{\Delta}{\Delta t}(m \cdot V) = \Delta m \cdot \frac{\Delta v}{\Delta t}$$

This works out really well with what we observe in nature. Thus, we could alternatively express Newton's second law by stating that the external forces on an object (required to produce a change in motion) are equal to the time rate of change of momentum. For our purposes though, we'll assume that the mass of our object does not change with respect to time, and note that the change in velocity of our object with respect to time is simply equal to its acceleration. We can then write the final result as follows,

$$F_{EXT} = m \cdot \frac{\Delta v}{\Delta t} = m \cdot a$$

Or, *force equals mass times acceleration.* This simple expression states that force is proportional to acceleration, and is equal to acceleration times the universal *fudge-factor* known as mass. We now have a way to interchangeably work with force or acceleration in our exploration of physics, by knowing that the two are similar. As with our previous notation, since mass is a scalar and acceleration is a vector, force is also a vector quantity.

Example 5.1: A one-kilogram object falls under the influence of gravity at a constant acceleration of 9.81 m/s^2. What is the force acting on the object?

The acceleration due to gravity is 9.81 m/s^2. Using Newton's second law we find that the force on the object would be,

$$F_{EXT} = m \cdot a$$

$$= 1\,\text{kg} \cdot 9.81\,\frac{\text{m}}{\text{s}^2} = 9.81\,\frac{\text{kg}\cdot\text{m}}{\text{s}^2} = \mathbf{9.81\,N}$$

A one-kilogram object that falls under the action of gravity has a force of 9.81 N acting on it. Let's look at another problem involving a gravitational force.

Example 5.2: You have three masses of equal volume. All are equal to 1 cm^3. One is made of copper (density = 8.96 g/cm^3), one is made of magnesium (1.74 g/cm^3) and the last one is made of tungsten (19.35 g/cm^3). You drop them from a height of 20 meters and time how long it takes for them to hit the ground. What is the gravitational force acting on each mass? Which one hits the ground first?

Applying Newton's second law, and noting that mass is equal to the density times the volume, we can calculate the gravitational forces on each mass as follows.

$$F_{\text{copper}} = \rho_{\text{copper}} \cdot V_{\text{copper}} \cdot a_{\text{gravity}} = 8.96\,\frac{\text{g}}{\text{cm}^3} \cdot 1\,\text{cm}^3 \cdot 9.81\,\frac{\text{m}}{\text{s}^2} = 0.0878\,\text{N}$$

$$F_{\text{magnesium}} = \rho_{\text{magnesium}} \cdot V_{\text{magnesium}} \cdot a_{\text{gravity}} = 1.74\,\frac{\text{g}}{\text{cm}^3} \cdot 1\,\text{cm}^3 \cdot 9.81\,\frac{\text{m}}{\text{s}^2} = 0.0171\,\text{N}$$

$$F_{\text{tungsten}} = \rho_{\text{tungsten}} \cdot V_{\text{tungsten}} \cdot a_{\text{gravity}} = 19.35\,\frac{\text{g}}{\text{cm}^3} \cdot 1\,\text{cm}^3 \cdot 9.81\,\frac{\text{m}}{\text{s}^2} = 0.1898\,\text{N}$$

Even though the forces due to gravity are different (meaning that they do indeed have differing weights pulling them towards the ground), the acceleration due to gravity is still a constant in each case. Recalling our previous kinematics expressions for the time it takes an object to fall under gravity from a height of 20 m we have,

$$y = y_0 + v_{y0} \cdot t + \frac{1}{2} g \cdot t^2$$

Substituting into the above equation we get,

$$20\ \cancel{m} = \frac{1}{2}(9.81\ \frac{\cancel{m}}{s^2}) \cdot t^2$$

$$\frac{40}{9.81}\ s^2 = t^2$$

t = 2.02 s

So the time required for the weights to fall 20 m would be 2.02 seconds each, regardless of how much they actually weigh.

The Guru Says...

The great philosopher and scientist Galileo Galilei published the startling finding that objects fall at the same rate regardless of their weight in 1634, completely erasing the 'well known' statement of Aristotle to the contrary.

Despite this scientific fact being known for close to 400 years, you can still encounter individuals today that profess (i.e., argue) otherwise.

Newton's Third Law – Accelerating Objects Hit Back

Lastly, Newton stated his third law of motion as, *"To every action there is always opposed an equal reaction: or, the mutual actions of two bodies upon each other are always equal, and directed to contrary parts."*

Simply stated, for every action {force} there is an equal and opposite reaction {force}. This is an intuitive law since we all know that if we hit an object, it will hit back. As we walk across the floor, the floor pushes back on every step we make with an equal force, *otherwise we'd fall {accelerate} through the floor – right?* The third law is a law of *reciprocity* in that it keeps forces in balance on interacting bodies. Imagine what a strange place the universe would be without this very fundamental physical principle.

Example 5.3: An interesting mechanical device that has been known about for a long time is called the pulley. A pulley is a wheel on an axle that transfers a force in a different direction. A pulley alone cannot modify the magnitude of a force; it can only change its direction.

The system shown below consists of three pulleys (two bolted to the ceiling, and one bolted to the floor). A steel cable runs through all the pulleys, as shown, and is connected to a 100 N weight.

Apply Newton's third law of motion and determine if there is any mechanical advantage to be gained in building a contraption like the one shown. What force 'P' would be required to move {accelerate} the mass shown? What happens as we add additional pulleys to the system? What happens when we remove pulleys or move the mass?

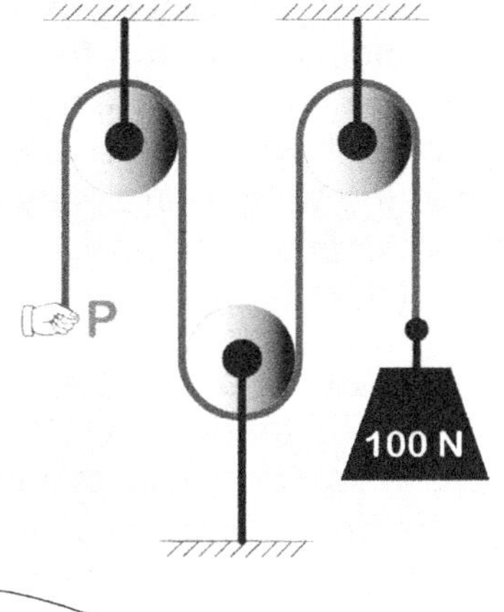

This complex looking problem is actually quite simple. Since a pulley cannot modify the magnitude of a force (it can only change the direction of the applied force vector), the 100 N force is pulled from the weight, over the three pulleys, and to the person's hand. Thus we can state that the *tension* in the cable is the same everywhere and the applied force 'P' is also 100 N. Adding or subtracting pulleys has no effect on this system, which has no mechanical advantage.

The force 'P' is the force required to hold the mass in place – a concept that engineers call *statics*. If the forces were imbalanced, say by the hand applying more or less force to the cable, the mass would accelerate up or down – a concept known in engineering as *dynamics*. Thus, we know that an imbalance of forces in a system only has one possible answer – the system *must accelerate* according to Newton's second law.

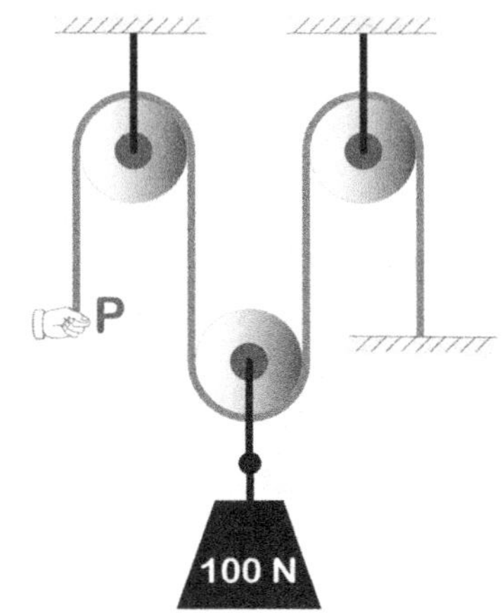

Figure 5.2: A Pulley System That Actually Does Something Meaningful

Now, let's look at Figure 5.2, a slight variant of our previous example, where the Guru switched the weight and the anchor point of the pulleys. What do you think the required force 'P' is to maintain a static condition in this case?

To analyze a pulley problem like this, the only thing you have to remember is that the force in the cable 'P' is going to be the same everywhere just as long as there are no breaks in it. Looking at the pulley closest to the hand pulling down we can see that it merely rotates the force vector into a downward direction with no change in its magnitude. The same applies for the pulley to the far right which only again rotates the force vector to point it towards the anchor point on the floor. This must mean that if anything important is going to happen, it must occur where the 100 N weight is attached to the axle in the middle of the pulley.

Looking at Figure 5.3, we can draw what is known in mechanics as a *free-body diagram*, which is a picture of the object in question, isolated from the rest of the system, showing the applicable forces involved. We know that the weight produces a downward action equal to 100 N because of gravity, and that this force acts on the axle of the pulley.

We also know by Newton's third law that for this 100 N action, there must be an equal and opposite reaction, which because of the physical nature of how pulleys are constructed, is divided across the two lengths of the cable leaving this pulley.

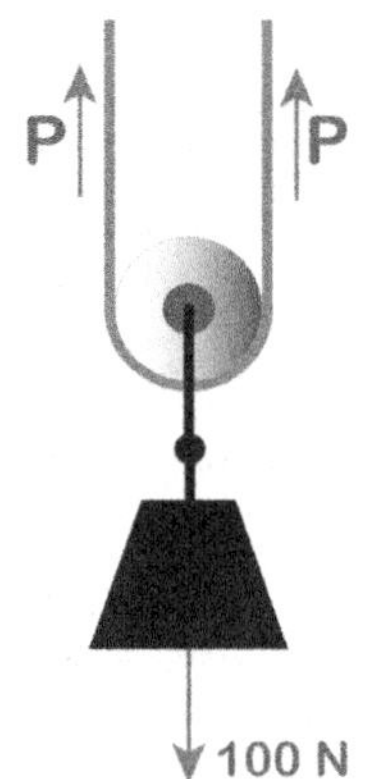

Figure 5.3: A *Free-body* Diagram of the Pulley Problem.

Since we're now experts at vectors, we know that we have two 'P' forces moving upward and one 100 N force moving downward, and that these forces must sum to zero if we're to maintain a static condition. Thus we can write,

$$P + P - 100 = 0$$

$$2P = 100$$

$$\mathbf{P = 50\ N}$$

So the force required to maintain the 100 N weight is now only 50 N. Our mechanical advantage for this system would be '2' and we could conclude that pulleys are a pretty good thing for manipulating forces. In fact, every time we add another cable segment across a pulley, we reduce the load by that fractional amount (i.e., for three segments the load would be P/3, four segments would be P/4, and so on). Simple machines of this type are typically called *block and tackles*, and they all work because of Newton's third law.

The Direction of Newton's Travels

Finally, we would be remiss if we did not reiterate that all of Newton's findings are vectors; thus they follow all the rules that we've previously outlined for such mathematical quantities.

Newton's first law of motion, or the law of inertia, always acts along a perfectly straight line, called the *'line of action'*. When we apply a force to an object to get it moving (or place it at rest) we automatically discover that we have to apply that force along the direction of motion – the line of action. This is not only true for single objects, but also for systems of objects (such as billiard balls on a pool table) whose individual motion vectors can always be resolved into a single resultant vector that always defines the line of action of the desired motion.

Newton's second law of motion is a no-brainer for vectors. We already know that acceleration is a vector and that multiplication of a scalar, like mass, times a vector (acceleration) also mathematically yields a vector (force); so this is pretty self-evident. Since the mass of a system merely provides the inertia and does not alter the direction of the acceleration, we therefore know that the direction of the force vector has to be in the same direction as the acceleration, and vice versa.

The Guru Says...

The pulley is one of the six ancient simple machines known by mankind for thousands of years to modify force and allow heavy objects to be moved with relative ease. The other five devices were: *the wedge, the lever, the inclined plane, the screw, and the wheel.*

The third law of motion is probably the most interesting in terms of how we perceive the universe, since typically we never really experience the actual force on something; rather we always feel the reaction force. *Confused* – you won't be after this next example.

Example 5.4: Pretend you're sitting in an automobile at a stoplight. When the light turns green, you step on the gas and take off like a speed demon. You are clearly accelerating forward, and we know that in order to accelerate forward the force vector must also be pointing forward (otherwise you'd in reverse). So, why are you then pushed backwards into the seat?

Newton's third law produces, as a direct consequence, a phenomenon known as a *reverse inertia force*. This force has been termed as a *d'Alembert force* after the 18th century French mathematician Jean le Rond d'Alembert who first commented on it.

The d'Alembert force is the inertial reaction force your body feels as you are being accelerated from a stop (remember, mass does not like to accelerate and will fight you all the way). The inertial force is identically equal to the negative of Newton's second law, or,

$$F_{\text{d'Alembert}} = F_{Inertial} = -m \cdot \frac{\Delta v}{\Delta t} = -m \cdot a$$

So, even though you clearly are accelerating in one direction, *you will always feel the inertial force in the opposite direction due to the simple fact that you have mass*. We take this for granted every day, which makes it all the more stranger when you stop to think about which way the vectors are actually pointing in space.

Inertial forces, such as the d'Alembert force, are *fictitious forces* in physics, since they occur as a consequence of the third law of motion, and don't really exist as a real force by themselves. There are many different types of fictitious forces that arise in nature by virtue of an object having mass (inertia), with most of these occurring for the rotating systems that we will explore in Chapter 8. These phantom forces include the centrifugal force and the Coriolis force.

Chapter 6: Work and Energy

How the concepts of work and energy drive the universe

The famous American author and humorist Mark Twain once said that, *"Work and play are words used to describe the same thing under differing conditions."* The same thing can be said about the physical concepts of *work and energy,* since work is defined as the ability to apply a force over a distance to do accomplish some type of task, and energy is pretty much stored work.

There are numerous types of energy storage and conveyance mechanisms in the universe, all being completely interchangeable with work from both a mathematical as well as a physical standpoint. These sources take the forms of potential energy; kinetic energy; thermal, chemical, electrical, and electrochemical energies; electromagnetic, acoustic, and wave energy; and nuclear energy.

The best part is that work and energy are scalar quantities, which means they have no direction associated with them – *only magnitude*. This makes the analysis of work and energy quite simple actually from a mathematical standpoint.

Work Produced by a Force – *Whether You Like it or Not*

We've already seen from some of our previous examples that the work produced by a single force is equal to that force times the distance over which you apply said force. Since we know that both a force 'F', and distance 'd' are vector quantities, but work is a scalar, this implies that the 'times' part of the equation must be a dot product, or,

$$W = \bar{F} \bullet \bar{d} = F \cdot d$$

Looking at Figure 6.1, when we talk about the amount of work an engine performs, we always measure the force each piston receives times the length that the piston travels. This quantity, force times distance, determines how much work you can get out of the engine. If you want an engine to do more work, you need to apply more force to the piston or apply the same force over a longer distance. In this context, the work being performed by the Guru lifting the weights is mechanically similar to the piston in the engine (although the engine doesn't get nearly as tired from the experience).

We also know from our previous sneaky Example 2.6 that we must apply a force in the direction of the motion (i.e., along its *line-of-action*), in order to make anything meaningful happen in the universe. All work must be defined in this manner, if it is actually going to affect objects in the way that Newton's Law's likes.

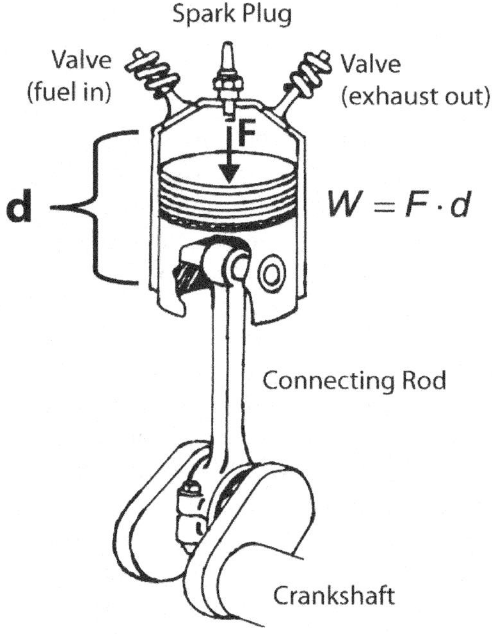

Figure 6.1: A Simple Gasoline Engine Cylinder Doing Work.

This concept also applies if we have multiple forces acting on an object in different directions. In this case, we merely reduce (find the resultant of) all the different force vectors, and apply this new single vector along the line of action of our motion.

The important point to take away from the concept of 'work' is an *absolute* one, which is interchangeable with a relative amount of force and/or distance we choose to apply. If we have a force (or weight) that we want to move a certain distance, then we know absolutely how much work we need to do to accomplish the task. Consequently, if someone hands us a fixed quantity of work, we then have an infinite number of combinations of force and distance we can use to accomplish the job. Think of work in absolute terms, and how much force or distance as relative to that quantity.

Potential Energy and Saving Work for Later On

It's one thing to raise a weight up or push on a piston, but what if you wanted to store some of that work and use it later on when you don't feel like doing as much physical effort? This is the concept behind *potential energy*.

Potential energy, as shown in Figure 6.2, is *'the energy of a stored force'*. When you hear a term using the word 'potential' involving mechanical systems and weights or masses – *think gravity*. Potential energy in this form is the energy stored by gravity. The work that this energy can produce would therefore be related to the distance that a mass or weight can fall under the action of gravity (i.e., force times distance).

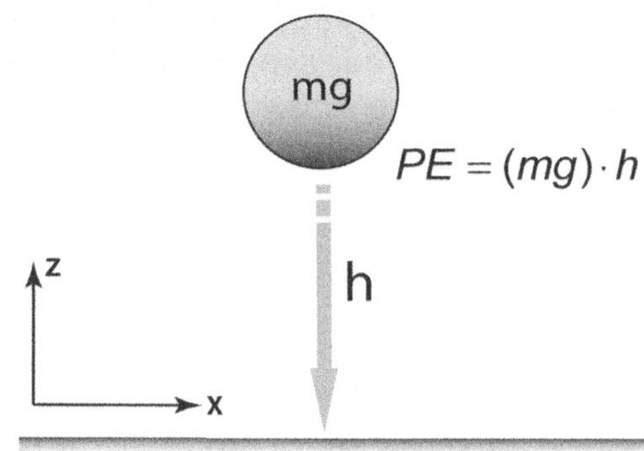

Figure 6.2: Example of Potential Energy Due to Gravity.

There are many mechanisms that employ this age-old technology of a falling mass, including grandfather clocks and counterweight systems in elevators and hoists.

Mathematically, we can define the potential energy 'PE' of a mass 'm' at a certain height 'h' above the ground as,

$$PE = (mg) \cdot h$$

We can easily derive this expression by using what we've already learned. We know that work equals energy, and this work equals force times distance. Newton's second law gives us the force part of the expression, since force equals mass times acceleration, and the acceleration in this case is due to gravity (g). Thus, our final expression for stored work in a heavy object lifted above the ground (i.e., potential energy), is equal to the force due to gravity (mg), times the distance we raise the object (h). Potential energy has the same units as work and is fully interchangeable with any work expression, since they are scalars.

Potential energy conversion into work is not just limited to anvils, rocks, or pianos falling on cartoon characters; there are also mechanical and electrical ways of doing the same thing. Universally, the best way to store mechanical energy is using an ancient device known as a *spring*. Any device (such as a wristwatch or pinball machine), which needs to temporarily store mechanical energy for later use, will use a spring to accomplish the task.

The Guru Says...

Both work and energy are measured using the unit of the Newton for force, and the unit of meters for distance, so work and energy will always have the derived unit of the Newton-meter (N·m or Joule 'J') with base units consisting of $m^2 \cdot kg \cdot s^{-2}$).

A spring is a linear device that stores energy in a material by slightly deforming it. Thus, from an atomic standpoint, a spring stores energy by pushing and pulling on the atoms in the material causing the electrons to want to push back (yes, Newton's third law works at this level as well). Macroscopically, this electron resistance is known as a *spring constant or spring stiffness* (k), and has units of *force per distance* (i.e., the more you pull or push on the spring, the more reaction force it exhibits). Different types of springs will have different types of spring constants, and they *always* resist motion opposite their deflection direction, so they are analogous in this regard to mass in Newton's second law, and can be treated as such. Mathematically, we can define the potential energy 'PE' of a spring 'k' being stretched a distance 'x' as shown in Figure 6.3 as,

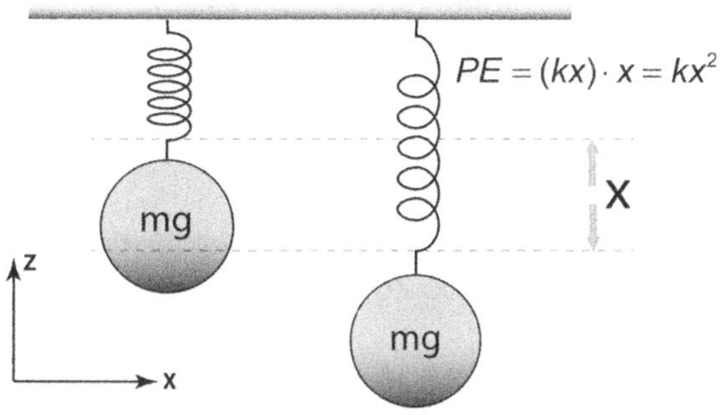

Figure 6.3: Example of Potential Energy due to a Spring Deflecting

$$PE = (kx) \cdot x = kx^2$$

The derivation for the potential energy due to the stretching of a spring follows the same logic as before. We know that work equals energy and this equals force times distance. Newton's second law gives us the force part and that equals the spring constant times the displacement. Thus our final expression for stored work in a stretched spring equals the force ($k \cdot x$), times the distance we stretch the spring (x). As before, this energy has the same units as work and is fully interchangeable with our other expressions.

There are many different ways in which to store potential energy in the universe. We have only looked at the two of the most common methods seen on Earth, namely *elastic potential energy* (e.g., springs, rubber bands, etc.) and *gravitational potential energy* (e.g., masses, weights, hydroelectric dams, etc.). Potential energy can also be stored in electrostatic forms (such as batteries or capacitors), in magnetic forms (like chokes and inductors), and in atomic and nuclear form (including fusion, fission, and chemical reactions) just to name a few.

Let's look at a quick example of how we can apply these two different forms of potential energy we have developed.

Example 6.1: The Guru drops a mass weighing 50 kg onto a spring-loaded plate. If the stiffness of the spring is 10 kN/m (kilo-Newtons per meter) and the mass can drop 5.0 meters, how much will the spring deflect?

Based upon the problem description, we have something going on that looks like the figure below. We have a 50 kg mass free-falling a maximum distance of 5.0 meters onto a spring-loaded plate. The first thing we have to do is calculate how much potential energy our falling mass contributes to the system.

A mass that can fall from a height of 5 meters has a potential energy of,

$$PE_{mass} = (mg) \cdot h$$

$$= (50\text{ kg} \cdot 9.81\frac{\text{m}}{\text{s}^2}) \cdot 5\text{ m}$$

$$= 2452.5\text{ N} \cdot \text{m}$$

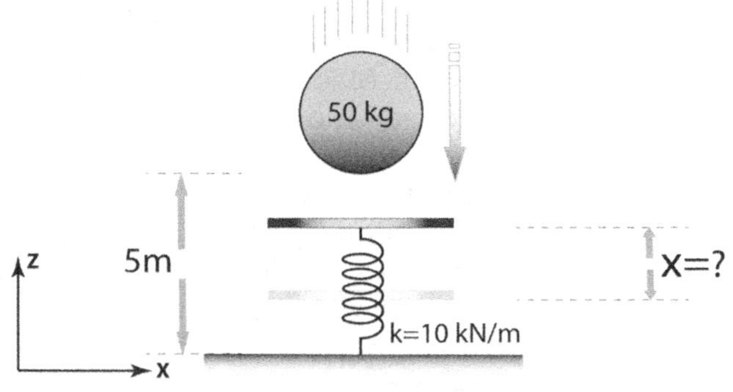

Now, we have previously found that the energy a linear spring can store is defined as,

$$PE_{spring} = (kx) \cdot x = kx^2$$

Equating the potential energy for the spring to the potential energy for the falling mass we can write the following *equilibrium equation* and solve for the maximum deflection of the spring assuming that all the energy is transferred.

Thus,

$$PE_{mass} = PE_{spring}$$

$$2452.5\ \text{N} \cdot \text{m} = (10\ \text{kN/m})x^2$$

$$x^2 = \frac{2452.5\ \cancel{\text{N}} \cdot \text{m}^2}{10000\ \frac{\cancel{\text{N}}}{\cancel{\text{m}}}}$$

$$\mathbf{x = 0.4952\ m = 495.2\ mm}$$

So the spring would deflect about one-half meter before it would come to a stop and reach equilibrium. Using this example you can now model many different types of mechanical systems that use an interrelationship between elastic and gravitational potential energy to accomplish some sort of task.

Now that we have seen how to store energy in something that stands still, let's look at how to store energy in something that is moving.

Kinetic Energy – Creating Work from Moving Objects

On the opposite end of the spectrum, physics tells us that we can also store energy in a moving object, since this is also consistent with Newton's first law of motion and how inertia behaves. This *energy of motion* (or energy of momentum) is known as *kinetic energy*, and is typically denoted by the symbol *'KE'.*

As you would expect, kinetic energy is going to have terms involving mass and velocity in order to make its units consistent with those for work, and our potential energy expressions.

Following the same logic we used to derive our expressions for potential energy, we can write our generalized expression for kinetic energy by noting that the energy produced equals a force times a distance, and force equals mass times acceleration, which in turn equals mass times the change in momentum over time - *whew.*

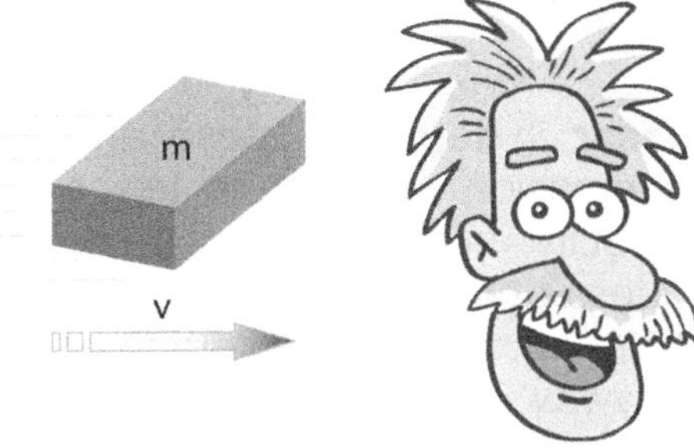

Figure 6.4: Kinetic Energy is the Energy of Motion – *Never, Ever, Forget That*

So, taking a look at Figure 6.4, in which the Guru is about to experience first hand the effects of kinetic energy, we can write,

$$KE_{brick} = \text{Work}_{brick} = \text{Force}_{brick} \cdot \text{Distance}_{brick\ travels}$$

By Newton's second law, we know that the force produced by the brick is equal to its mass times its acceleration. So, we can start our derivation of kinetic energy by making this fundamental substitution as,

$$KE_{brick} = \text{mass}_{brick} \cdot \text{acceleration}_{brick} \cdot \text{Distance}_{brick\ travels}$$

Since we are trying to avoid the use of the Calculus in the Guru's Guide, we will make a simplifying assumption by stating that all of our problems have *constant acceleration* (i.e., the acceleration of the object being examined does not change with time and simply has starting and ending velocities of v_1 and v_2, with a change of Δv). Thus, the acceleration of the object (defined as velocity divided by time) is simply the change in velocity over the change in time (or $\Delta v / \Delta t$).

Similarly, if we have constant acceleration, then the distance that the object travels is equal to the average change in velocity times the duration of travel.

$$\text{distance} = \frac{\Delta v}{2} \cdot \Delta t$$

Performing our substitution, we can write the expression for the kinetic energy of an object as follows,

$$KE = \text{mass} \cdot \text{acceleration} \cdot \text{distance} = \text{m} \cdot \left(\frac{\Delta v}{\cancel{\Delta t}} \right) \cdot \left(\frac{\Delta v \cdot \cancel{\Delta t}}{2} \right) = \frac{1}{2} m \Delta v^2$$

$$KE = \frac{1}{2} m v^2$$

So the amount of 'work' the Guru's head experiences due to the flying brick is equal to one-half the mass of the brick times the square of its velocity.

Kinetic energy is an impressive quantity, since unlike momentum which was linear in terms of mass and velocity (i.e., all terms had the same exponent), this expression is linear in terms of mass, but squared (nonlinear) with respect to the velocity. In retrospect, we can also see that our assumption of constant velocity really doesn't change our problem at all, since if we assume a large enough time interval we wind up taking the average of the velocity and the time and get the exact same result anyway.

So, a doubling of mass would equate to a doubling of kinetic energy (or momentum for that matter), but a doubling of velocity would equal a *four-fold increase* in the energy of the object (or the work required to effect this change in velocity). Thus we can state that similar objects that have the same kinetic energies, need not necessarily have the same momentum. A stationary object cannot have kinetic energy.

The Guru Says...

With some rearrangement and grouping of terms, the kinetic energy of an object is related to its momentum via the following relationship: $KE = \frac{(m \cdot v)^2}{2m}$, where 'KE' is the kinetic energy and the product of $m \cdot v$ is the momentum. Clearly kinetic energy and momentum are not the same thing.

With that in mind, take a look at Table 6.1 and notice the different levels of kinetic and potential energy possessed by certain everyday objects (even ones that would have similar amounts of momentum).

Table 6.1: Some Interesting Potential and Kinetic Energies of Objects by Increasing Mass

System Relative to Your Inertial Coordinate System	Energy Types Involved	Mass of Object (kg)	How High Up (m)	How Fast Moving (m/s)	Potential Energy, PE (J)	Kinetic Energy, KE (J)
M16 Rifle Bullet	Kinetic	0.0036	- -	980	- -	1,729
Movement of a Snail	Kinetic	0.012	- -	<0.01	- -	0
Golf Ball off the Tee	Kinetic	0.045	- -	80	- -	144
Meteor Entering Earth's Atmosphere	Both	0.1	120,000	10,000	117,720	5,000,000
Baseball Fast Pitch	Kinetic	0.15	- -	40	- -	120
Hitting Something with a Hammer	Kinetic	2.5	- -	9	- -	101
Anvil on Cliff Above Genius Cartoon Coyote	Potential	23	200		45,126	
Anvil as it Hits Genius Cartoon Coyote	Kinetic	23	- -	62	- -	45,126
Human Walking	Kinetic	82	- -	2	- -	92
Satellite in Low Earth Orbit	Both	500	2,000,000	7,800	9.81×10^{9}	1.52×10^{10}
Driving at Highway Speed	Kinetic	1,500	- -	24	- -	432,000
Lockheed SR71 Blackbird	Both	77,100	1,100	27,500	8.32×10^{8}	2.92×10^{13}
Space Shuttle at Main Engine Cutoff	Both	80,000	120,000	8,000	9.42×10^{10}	2.56×10^{12}
747 Jet at Cruising Altitude	Both	81,600	12,000	920	9.61×10^{9}	3.45×10^{10}
A Typical Freight Train	Kinetic	7,257,477	- -	20	- -	1.451×10^{9}
The Moon	Both	7.35×10^{22}	3.85×10^{8}	1,010	2.78×10^{32}	3.75×10^{28}

So clearly it takes a lot of kinetic energy to keep large objects aloft like the moon and even small objects like a satellite can have more 'punch' than a fully loaded freight train. Let's now look at an example comparing the kinetic energies of two different objects keeping in mind that the work performed by each is physically identical.

Example 6.2: A cowboy in the Wild West shoots his Winchester .30-30 rifle at a large boulder. If the bullet weights 10 g and travels at 750 m/s when it hits its target, how much energy is transferred to the boulder? How fast would someone have to smash a small 800 kg car into the boulder to achieve the same level of energy transfer?

(Neglect any time travel problems associated with having an automobile in the Wild West)

To solve this, we would calculate the energy of the rifle bullet by applying our kinetic energy equation, and then equate this result to the energy of the more massive car to figure out how fast it must move to produce the same physical result.

Thus, we have,

$$\begin{aligned} KE_{bullet} &= \frac{1}{2}\text{mass}_{bullet} \cdot \text{velocity}^2_{bullet} \\ &= \frac{1}{2}(0.01\,\text{kg})(750\,\frac{\text{m}}{\text{s}})^2 \\ &= \frac{1}{2}(0.01)(750)^2\,\frac{\text{kg}\cdot\text{m}^2}{\text{s}^2} = \mathbf{2812.5\,N\cdot m} \end{aligned}$$

The bullet would hit the boulder with energy equal to 2,812.5 Joules (J). To figure out how fast the car would have to travel, we merely equate the energy of the car to that of the bullet and solve for the velocity.

$$KE_{car} = KE_{bullet} = 2812.5\,\text{N}\cdot\text{m}$$

$$\frac{1}{2}(800\,\text{kg})(v_{car}\,\frac{\text{m}}{\text{s}})^2 = 2812.5\,\text{N}\cdot\text{m}$$

$$v^2_{car} = \frac{2\cdot 2812.5\,\text{N}\cdot\text{m}}{800\,\text{kg}} = 7.03\,\frac{\text{m}^2}{\text{s}^2}$$

$$\mathbf{v_{car} = 2.65\,\frac{m}{s}}$$

Thus, the energy imparted to the boulder from the teeny-tiny 10-gram bullet is equivalent to a small car ramming into the same boulder at parking lot driving speeds. In either case, damage is going to occur.

The Work-Energy Theorem – Newton's Laws in One Tidy Package

Now that we have a good idea about how work and energy relate to each other, let's take another look at our derivation for kinetic energy, and remember that the quantity of mass times acceleration times distance also equals our expression for work. We could therefore write that,

$$KE = \text{mass}\cdot\text{acceleration}\cdot\text{distance} = Work$$

Also recall that our previous Δv is nothing more than the change in velocity from a starting point (let's call it point '1') to an ending point '2' (as in v_2 minus v_1). Given this, we can write,

$$Work = \text{m}\cdot\left(\frac{\Delta v}{\cancel{\Delta t}}\right)\cdot\left(\frac{\Delta v\cdot\cancel{\Delta t}}{2}\right) = \frac{1}{2}m\Delta v^2 = \frac{1}{2}mv_2^2 - \frac{1}{2}mv_1^2$$

Thus, the work performed on an object is really nothing more than a measure of the change in kinetic energy of the same object. *This concept is known as the work-energy principle of classical mechanics.*

Written in a compact form we can state,

$$Work = KE_2 - KE_1 = \Delta KE$$

As a direct consequence of this we can see that if there is no change in the velocity of an object of constant mass, there is no change in the kinetic energy of the object, and thus no work is being performed. No work implies 'no external forces' and therefore we satisfy Newton's first law of motion.

Also note that in order for the above scalar relationship to be satisfied, we have to make sure that the vectors that indirectly contribute to the final result are also satisfied. Thus, only resultant forces that act along the line of action of the motion contribute to the final change in velocity and work. Motion of objects at right angles (such as what we saw in Example 2.6) produces no work.

The Guru Says...

Generally speaking, any motion on an object (whether it is a tetherball on a string or a planet revolving around a star) that produces no change in the magnitude of the velocity {momentum} produces no work and is effectively 'static' from the standpoint of Newton's first law.

We now have a very powerful tool to avoid many of the pitfalls and traps that people fall into when trying to understand problems in physics. But we're not quite done with this whole work-energy theorem thing yet.

If we look at the case where the kinetic energy of a particle decreases (i.e., KE_2 is some finite value of kinetic energy less than the starting point KE_1 in our previous expression), this requires that the external force must be in a direction opposite the direction of motion. The work must also be in a negative sense to satisfy the resultant force vector floating around in the second law of motion. Thus, we can say that the work done on an object by a resultant force is equal to the negative of the work that the object does back on the force that it is interacting with (i.e., Newton's third law of motion).

This intuitively makes sense since if we have a tennis ball moving at a speed of 'x' in one direction and we want to slow it down, we need to apply some work to it in the opposite direction to reduce its energy of motion (kinetic energy). We can therefore make the general statement that *the total kinetic energy of an object is equal to the work that the object would do if we bring it to a resting state*. This fundamental concept of how matter moves in the universe is hidden inside the first law of motion, since this law defines the two extremes of the no-work condition (i.e., *uniformly unaccelerated motion* or *no motion* at all). Everything else in between is the interplay of work and energy. We are indeed making good progress in understanding how the universe works.

Conservative Forces, Friction, and the Conservation of Energy

There are two types of mechanical forces in the universe, those that are *conservative* and those that aren't. Unlike the common political usage of the term 'conservative', the definition of conservative in physics is something entirely different, and extremely rigid and unwavering in its application.

A *conservative force* is a force that is capable of recovering all of its energy over the course of its motion. This means that if the Guru moves from 'A' to 'B', energy is expended, but a subsequent motion from 'B' back to 'A' would return all the energy back.

A good example of a conservative force is gravity. If we lift an object up to a certain height, the object gains energy at our expense, which is stored as gravitational potential. If we drop the object, we get that energy back again. In a vacuum, and in the absence of any other external forces, we could repeat this process over and over until the end of time and always recover back the same amount of energy that we put into the object. Mathematically, we are stating the following regarding a conservative system,

$$\Delta KE + \Delta PE = 0; \quad \Delta KE = -\Delta PE$$

Since we know that work is equal to the change in kinetic energy, and that the kinetic energy of an object is equal to the work that the object can do if we brought it to a rest state, we can also write the following,

$$Work = W = \Delta KE = -\Delta PE$$

Suppose you wanted to siphon some water out of a tank and into a smaller container using a garden hose. How would you do that? Well, you would start the water flowing in the hose, and then make sure that the output end of the hose is lower than the water level in the tank. Since gravity is moving the water from the tank to the container, you really do not need to concern yourself with what path the garden hose takes. The only requirement is that the starting point of the hose is higher (i.e., has a greater gravitational potential energy) than the exit point. Imagine how cumbersome this simple task would be if gravity was not path independent.

Although conservative forces simplify the computations immensely, since we only need to know about the starting and ending points, they do not strictly exist in nature since somewhere there is always a little bit of friction around to slow things down. Friction is therefore a *non-conservative force*.

The Guru Says...

Thus we can say that the work performed by a potential {conservative} force (such as gravity) really only depends on the starting and ending points of the motion. Everything else in between is unimportant from the standpoint of doing meaningful work within the universe. Motion of this type is termed to be 'independent of path', or having 'path independence' in physics. Other conservative forces include the forces produced by an electric field and those produced by a magnetic field (although some physicists still argue about that last one).

In our previous siphon example we stated that we really do not care about what the hose did just as long as the output was lower than the input, so that the gravitational potential energy was downwardly positive and the water moved into the container. But, if we made the garden hose, say, 10 kilometers long, then we probably would not get any water flow since all of our gravitational potential would be consumed by the frictional losses in the hose. For most hardware store products this is not a significant problem, but we should be aware of it in any case. Let's explore the interaction of non-conservative forces using a noteworthy example.

Example 6.3: You're driving down a wooded interstate traveling at 60 km/h. Suddenly Bigfoot jumps out in front of your car 15 m ahead. You instantly slam on the brakes. Do you hit the clearly oblivious creature? Assume dry pavement conditions and a vehicle weight of 7500 N.

The mass of your vehicle is equal to the weight of the vehicle (7,500 N) divided by the acceleration of gravity (9.81 m/s^2). Thus, the mass would be,

$$m = \frac{7500\text{ N}}{9.81\frac{\text{m}}{\text{s}^2}} = 764.5\text{ kg}$$

The kinetic energy of your car would therefore be equal to one-half the mass times the velocity-squared, thus,

$$KE_{car} = \frac{1}{2}m_{car}\cdot v_{car}^2$$

$$= \frac{1}{2}(764.5\text{ kg})\cdot(60\frac{\cancel{\text{km}}}{\cancel{\text{hr}}}\cdot\frac{1\ \cancel{\text{hr}}}{3600\text{ s}}\cdot\frac{1000\text{ m}}{1\ \cancel{\text{km}}})^2 = \frac{1}{2}(764.5\text{ kg})\cdot(16.7\ \frac{\text{m}}{\text{s}})^2$$

$$= \mathbf{106,605.7\ J}$$

Looking back at Table 4.1, the coefficient of kinetic friction for a tire on dry asphalt is 0.8. Thus, the work produced by the 7,500 N car on the road is,

$$\text{Friction Force} = \text{Kinetic Friction Coefficient}\cdot\text{Weight of Car}$$

$$F_f = \mu_k\cdot W$$

$$= 0.8\cdot 7500\text{ N} = 6,000\text{ N}$$

Over a distance of 15 m, the work performed to slow the car down would be,

$$W = (6000\text{ N})\cdot(15\text{ m}) = 90000\text{ N}\cdot\text{m} = \mathbf{90000\ J}$$

This work is insufficient to remove all the kinetic energy from the vehicle in accordance with what we know about the work-energy theorem. **The excess 16,605 J of kinetic energy clobbers Bigfoot**. On the upside though, it's apparently not enough to severely injure him, and you manage to snap a couple of blurry photographs of the creature running away that you can sell to a tabloid magazine.

So, just for curiosity sake, how much more distance would we need to stop and miss Bigfoot entirely? This would be 16,605 J / 6,000 N or 2.8 m. Working this problem backwards you can now figure out how fast a car was traveling based upon the length of the skid marks – and catch some of the mistakes made on television crime shows.

Finally, it should be noted that from the expression $Work = W = \Delta KE = -\Delta PE$ for a conservative system, we can now generalize the expression as $\Delta KE + \Delta PE - W = 0$ or simply $KE + PE - W = 0$, to handle a non-conservative system. This states that the sum of the kinetic and potential energies of a system minus any work you can get out of the system is equal to zero – namely, *it is all a 'wash'*.

In other words, energy may be transformed into other kinds of energy, or work such as frictional heat or usable work, but the total of this quantity cannot be created or destroyed. The total energy of a system is always conserved. This is known as the *Principle of the Conservation of Energy*, and is a big deal in keeping the universe working.

Power and Mechanical Efficiency in the Universe

Work and energy are fine-and-dandy quantities, but in the real world *'power is king'*. While a small electric motor can provide the work to pump 1,000 liters of water, a larger motor can do the job in a much shorter period of time. Since humans are an impatient lot, we're typically more interested in how much work we can do in a given amount of time, rather than the overall quantity of work being done (which would place no physical restrictions on the device actually doing the work).

Enter the new physical quantity of *power,* whose unit is defined in the SI system by the *Watt* (W), with derived units of a *Joule per second* $(\mathrm{J/s})$, or *Newton-meter per second* $(\mathrm{N \cdot m/s})$.

Power is equal to the time rate of change of work. Since work is defined as force times distance the expression for power can be represented as,

$$Power = P = \frac{\text{Work}}{\text{time}} = \frac{\text{Force} \cdot \text{distance}}{\text{time}} = \text{Force} \cdot \frac{\text{distance}}{\text{time}} = Force \cdot \text{velocity}$$

$$\boxed{Power = P = Force \cdot \text{velocity} = F \cdot v}$$

We previously stated the above finding without proof, and now we know why it's true. We'll be using this simple scalar expression throughout the remainder of the Guru's Guide to tackle some fairly complex problems.

Example 6.4: In our previous Bigfoot example, how much power do the brakes dissipate when we first apply them?

If our braking force is 6,000 N, then our *effective power dissipation* by the tires at the moment we first hit the brakes would be,

$$(6000\ \mathrm{N}) \cdot (16.7\ \mathrm{m/s}) = 100{,}200\ \mathrm{W} = 100.2\ \mathrm{kW}$$

Or slightly over 100 kilowatts of power when we first apply the brakes. Where do you think the most amount of rubber from a tire in a skid mark occurs – *at the start of the skid, or at the end?*

Finally, the mechanical efficiency of a system is defined as the amount of output work you get versus the amount of input work. If we divide both the output and input by time, we can also state that the mechanical efficiency is the ratio of the output power of a system to its input power. Mathematically we can write this expression as,

$$efficiency = \eta = \frac{work_{out}}{work_{in}} = \frac{power_{out}}{power_{in}}$$

This is a very useful quantity when we start looking at how well a system converts energy into a usable form that we humans like.

Example 6.5: An electrical motor is lifting a 1000 N elevator car to a height of 100 m. Each cycle of the elevator consumes 100 watt-hours (Wh) of energy. How efficient is this activity?

To lift a 1000 N elevator car 100 m, we require the following amount of mechanical work from the motor,

$$W = 1000\ \text{N} \cdot 100\ \text{m} = 100000\ \text{N} \cdot \text{m} = 100000\ \text{J} = 100\ \text{kJ}$$

If the electric motor consumes 100 Watt-hours (Wh) of energy 'E' to accomplish this task have,

$$E = 100\ \text{Wh} = 100\ \text{W} \cdot 3600\ \text{s} = 100\ \frac{\text{J}}{\cancel{\text{s}}} \cdot 3600\ \cancel{\text{s}} = 360000\ \text{J} = 360\ \text{kJ}$$

Thus, the mechanical efficiency of the elevator system would be given by the following expression,

$$\eta = \frac{work_{out}}{work_{in}} = \frac{100\ \text{kJ}}{360\ \text{kJ}} = 0.2778 \approx \mathbf{27.8\%}$$

Which is pretty awful, and we should shop for a new electrical motor before this one catches on fire.

The Guru Says...

There is no difference between mechanical power and electrical power since – *power is power*. For that matter, *work is work*, and *energy is energy*, as these quantities are universal and independent of what source is generating or consuming them.

So, whether you see a unit expressed as part of a mechanics problem, or an electrical problem, or a problem dealing with heat or temperature, the same concepts of work and energy hold true no matter what the particular problem is under examination.

Chapter 7: Collisions

Things that go bam, wham, smash, crash and thud

The Guru is not paying attention to his driving and is about to experience one of the best kinetic energy transfer mechanisms in the universe – *a collision*. Collisions occur between objects large and small. An inattentive driver might collide with a tree; asteroids and comets occasionally smash into planets; electrons collide with objects in order to make a scanning electron microscope work; nuclear fission occurs due to neutrons colliding with other atoms; and the pressure of any gas exists solely as a result of random collisions of the gas molecules pushing against their container. If you think about it, collisions occur everywhere, *and their sole purpose is to transfer energy*.

You'll be happy to know that the topic of collisions is, for the most part, a simple extension of what we have already learned in our discussions of Newton's Laws and the concepts of work and energy. In fact, the fundamental basis for analyzing the forces due to collisions (and culminating with the Guru's favorite equation in physics) is derived entirely from Newton's second law.

When a collision occurs between two objects, a tremendous amount of force can be transferred over a relatively short period of time. The magnitude of this *'impulsive'* force can be so great that it can literally deform the physical shape of the impacting objects over the time it takes for the impulse to occur. We'll start by examining the kinematics of collisions, and then conclude by discussing the kinetics and the forces involved when things go smashing into each other.

One Dimensional Collisions

Let's start by looking at a simple example in Figure 7.1, where we will assume that both balls are the same size and mass, and that mass m_1 is moving with initial velocity v_1, and mass m_2 is stationary. Upon completion of the collision, we would expect m_1 to then be stationary, and mass m_2 to now be traveling at a velocity v_2', which should be somewhere in the neighborhood of v_1 for this particular example.

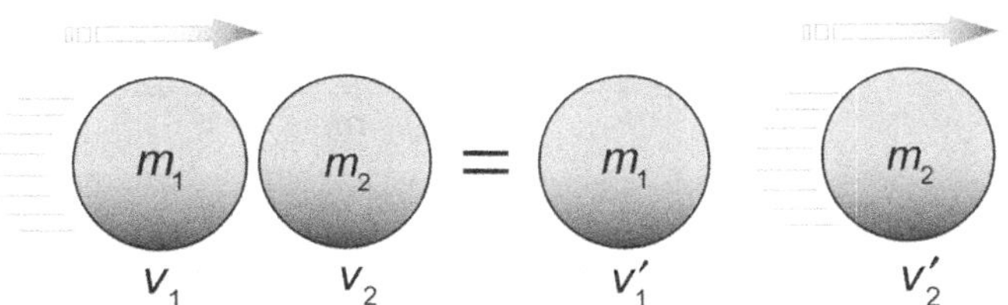

Figure 7.1: An Example of an Elastic Collision Where $m_1 \simeq m_2$ and $v_2' \simeq v_1$.

In the more general case where m_1 and m_2 are not equal in mass (let's make one mass out of magnesium and the other out of tungsten), we would intuitively expect that the resultant velocities v_1 and v_2 would be different, since the momentum $(m \cdot v)$ imparted by the different masses would be different. Mathematically we could write this general expression in the following form.

$$m_1 v_1 + m_2 v_2 = m_1 v_1' + m_2 v_2'$$

This states that both before and after the collision, we still have masses m_1 and m_2, but the transfer of momentum between them has now produced a new set of final unknown velocities v_1' and v_2'. We also know that since both mass and momentum is conserved in the universe, the change in velocities (i.e., v_1' and v_2' as well as v_1' and v_2') both before and after the impact, must be related to each other. Thus, we can also write the following intuitive expression describing the collision as,

$$e \cdot (v_1 + v_2) = v_1' + v_2'$$

Where the *fudge factor* 'e' is known as the *coefficient of restitution* and is a numerical description of just how 'good' the impact between the two objects was.

The coefficient of restitution ranges from zero to one, as can be seen in Table 7.1, and is a measure of how much energy is lost due to mechanical deformation, heat transfer, or the sound of the objects hitting each other when the collision occurs.

Table 7.1 Typical Coefficients of Restitution

OBJECT DOING THE HITTING	OBJECT GETTING HIT	COEFFICIENT OF RESTITUTION (e)
Billiard Ball	Billiard Ball	0.80
Any Wooden Object	Croquet Ball	0.60
Any Steel Object	Golf Ball	0.85
Any Wooden Object	Hockey Puck	0.39
Glass Marble	Glass Marble	0.66
Any Wooden Object	Ping Pong Ball	0.74
Tennis Racket	Tennis Ball	0.71

Solving the previous expression for v_1' and v_2' respectively, and substituting into our initial momentum expression for the collision, yields {after a lot of algebra} the following two independent equations describing the final velocities v_1' and v_2'.

$$v_1' = \frac{m_1 v_1 + m_2 v_2 + m_2 \cdot e(v_2 - v_1)}{m_1 + m_2}; \quad v_2' = \frac{m_1 v_1 + m_2 v_2 + m_1 \cdot e(v_1 - v_2)}{m_1 + m_2}$$

With these two kinematic expressions describing the final velocities of any two impacting objects in terms of their initial momenta, we can now analyze practically any impact that we could ever encounter. Even off center collisions, which do not occur along the line of action of the motion shown in Figure 7.1, can still be resolved into vectors in the correct directions and solved the same way. These types of impacts are called *oblique collisions*.

Elastic and Plastic Collisions

There are two special classifications of collisions, which arrive as a consequence of our solutions for the final velocities v_1' and v_2', namely the extreme cases where the restitution coefficient (e) is either '0' or '1'. Collisions that occur between really hard objects, like hardened steel ball bearings, and those that occur between soft, gushy materials, like two lumps of clay hitting each other, are good examples of these two extremes of restitution.

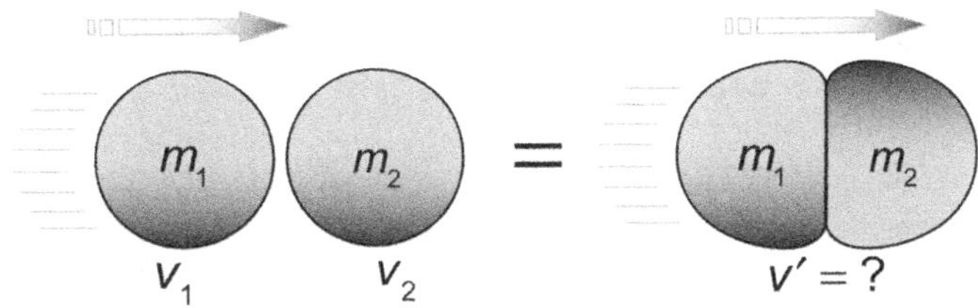

Figure 7.2: An Example of a Fully Plastic Collision.

We would term the first collision a *perfectly elastic collision* (shown in Figure 7.1) where $e = 1$, and the second a *non-elastic (or perfectly plastic) collision* (shown in Figure 7.2) where $e = 0$. Thus, for a perfectly elastic collision ($e = 1$) our restitution expression reduces to $v_1 + v_2 = v_1' + v_2'$ and we can write the final velocities as,

$$v_1' = \frac{m_1 v_1 + m_2 v_2 + m_2 \cdot (v_2 - v_1)}{m_1 + m_2}; \quad v_2' = \frac{m_1 v_1 + m_2 v_2 + m_1 \cdot (v_1 - v_2)}{m_1 + m_2}$$

For a perfectly plastic collision (i.e., $e = 0$) our restitution expression reduces to $0 = v_1' + v_2'$ or $v_1' = -v_2'$ since both masses are effectively fused together into one object and we can write the following final velocity solution as,

$$v_1' = v_2' = \frac{m_1 v_1 + m_2 v_2}{m_1 + m_2}$$

Example 7.1: If we modify our Newton's Cradle example from Chapter 4 to only have two masses, and then let Ball #1 fall, what is the velocity of Ball #2 under the following circumstances: a perfectly elastic collision, a perfectly plastic collision, and a collision where both masses are made out of glass marbles?

In a Newton's Cradle device, we know that all the masses are of equal size, so for the case where we have a perfectly elastic collision we can directly write the velocity of Ball #2 (v_2') in terms of the velocity of Ball #1 (v_1') as,

$$v_2' = \frac{m v_1 + m v_2 + m \cdot (v_1 - v_2)}{2m}$$

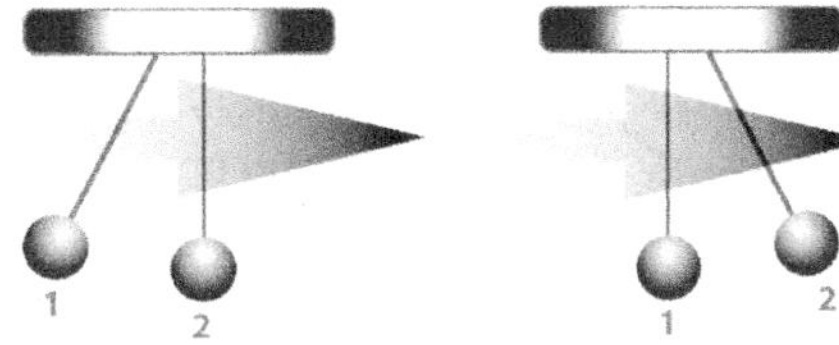

Performing the math and cleaning things up a bit, we find that the final velocity of Ball #2 is equal to the initial velocity of Ball #1. Momentum is completely conserved in this example. This is exactly what we would expect from observing the behavior of a Newton's Cradle.

$$v_2' = \frac{\cancel{m} \cdot (v_1 + v_2 + v_1 - v_2)}{2\cancel{m}} = \frac{2v_1}{2} = v_1$$

Now, what happens when Ball #1 collides into Ball #2 and they stick together? We start by writing the expression for the final velocity in a perfectly plastic collision with equal masses and $v_2 = 0$ as,

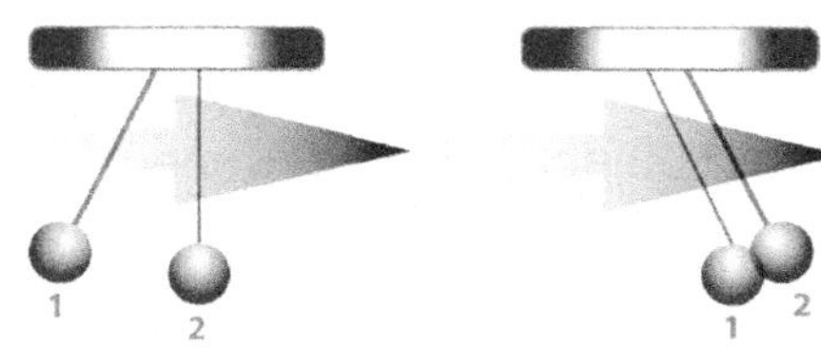

$$v_2' = \frac{m_1 v_1 + m_2 v_2}{m_1 + m_2} = \frac{m v_1 + m v_2}{2m} = \frac{\cancel{m}(v_1 + v_2)}{2\cancel{m}} = \frac{(v_1 + 0)}{2} = \frac{v_1}{2}$$

Thus the final velocity of Ball #2 would be only one half that of the initial velocity, since we lost velocity and effectively gained mass through the plastic collision. Finally, let's change the materials used in the Newton's cradle and see what our expressions would look like. Grabbing the generalized equations for the final velocities and substituting the restitution coefficient for glass marbles, we find,

$$v_1' = \frac{\cancel{m} \cdot (v_1 + 0 + 0.66 \cdot (0 - v_1))}{2\cancel{m}} = \frac{(v_1 - 0.66 \cdot v_1)}{2} = 0.17 \cdot v_1$$

$$v_2' = \frac{\cancel{m}(v_1 + 0 + 0.66 \cdot (v_1 - 0))}{2\cancel{m}} = \frac{(v_1 + 0.66 \cdot v_1)}{2} = 0.83 \cdot v_1$$

So in the case where we have neither a perfectly plastic, nor perfectly elastic collision, we can see that the amount of momentum transferred to each object becomes highly dependent on the restitution coefficient. In this case 83-percent of the momentum is transferred to the Ball #2, while 17-percent remains in Ball #1.

The Guru Says...

To find the coefficient of restitution of two objects we use a similar approach to how we found the coefficients of static and kinetic friction.

We start by taking the two objects and let one fall and hit the other (which one you drop is not important – remember Newton's third law). Measure the speed of the {first} falling object just before it hits the second {stationary} object, and immediately after it {the first object} bounces back from the surface. The ratio of the falling speed to the rebound speed is equal to the coefficient of restitution.

The Energy of Collisions

Recalling that our expression for a simple one dimensional collision is $m_1v_1 + m_2v_2 = m_1v_1' + m_2v_2'$ and that the kinetic energy of a particle is defined in terms of its momentum via the following expression,

$$KE = \frac{1}{2} \cdot m \cdot v^2 = \frac{(m \cdot v)^2}{2m}$$

We can rearrange our momentum expression to get all the v_1's on one side of the equation and all the v_2's on the other, and performing the above kinetic energy transformation we get,

$$m_1v_1 - m_1v_1' = m_2v_2' - m_2v_2; \quad \frac{(m_1v_1 - m_1v_1')^2}{2m} = \frac{(m_2v_2' - m_2v_2)^2}{2m}$$

Expanding terms gives the following ugly expression,

$$\frac{(m_1v_1 - m_1v_1')^2}{2m} = \frac{(m_2v_2' - m_2v_2)^2}{2m}$$

$$\frac{(m_1v_1 - m_1v_1') \cdot (m_1v_1 - m_1v_1')}{2m} = \frac{(m_2v_2' - m_2v_2) \cdot (m_2v_2' - m_2v_2)}{2m}$$

$$\frac{(m_1v_1)^2 - 2 \cdot (m_1v_1)(m_1v_1') + (m_1v_1')^2}{2m} = \frac{(m_2v_2')^2 - 2 \cdot (m_2v_2')(m_2v_2) + (m_2v_2)^2}{2m}$$

$$\frac{1}{2}m_1v_1^2 - (m_1v_1)(m_1v_1') + \frac{1}{2}m_1v_1'^2 = \frac{1}{2}m_2v_2'^2 - (m_2v_2')(m_2v_2) + \frac{1}{2}m_2v_2^2$$

Again rearranging terms so that they are in the same order as the previous expression, and noting that for an elastic collision the products $(m_1v_1)(m_1v_1')$ and $(m_2v_2')(m_2v_2)$ are zero, since before the collision v_1 is finite and v_2 is zero, and after the collision v_2' is finite and v_1' is zero, we get,

$$\frac{1}{2}m_1v_1^2 = \frac{1}{2}m_2v_2'^2$$

$$KE_1 = KE_2$$

So the kinetic energy of the system is maintained during the collision, *but only for the case where the collisions are perfectly elastic.* In all other cases, there is a loss in kinetic energy and after a couple of collisions the system will come to rest with all the kinetic energy being converted to heat and sound.

Impulse and Momentum – *Friend or Foe?*

Now we get to the really good stuff – *the kinetics of collisions*. We've all experienced the force of a collision. Have you ever stubbed your toe on something? Run into a door or wall by accident? Fallen down while ice skating or running? We know from personal experience just how much force can be experienced during these small *impulsive motions*, and how most of them really do not feel all that good.

To analyze just how much force is generated during the collision of two objects, like your toe on a heavy piece of furniture, we need look no further than Newton's second law which relates the acceleration of an object (your toe) to the force of what it is hitting. Let's start by going back to our original derivation point in Chapter 5 where we defined the force on an object as its time rate of change of momentum.

$$Force = F = \frac{\Delta(m \cdot V)}{\Delta t}$$

We'll rewrite our expression by multiplying both sides of the equation by the change in time (Δt), so that we have the following expression relating force and the change in momentum of the two objects,

$$F \cdot \Delta t = \frac{\Delta(m \cdot V)}{\cancel{\Delta t}} \cdot \cancel{\Delta t} = \Delta(m \cdot V)$$

Which basically states that the force experienced by the object doing the 'hitting' times the time increment (Δt) over which the 'hitting occurs', is equal to the change in momentum of the 'object doing the hitting'. The quantity $F \cdot \Delta t$ is referred to as the *impulse* of the system, and is what determines whether or not an impact is minor or severe, but more on that in a minute.

Since we know that $\Delta(m \cdot V)$ is nothing more than the change in momentum from some starting point '1' ($m_1 v_1$) to a final state '2' ($m_2 v_2$), let's write it as such. Thus, our revised expression relating the impulsive force to the change in momentum would be,

$$F \cdot \Delta t = \Delta(m \cdot V) = m_2 v_2 - m_1 v_1$$

$$m_1 v_1 + F \cdot \Delta t = m_2 v_2$$

Which renders the so-called *impulse and momentum equation* of classical physics. This simple looking equation has more applications in the universe than we could write about all day long, mostly due to the fact that it relates mass, velocity, force, and time all in one nice neat package. Anything that moves in the universe is subject to the constraints of this harmless looking equation.

But, *the cute equation has a sinister dark side that we'll see in a moment.*

Reading through this equation from left to right, it states that the starting momentum of an object (m_1v_1) plus a force times a time interval equals the final momentum of the object (m_2v_2). This would imply that the quantity of force 'times' time also has units consistent with those of momentum.

So, for a conserved system, the change in momentum during an impact, like that shown in Figure 7.3, is just equal to a numerical value. We'll call this number *'wham'*.

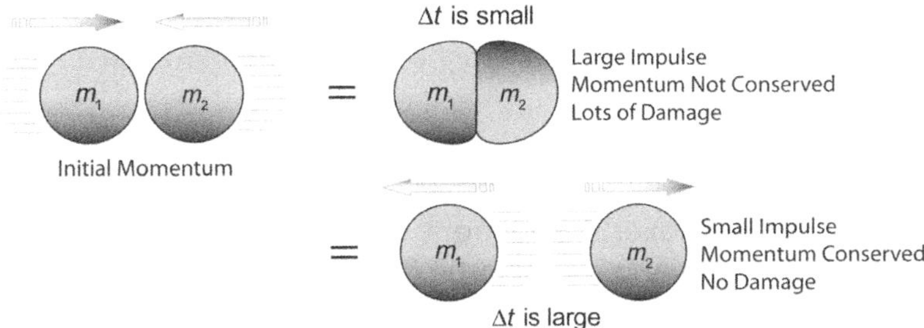

Figure 7.3: Examples of Impulses Applied to Moving Particles.

Thus for any collision, the following statement regarding the impulsive force and the time interval over which the impulse is applied must hold true,

$$F \cdot \Delta t = wham$$

There is only one mathematical way that you can adjust $F \cdot \Delta t$ and still have *'wham'* equal to a constant. *That is, if F goes up, Δt must go down, and if Δt goes up, F must go down.* This must occur in a fixed ratio so that we always get the same value of *'wham'* each and every time. *And here is the sinister part…*

For any given level of momentum exchange, the impulsive force generated by the impact can either be deadly great, or almost nothing at all, based entirely upon Δt – *the time of contact of the impulsive force*.

So, utilizing this newfound knowledge, have you ever wondered why getting hit by a brick is not as much fun as getting hit by a pillow? We can easily make the momenta of the two objects the same, yet the Guru bets that you'll always like the pillow more. This thought process is the same reason we wear helmets when riding a motorcycle and pad the dashboards of automobiles. Thus, if we *extend the time of contact between two objects* hitting each other, *we reduce the overall force of the collision.* Let's look at a couple of examples that take advantage of this point.

The Guru Says…

In mechanics problems, you typically do not know the time of contact (Δt) between two objects to a high level of accuracy without testing the exact problem in a laboratory.

The typical 'workaround' to this is to calculate the final answer in terms of the impulse ($F \cdot \Delta t$) since we always know the exact solution for the change in momentum of the system.

Example 7.2: Boxers use a technique known as 'riding with the punch' to reduce the force of impact from their opponent's fist. If a typical boxer's fist has a mass of approximately 4 kg and a punching speed of 9 m/s, what is the specific impulse to the opponent's head? If the boxer does not try to move out of the way and the time of contact is ($\Delta t = 0.1\,\text{s}$), what force does he feel, and how fast will his 7 kg head move backwards? Suppose the boxer rides with the punch for one-second, what is the force of the blow then?

If the boxer's opponent does not move, we have the condition where the final velocity is zero and he absorbs the entire blow. Thus,

$$m_1 v_1 + F \cdot \Delta t = m_2 v_2$$

$$F \cdot \Delta t = m_2 \cdot (0) - m_1 v_1$$

$$= 0 - (4\text{ kg}) \cdot (9\frac{\text{m}}{\text{s}}) = \mathbf{-36}\frac{\mathbf{kg \cdot m}}{\mathbf{s}}$$

Thus the impulse felt by the boxer would be -36 kg-m/s (it's negative because impulses are always felt opposite the acceleration due to Newton's third law). If we apply this impulse over a period of 0.1 seconds we get a resultant force of **360 N**, which is similar to having an anvil smack into his face. The final velocity of the boxer's head can be determined by setting the specific impulse we just determined equal to the momentum and solving for the velocity.

$$7\,\cancel{\text{kg}} \cdot v_2 = -36\frac{\cancel{\text{kg}} \cdot \text{m}}{\text{s}}$$

$$v_2 = \mathbf{-5.1}\frac{\mathbf{m}}{\mathbf{s}}$$

If the boxer rides with the punch for one second, the force on his head is reduced from 360 N to **36 N**, which is about the equivalent of a slap.

Example 7.3: A stuntman jumps off the roof of a 20 m building into a 5 m thick airbag. The airbag is designed to deflate and bring him to a stop over a period of 5 seconds. What is the force on the stuntman? What is the force if he misses the airbag such that he comes to a stop in only 0.2 seconds? Assume the stuntman has a mass of 80 kg.

We first need to determine how fast the stuntman is moving just before he hits the airbag. Since the total fall is 20 m and the airbag is 5 m above the ground, the distance from the roof to the airbag is therefore 15 m.

Using our kinematics expression for a freely falling object (neglecting air friction and with no starting velocity) we have,

$$y = -\frac{1}{2} g \cdot t^2$$

$$15\text{ m} = -\frac{1}{2}(9.81\frac{\text{m}}{\text{s}^2}) \cdot t^2$$

Where we chose 'down' as the positive direction to simplify the math. Solving for 't' we get,

$$\frac{1}{2}(9.81\frac{\text{m}}{\text{s}^2}) \cdot t^2 = 15\text{ m}$$

$$t^2 = \frac{15}{4.9}\text{s}^2$$

$$t = \sqrt{3.06}\text{ s} = 1.74\text{ s}$$

Thus the velocity of the stuntman is equal to the acceleration of gravity times the time he falls, which is 9.81 m/s^2 'times' 1.72 seconds, or 16.9 m/s. The force on the stuntman can therefore be determined as,

$$m_1 v_1 + F \cdot \Delta t = m_2 v_2$$

$$F = \frac{m_2 v_2 - m_1 v_1}{\Delta t} = \frac{0 - (80\text{ kg}) \cdot (16.9\frac{\text{m}}{\text{s}})}{5\text{ s}} = -270.4\frac{\text{kg} \cdot \text{m}}{\text{s}^2} = \mathbf{-270.4\ N}$$

This is a soft landing when distributed throughout his whole body. If the stuntman misses the airbag, he would fall the full 20 meters to the ground, making his final velocity 19.8 m/s at impact. More significantly, however, we need to replace the five seconds in the above denominator with 0.2 seconds, and the result would be **-7,920 N**. This is equivalent to getting hit by an automobile moving at freeway speed.

So, the morale to the story is, *never take your eye off the airbag when jumping from the roof of a building (or just avoid jumping off of buildings whenever possible).*

Problem Solving Tip...

When analyzing impulse and momentum problems, you always have to calculate the momenta of the system just before the moment of impact and find the impulse due to that.

Once you know the impulse, you can then apply it to determine what the final velocities will be. It's a two-step process in most cases.

Chapter 8: Rotational Mechanics

A discussion on things that whirl, spin, twist, and swirl

So far, we've discussed things in the universe that move in a straight line or up, down, left, and right – a concept physicists call *translation*. This works well for things like falling objects and spaceships, race cars and people running, but is missing something when we analyze spinning planets and ice skaters.

We need to develop a corollary to what we know so far, for things that have a propensity to spin about an axis. Lucky for us, we already have the framework to tackle this task and will discover in the process that translation and rotation are linked by something called the *center of gravity*. We'll also find out that rotational objects like to *'daydream'* a lot, and invent imaginary forces that appear real but are indeed science fiction at its best.

Basic Rotational Definitions

In order to discuss rotational motion, we need a coordinate system like we did in Figure 2.1; but in this case, we need to define an *axis of rotation* and some type of rotational variable, to help us keep track of how much something has moved relative to this axis of rotation. Most physics books do this by drawing some sort of 'blob' shaped object rotating about a fixed axis to show how we quantify rotational motion.

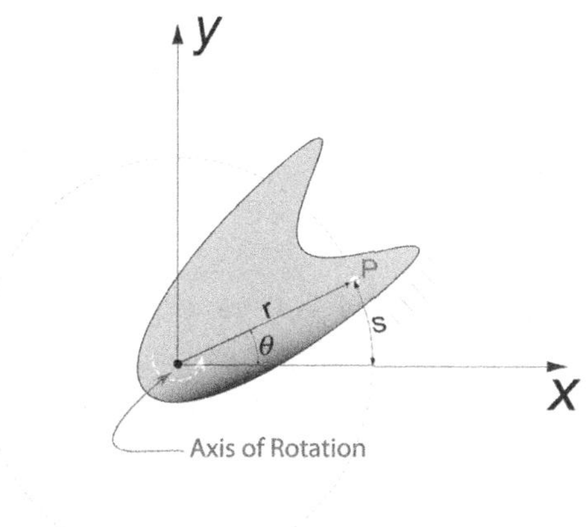

Figure 8.1: The Polar Coordinate System for *Positively Defined* Rotational Mechanics and the Obligatory 'Blob' of Physics.

Not to be one to alter tradition, the Guru has drawn a representative 'blob' in Figure 8.1, which is shown rotating counterclockwise through an angle θ.

Any point 'P' within the 'blob' will therefore rotate a distance from the 'x' axis, by an amount 's' called the arc length, which is defined by the following simple geometric expression relating the angle of rotation, to the distance 'r' between the rotational axis and any point 'P' as,

$$s = r \cdot \theta \qquad \theta = \frac{s}{r}$$

At this point, it is irrelevant how long it took for the 'blob' to rotate through the angle θ; it is only important that we define this rotation with the variables as shown, to completely describe the motion (i.e. using the *polar coordinate system* of 'r' and 'θ'). Looking back at the SI units in Table 1.6, the angle θ uses a non-dimensional unit called the *radian (rad)*.

Recalling the *right hand rule* in Chapter 2, we know that we could place our right hand along the 'x' axis, and curl our fingers towards the 'y' axis, producing a positive 'z' axis in the direction that our thumb is pointing, which for Figure 8.1 would be straight out of the page. We will also ascribe that positive rotations occur in the direction shown in Figure 8.1, or *counterclockwise positive*, making them consistent with the coordinate system axes. In this chapter we are only going to concern ourselves with rotation about a single 'z' axis, although we could extend this coordinate system logic to the other two rotational dimensions, just as long as we remember where our right hand is at all times.

Now that we have defined how an object rotates in a general sense, we are never too far away from determining the vector components of the motion as a function of the 'x' and 'y' coordinates (yes, rotational motion is also a vector quantity). Since the angle of rotation is always measured with respect to the 'x' axis, as shown in Figure 8.1, the motion in the 'x' vector direction is defined as $|r| \cdot \cos(\theta)$, while the motion in the 'y' direction is $|r| \cdot \sin(\theta)$. There is no motion defined on the 'z' axis since this is the axis of rotation. Thus, our vector describing the motion in a Cartesian reference frame would be,

$$\bar{r} = <|r| \cdot \cos(\theta), |r| \cdot \sin(\theta), 0>$$

We'll find that expressing rotating objects in terms of just the scalars 'r' and 'θ' is simpler than doing everything in a vector-driven Cartesian frame, but we now know a transformation to convert from one coordinate system to another just in case.

Knowing how much something rotated in totality is one thing, but often we are more interested in knowing *how fast something is rotating*. For this, we'll define the *angular speed* of an object by dividing the total rotational angle by how long it took to rotate through said angle. Thus, we have,

$$\omega = \text{angular speed} = \frac{\Delta\theta}{\Delta t}$$

Since the rotational angle has no units (i.e., the radian is equal to distance divided by distance, and thus is unitless), angular speed is simply equal to $1/\text{time}$ and would be expressed in revolutions per second, like an old record player, or more commonly in radians per second (rad/s).

Similarly, if we divide the above expression by the time interval again, we are dividing velocity by time, and arrive at what is known as the *angular acceleration*, which is now measured in units of $1/\text{time}^2$, or radians per second-squared (rad/sec^2).

We define this as,

$$\alpha = \text{angular acceleration} = \frac{\Delta\omega}{\Delta t} = \frac{\Delta\theta}{\Delta t^2}$$

We now have at our disposal all the fundamental quantities describing rotational motion. Let's put them together to generate some meaningful kinematic expressions.

Combined Translational and Rotational Kinematics

The choice of the rotational axis of a freely spinning 'blob' in space is not arbitrary by any means. Whereas an object with a fixed rotational axis, such as a tire on the axle spindle of an automobile, is constrained to move about a certain predefined location, the tire itself, disconnected from the car and whirling through space, will *always* rotate about a point mathematically equivalent to the *mass centroid*. This point, known appropriately as the *center of mass* or *center of gravity*, is a function of the geometry of the object and how the mass is distributed within it.

Thus, in a two-dimensional object, such as the one previously shown in Figure 8.1, the center of mass will lie somewhere along the surface of the object, and as we have shown, need not be located in the exact geometric center of the object. In fact, only in uniformly symmetrical objects (in both mass distribution and shape), like a circle or equilateral triangle, will the center of mass be located exactly in the geometric center of the object.

This is the reason that all physics books draw the obligatory 'blob', to show that the theories applied to rotational objects are not strictly applicable to uniformly symmetric objects, but are generally applicable to any weird looking shape having rotational motion through the center of gravity.

The Guru Says...

Think of the center of gravity as the point in space that the gravity vector would act at, if you were somehow able to take all the mass of an object and condense it down to a single point.

Despite its name, the center of gravity does not require a gravitational force in order to work. This concept applies with or without gravity; hence its alternate name 'center of mass'.

Clever engineers place a rotational axis of an object at the same location as the center of gravity to avoid unwanted horizontal force effects, called *eccentricity* or *eccentric mass effects*, from producing unwanted (and potentially dangerous) motion.

In a three-dimensional object, the center of gravity is located somewhere internal to the shape as shown in Figure 8.2. We have shown the two-dimensional and three-dimensional center of gravities to be in generally the same location (i.e., one on the surface and one internal to the object), although the reader is cautioned that this need not be the case.

Given this, an object would therefore have a natural inclination to rotate about this central point in a way that *is completely independent from any translational motion it may have*. Yes, you heard that correctly. The complete kinematic motion of a complex three-dimensional object (or two-dimensional one for that matter) moving through space is completely decoupled from a rotational and translational standpoint, when we examine the motion from the position of the center of mass. The only thing locking the two different motions together is again the time variable 't'. This concept is demonstrated in Figure 8.3.

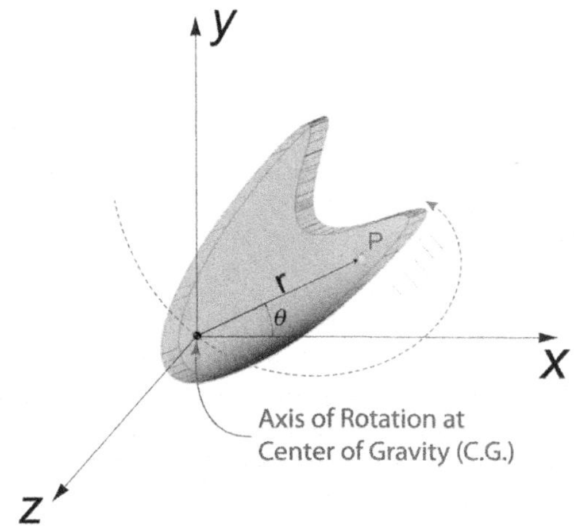

Figure 8.2: Our 'Blob' in Three-Dimensions Showing an Arbitrary Rotation about the Center of Gravity (C.G.).

The reason for this is firmly founded in Newton's first law of motion, since if we can treat the translation of a complex object as inertial motion of a simple particle, then the corresponding rotation must also go along for the ride and has no impact on the kinematic path being followed.

Anyway think about it, other than the effects of aerodynamic forces, and the like, on a twisting object, how confusing would the universe be if translation inertially depended on the final orientation of the object at any given time. The motion of objects in the universe would be completely different from what we know them to be.

Given that the design of this mathematical machine known as the universe is clever enough to decouple inertial motion, we should be able to exploit this fact by drawing on our previous knowledge and directly substituting our newfound expressions for rotation (θ), angular velocity (ω), and angular acceleration (α) into the previous kinematic expressions from Chapter 4, to arrive at a set of equivalent expressions for rotational motion.

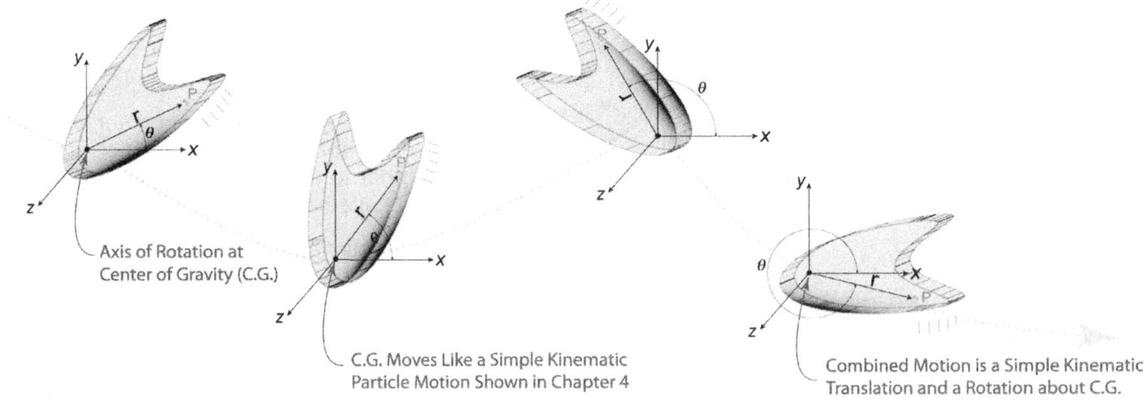

Figure 8.3: Combined Translation and Rotation of an Object in the Universe (the motion is a combination of simple particle motion and rotation about the C.G.)

Performing the substitution for the applicable variables, and assuming constant angular acceleration, we arrive at the following equations of motion for combined translation and rotation, which is pretty cool from a kinematics standpoint, since once we understand what's going on, we literally get two sets of equations for the price of one.

If the Translational Motion is Defined by…	*The Corresponding Rotational Motion is…*
$v = v_0 + a \cdot t$	$\omega = \omega_0 + \alpha \cdot t$
$x = x_0 + v_0 \cdot t + \frac{1}{2} a \cdot t^2$	$\theta = \theta_0 + \omega_0 \cdot t + \frac{1}{2} \alpha \cdot t^2$

Since the time variable appears in each of the expressions shown above, we can perform some additional substitution of one set of equations into the other and completely eliminate time to obtain one final alternate (and handy) relationship linking all three variables together.

This expression works out to be with some algebra,

$$v^2 = v_0^2 + 2 \cdot a \cdot x \qquad \omega^2 = \omega_0^2 + 2 \cdot \alpha \cdot \theta$$

Example 8.1: What is the angular speed of the second hand on your wristwatch? What is the angular speed of the minute hand? If the second hand is 13 mm long, how far does the tip travel in one day?

The second hand on a wristwatch completes one revolution every minute. In the units of angular measurement, one complete revolution is defined as $2 \cdot \pi$ radians since geometrically there are $2 \cdot \pi$ radians in a circle (this is also the unit we work in when analyzing things that spin about an axis). Thus for every minute (60 seconds) the second hand moves according to the following unit conversion.

$$\text{Angular speed}_{\text{second hand}} = \frac{1\ \text{rev}}{\text{min}} \cdot \frac{1\ \text{min}}{60\ \text{sec}} \cdot \frac{2\pi\ \text{rad}}{1\ \text{rev}} = \frac{2\pi}{60}\ \text{rad/sec} = \mathbf{0.105\ rad / sec}$$

The corresponding movement of the minute hand can be accomplished using the same process. Thus we can write,

$$\text{Angular speed}_{\text{minute hand}} = \frac{1\ \text{rev}}{\text{hour}} \cdot \frac{1\ \text{hour}}{3600\ \text{sec}} \cdot \frac{2\pi\ \text{rad}}{1\ \text{rev}} = \frac{2\pi}{3600}\ \text{rad/sec}$$

$$= 0.001745\ \text{rad/sec} = \mathbf{1.745 \times 10^{-3}\ rad / sec}$$

Finally, if we start at midnight with the second hand position at $\theta_0 = 0$, we can express the total angular rotation over a day (86,400 sec) as,

$$\theta = 0 + \omega_0 \cdot t = 0.105 \frac{\text{rad}}{\text{sec}} \cdot (86400\,\text{sec}) = 9{,}072\ \text{rad}$$

Remember that any angular motion is related to the linear motion of any point 'P' by the relationship $s = r \cdot \theta$. Since the unit of the radian is unitless, we can therefore calculate the total linear distance moved by multiplying the total angular measurement by the distance from the axis of rotation.

Thus,

$$s = r \cdot \theta$$
$$= (13\text{ mm}) \cdot (9072\text{ rad})$$
$$= 117{,}936\text{ mm} = \mathbf{117.94\ m}$$

Problem Solving Tip...

Remember, the 'glue' that holds translational and rotational motion together is 'time'. Both motions are inertially independent, but are time synchronized at every moment along an object's path.

This is slightly over the length of a football field each and every day. Considering what we know about the physical concept of work, is it any wonder that the batteries wear down over time?

Torque, Energy, and Momentum Corollaries of Rotating Objects

We have learned that when it comes to analyzing the rotational mechanics of an object, we can use the same type of equations that we developed for translational kinematics, and solve away. We also discovered that complex motion in space, containing rotation and translation, could be examined separately, since Newton's first law treats translational inertia and rotational inertia as different phenomena.

So if rotational inertia is its own separate phenomenon from its translational counterpart, we should probably expect that the forces that cause things to whirl, spin, twist, and swirl would also somehow be separately described as well. We can't just sum these two different forces together, since a 'force' in the translational sense implies a vector pushing or pulling on an object along its line of action, and this motion cannot produce any rotational acceleration. Thus, we need a corollary to Newton's second law to separately treat the forces producing rotation. This brings us to the concept of *torque*.

The Guru Says...

The moon is what is called an *'eccentric mass'*. It is heavier on one side than the other, and is strongly influenced by the gravity of objects like Earth. This heavier side (which faces us) produces an inertial force that locks its rotation around Earth to a period equal to Earth's daily rotation (a phenomenon known as co-rotation). This is why we only see one side of the moon (the 'man on the moon') and cannot see its 'dark side' without the use of a spacecraft.

The torque on an object is in some ways a representation of how much 'rotational work' an object does. We've all experienced a rotational work problem. How many times have you encountered the lid of a jar or a stubborn bolt that required some extra *'umph'* to unloosen? That 'umph' is torque, and based upon how much of it you had to apply, you would definitely call it work.

Torque is defined as a *distance multiplied by a force*, which if you think about it is backwards from what our previous definition of what constituted 'work'. Technically we could have defined work as distance times the dot product of force, since it really doesn't matter which way we do the dot product operation. Here, however, it matters because the physical quantity of torque is a vector and the order of operation is something that we have to keep an eye on.

Mathematically, we define the torque on an object 'τ', to be equal to the following vector cross product,

$$\tau = \bar{r} \otimes \bar{F}$$

This is shown pictorially in Figure 8.4 in both a mathematical form as well as from the standpoint of our rusted bolt analogy. Both pictures in Figure 8.4 are identical in context.

We can see that it naturally follows that if we want more torque applied to an object, we could either increase the applied force '$\bar{F}$', or we could increase the *lever arm* '$\bar{r}$'. In fact, as we keep increasing the lever arm, we can correspondingly reduce the applied force making our part of the work less.

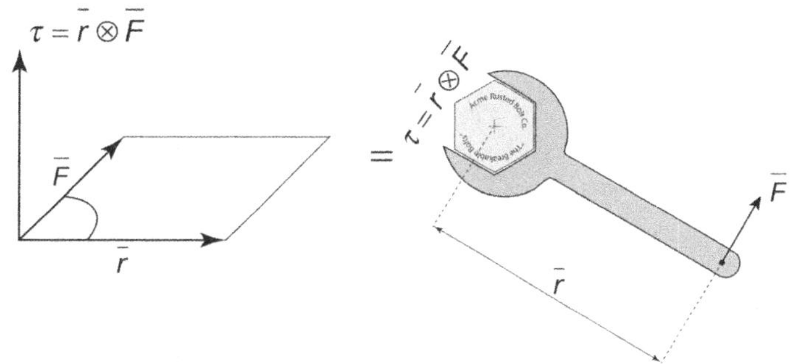

Figure 8.4: The Mathematical Definition of the Torque on an Object, and an Example.

Recalling from Chapter 2 that we could expand a vector cross product into its scalar components, we can also write our equation for torque in the following manner.

$$\tau = \bar{r} \otimes \bar{F} = <(r_y F_z - r_z F_y),(r_z F_x - r_x F_z),(r_x F_y - r_y F_x)>$$

A lever is one of the ancient machines used to modify force in a way that allows humans to move objects far more massive than could be moved with bare hands alone.

Example 8.2: Examining the wrench problem shown in Figure 8.4, if the effective lever arm is 50 cm, what is the applied torque on the bolt when we apply 20 N of force?

The Guru has redrawn the problem below where we have placed the origin of the coordinate system at the point where the bolt would rotate (Hint: this always makes the math easier). As shown in this figure, this is the correct positive mechanical sense for removing a stubborn bolt. Thus, we have a force of 20 N applied in the 'x' direction, or <20,0,0> N and a lever arm of 50 cm along the negative 'y' axis, or <0,-50,0> cm.

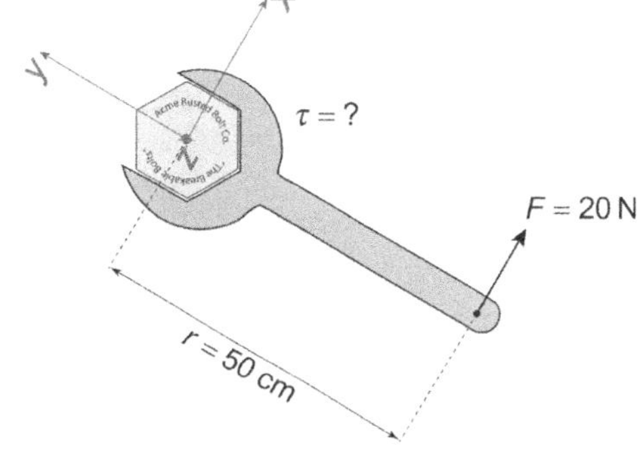

Since we are performing a vector cross product operation, we know that the resultant will be yet another vector which is perpendicular (i.e., orthogonal) to both the applied force and the lever arm, and this operation will automatically add another dimension to our final answer. Thus we would expect the torque to be in the 'z' direction and pointing out of the page (remember, the right hand rule applies for all cross products). Applying our cross product formula we can write the expression for the torque on the bolt due to the wrench by multiplying each of the vector terms by their applicable pairs.

Doing this we have,

$$\begin{aligned}\tau &= \bar{r} \otimes \bar{F} \\ &= <0,-50,0> \otimes <20,0,0> \\ &= <(-50)\cdot 0 - 0\cdot 0>, <0\cdot 20 - 0\cdot 0>, <0\cdot 0 - (-50)\cdot 20> \\ &= <0,0,1000> \text{N}\cdot\text{cm} = \mathbf{<0,0,10>\ N\cdot m}\end{aligned}$$

So our applied torque to the bolt is a positive 10 $\text{N}\cdot\text{m}$. We can see that if we want to increase the torque, we could either apply more force, or make the lever arm longer; either one would increase the desired twist on the bolt. In this case, the mechanical advantage is proportional to the length of the lever arm (double it, and we double the applied torque, and so on).

Moving on to a slightly different topic, we know that the kinetic energy of an object is physically defined as,

$$KE = \text{mass}\cdot\text{acceleration}\cdot\text{distance} = \frac{1}{2}m\cdot v^2$$

The linear velocity '*v*', shown in Figure 8.5, of any point 'P' on a rotating object is simply defined as the distance from the axis of rotation to point 'P' times the rotational speed 'ω' of the object, or $v = r\cdot\omega$. Performing this substitution into our energy expression above we can directly obtain a new rotational form of the kinetic energy equation dealing solely with rotational motion. This expression is given as,

$$\begin{aligned}KE &= \frac{1}{2}m\cdot v^2 \quad v = r\cdot\omega \\ &= \frac{1}{2}m\cdot(r\cdot\omega)^2 \\ &= \frac{1}{2}(m\cdot r^2)\cdot\omega^2\end{aligned}$$

Problem Solving Tip...

We should also note that in a manner similar to what we did for the dot product of two vectors, we can also obtain a scalar representation of the cross product (effectively the magnitude of the resultant vector of the cross product operation) by multiplying the magnitude of the projection of the $\bar{r}$ vector in the direction of the $\bar{F}$ vector and taking the sine of the angle between them. Thus we can also write an alternative 'scalar' solution to our torque problem as,

$$\tau = \bar{r}\otimes\bar{F} = |\bar{r}||\bar{F}|\sin\theta$$

We just need to remember that even though we can express torque as a scalar, this physical quantity is nevertheless a vector and acts in a certain direction only.

$$KE = \frac{1}{2} I \cdot \omega^2 \quad I = m \cdot r^2$$

Where the new mysterious quantity 'I' is known as the *rotational inertia* or *moment of inertia* of the object.

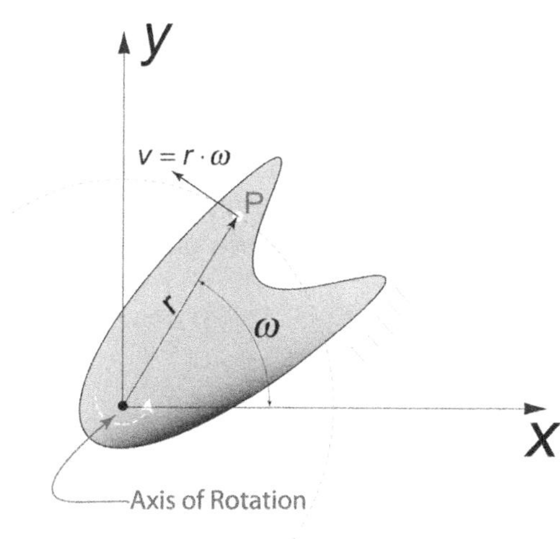

Figure 8.5: The Linear Velocity of Any Point on a Spinning Object is Equal to the Distance of the Point From the Axis 'r' Times the Rotational Speed 'ω'.

It is important to note that the moment of inertia of any object is highly dependent on where the axis of rotation is located. For example, the moments of inertia shown in Figure 8.6 are entirely different for each object shown even through we are using the same object to demonstrate this point. Thus it is vitally important to specify this quantity correctly based upon the axis of rotation.

The moment of inertia is a 'whole body' property in that it applies to the entire object; just in the same way 'mass' is a whole body property. In order to find out the moment of inertia for a particular object you need to either, a) cut the object into a bunch of little pieces and sum up all the masses times the squares of their distances to the rotational axis, since $I = m \cdot r^2$, or b) look up the moment of inertia of a particular object in an engineering textbook or on the internet. The second method is preferable for complex shapes.

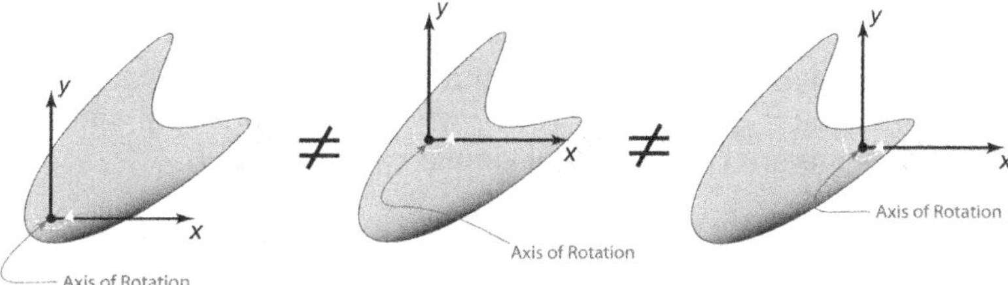

Figure 8.6: The Moment of Inertia is Different Depending upon where the Rotational Axis is Located. The Three Shapes Shown have Entirely Different Moments of Inertia.

Now, since kinetic energy is a scalar quantity (and we really like scalars a lot), we can express the total kinetic energy of an object that is in both translation and rotation as,

$$KE_{total} = \underbrace{\frac{1}{2} m \cdot v^2}_{\text{translational}} + \underbrace{\frac{1}{2} I \cdot \omega^2}_{\text{rotational}} \quad \text{with} \quad \underbrace{I = m \cdot r^2}_{\text{moment of inertia}}$$

We're almost done. We only need to figure out what the rotational equivalent of linear momentum would be, or the physical property of matter that keeps things spinning – something called *angular momentum*. The derivation of our expression for angular momentum is fairly straightforward and we'll borrow from what we've already done to make the derivation even easier.

Looking back at Figure 8.5 we note that every point 'P' in a spinning object, no matter how small it may be, has some sort of mass associated with it, and that mass is separated from the rotational axis by a finite value of 'r'. Thus, remembering that as the point 'P' moves in a circle it would have a linear momentum equal to $m \cdot v$ and a linear velocity of $v = r \cdot \omega$, we get the following,

$$L = r \cdot (m \cdot v_{\text{any point 'P'}}) = r \cdot m \cdot (r \cdot \omega) = m \cdot r^2 \cdot \omega$$

$$L = I \cdot \omega$$

Where 'L' is the physics abbreviation for angular momentum. Thus *the angular momentum of a system is simply equal to the moment of inertia of the object times its rotational speed about the axis of interest*. Angular momentum is what keeps a toy top spinning as well as what keeps the poles of a planet pointing in a fixed direction. Without angular momentum, the universe would be a very strange place indeed.

Now, that wasn't too painful, was it?

Example 8.3: A bowling ball is thrown down the alley at a speed of 10 m/s. If the ball weighs 7 kg and has a radius of 0.15 m, what is its rotational kinetic energy as it contacts the floor? What is the final translational kinetic energy?

(Assume the ball is uniform and that static friction applies everywhere)

We are told that the bowling ball is of uniform construction and that the rules of static friction apply. Thus, we can assume that the ball makes firm contact with the bowling alley and does not slide; it just starts to roll upon contact with the alley. Knowing this, we lookup the moment of inertia of a sphere and find it equal to,

y
r
x
z
$I = \frac{2mr^2}{5}$

$$I = \frac{2mr^2}{5}$$

Substituting the problem parameters into this expression, we find out that the moment of inertia of the bowling ball has to be about any central axis (since it is spherical),

$$I = \frac{2 \cdot (7\ \text{kg})(0.15\ \text{m})^2}{5} = 0.063\ \text{kg} \cdot \text{m}^2$$

We are told that when the ball hits the floor it is traveling at 10 m/s, thus we can calculate the angular velocity as,

$$v = r \cdot \omega$$

$$\omega = \frac{v}{r} = \frac{10 \text{ m/s}}{0.15 \text{ m}} = 66.7 \text{ rad/sec}$$

We also know that the total translational and rotational kinetic energy of the bowling ball is given by the following expression,

$$KE_{total} = \frac{1}{2} m \cdot v^2 + \frac{1}{2} I \cdot \omega^2$$

The total kinetic energy of our system is conserved, thus we start out entirely with translational kinetic energy and the ball moving through the air at 10 m/s. This gets converted to rotational kinetic energy as the ball starts rolling down the alley. Calculating the rotational component we find,

$$KE_{rotation} = \frac{1}{2} I \cdot \omega^2 = \frac{1}{2}\left(0.063 \text{ kg} \cdot \text{m}^2 \cdot (66.7 \text{ rad/sec})^2\right)$$

$$= \frac{1}{2}\left(4.2 \frac{\text{kg} \cdot \text{m}^2}{\text{s}^2}\right) = \mathbf{2.1 \ N \cdot m}$$

The final translational kinetic energy is going to be the difference between what we started with (pure translation) minus what is taken out by making the ball rotate (the pure rotational part). Thus we can write,

$$KE_{remaining} = \frac{1}{2} m \cdot v^2 - \frac{1}{2} I \cdot \omega^2 = \frac{1}{2}\left(m \cdot v^2 - I \cdot \omega^2\right)$$

$$= \frac{1}{2}\left(7 \text{ kg} \cdot (10 \text{ m/s})^2 - 0.063 \text{ kg} \cdot \text{m}^2 \cdot (66.7 \text{ rad/sec})^2\right)$$

$$= \frac{1}{2}\left(700 \frac{\text{kg} \cdot \text{m}^2}{\text{s}^2} - 4.2 \frac{\text{kg} \cdot \text{m}^2}{\text{s}^2}\right) = 347.9 \frac{\text{kg} \cdot \text{m}^2}{\text{s}^2}$$

$$= \mathbf{347.9 \ N \cdot m}$$

With the total kinetic energy of the two types of motion adding up to the initial value imparted by the bowler when the ball was thrown. So, the only thing we don't know how to do in this example is calculate the rotational force as a function of either the rotational speed or rotational acceleration, so that we can use what we already know about frictional forces to slow the ball down.

This is our final step, and the Guru has already laid all the groundwork for this problem when developing our expression for rotational kinetic energy. We'll do this by taking a brief side trip through the concepts of work and power in a rotating system to unlock an important physical property lurking inside.

Rotational Work and Power Concepts Reveal a Hidden Clue

In Chapter 6, we defined the translational work performed by a force as being equal to said force times the distance over which it acts. So we know,

$$W = F \cdot d$$

We also made the caveat that the force must act along the 'line of action' of the motion to produce non-zero work. The same rule applies to rotational work, and we need to make sure that any forces involved are applied in the direction of the object's rotation. Looking at our definition of work and referring to Figure 8.7 below, we know that the distance 'd' that the force 'F' is applied is equal to how far any given point 'P' travels along its circular path defined by the axis of rotation. Putting it another way, we will define the rotational work as how far the force 'F' can drag the point 'P' around the circle of radius 'r'.

From Figure 8.7 we can see that the change in *arc length* Δs that the point 'P' travels is equal to the distance that 'P' is from the rotational axis times the angle of rotation. Thus we can write that $\Delta s = r \cdot \Delta\theta$ and define the change in work of a rotating system as,

$$\Delta W = F \cdot \Delta s = F \cdot r \cdot \Delta\theta$$

But, as we have seen previously in this chapter, the force times the lever arm 'r' is equal to the applied torque 'τ' about the axis. We can therefore rewrite our expression for the change in work as,

$$\Delta W = \underbrace{F \cdot r}_{torque} \cdot \Delta\theta = \tau \cdot \Delta\theta$$

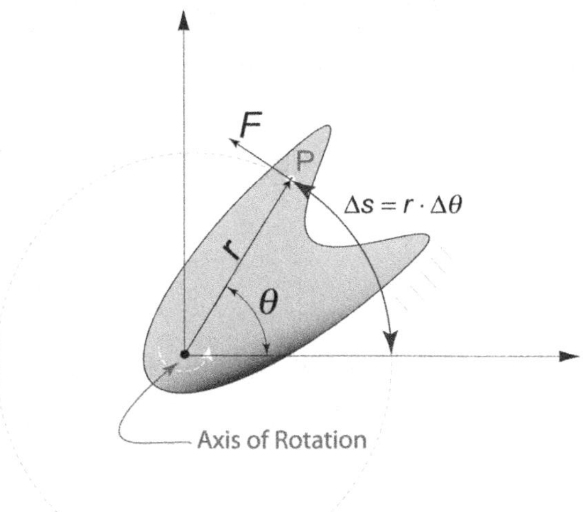

Figure 8.7: Representation of How Rotational Work is Defined

Now we know that the change in applied work on a rotating system is equal to the applied torque times how far the system rotates. We could divide both sides of the above equation by the change in time that it took for the angular rotation to occur and we would instantly have a handy relationship between power and torque.

$$\text{Change in Power} = \Delta P = \frac{\Delta W}{\Delta t} = F \cdot r \cdot \frac{\Delta\theta}{\Delta t}$$

$$= \tau \cdot \frac{\Delta\theta}{\Delta t} = \tau \cdot \omega$$

$$\boxed{P = \tau \cdot \omega}$$

So *the power generated by a rotational system is equal to the applied torque times how fast the object is spinning about an axis.*

Our expression makes sense from a physics standpoint, since in an earlier chapter we defined power as being equal to *force times velocity*. Since we know that 'ω' is the rotational velocity, it must therefore follow that the torque 'τ' must be equivalent to the force in this relationship.

We now know one of the variables in Newton's second law, namely that translational force 'F' is equivalent to rotational torque 'τ' – *a clue hidden within the work and power expressions of rotational mechanics.*

Rotational Kinetics – Newton's Second Law by Any Other Name

We could have started out this chapter using Newton's laws to develop a detailed vector derivation of rotational mechanics and the associated forces involved. This would have been a great ego builder for the Guru (at the expense of the reader), but you will conclude by the end of this section that it was infinitely easier, and far less painful, to do the energy and work derivations first, since they are scalar quantities, and then massage those findings to arrive at the final answer.

We know so far that the power produced by any spinning object (like an engine shaft, a flywheel, a windmill, etc.) is equal to the torque times the rotational speed $P = \tau \cdot \omega$. We also know that the kinetic energy of the same spinning object must also be equal to the following expression,

$$KE_{rotation} = \frac{1}{2} I \cdot \omega^2 .$$

Since power times some time increment Δt equals energy, it follows that we should be able to multiply power times Δt and equate this 'chunk' of energy to the rotational kinetic energy of the system. If we do this we wind up with the following expression,

The Guru Says...

Physical quantities that have the same SI units can be equated to each other regardless of their source. Therefore it is vitally important in physics to make sure that your units are correct and use them to guide you to the correct answer.

We can always multiply both sides of any physical expression by the same unit and produce no net change in the physical behavior of the problem. In fact, the units of 'time' and 'length' love to be multiplied and divided throughout equations to give other new expressions.

$$\underbrace{P \cdot \Delta t}_{\text{energy}} = \underbrace{\tau \cdot \omega \cdot \Delta t}_{\text{also energy}} = \underbrace{\frac{1}{2} I \cdot \omega^2}_{\text{still more energy}}$$

Now performing the requisite '*massaging of the equations*' gives an expression very close to what we are looking for.

$$P \cdot \Delta t = \tau \cdot \omega \cdot \Delta t = \frac{1}{2} I \cdot \omega^2$$

$$2 \cdot \tau \cdot \cancel{\omega} \cdot \Delta t = I \cdot \omega^{\cancel{2}}$$

$$2 \cdot \Delta t \cdot \tau = I \cdot \omega$$

A quantity of $2 \cdot \Delta t$ is still for all practical purposes equal to Δt, since it is assumed to be an arbitrary quantity no different than if it were equal to $3 \cdot \Delta t$ or $9.43 \cdot \Delta t$, etc.

If we were performing this derivation using differential Calculus, it would be a simple matter to show how the number '2' disappears when multiplied by our arbitrary Δt. However, since the Guru is avoiding using Calculus in this book, we have to approach the issue from a pragmatic standpoint rather than a wholly mathematical one. So, knowing this, we can clean up our answer from above to yield,

$$\Delta t \cdot \tau = I \cdot \omega$$

$$\tau = I \cdot \frac{\omega}{\Delta t} = I \cdot \alpha$$

$$\boxed{\tau = I \cdot \alpha}$$

Thus, *the torque on a rotating system is equal to the moment of inertia of the system times the rotational acceleration.*

This expression is Newton's second law for rotational motion. Notice how this expression follows the translational form of the equation $F = m \cdot a$. Physics is like that, and some of the best expressions also turn out to be the simplest.

Remember though, that we cannot add the rotational forces to the translational forces in a manner similar to what we did for the kinetic energies, since forces are vector quantities with very different lines of action.

With this final expression, we now have all the pieces of the puzzle to solve any problem in rotational mechanics. Time to look at an example of what we can do with it.

Problem Solving Tip...

Let's ponder this derivation for a moment. If we have a 'chunk' of arbitrary time Δt and we multiply or divide it by any number, we still have an arbitrary value – right? Arbitrary in this sense means that we really don't know or care about specifically how long the time interval is, only that it is finite and we can measure it. So if Δt were a really, really big number, $2 \cdot \Delta t$ would also be a really, really big number.

Consequently, if Δt were a really, really small number, $2 \cdot \Delta t$ would also be a really, really small number. This latter concept is the logic employed in Calculus to derive the limit of something.

Example 8.4: A four-cylinder automobile engine produces 100 kW of power at 2,000 RPM. What is the torque developed by the crankshaft? If the moment of inertia of the flywheel and crankshaft about the 'z' axis is $12.5 \text{ kg} \cdot \text{m}^2$, and the distance that each piston travels is 15 cm, how much downward force 'F' is required by each piston's connecting rod to accomplish this?

We first need to convert the rotational speed 'ω' of the crankshaft, in revolutions per minute (RPM), to the required SI units of rad/sec. One revolution is equal to $2\cdot\pi$ radians. Thus one revolution per second equals $2\cdot\pi$ rad/sec. For a 100 kW power output occurring at 2,000 RPM from the engine we can write,

$$P = \tau \cdot \omega$$

$$\tau = \frac{P}{\omega} = \frac{100{,}000\ \text{W}}{\frac{2000}{60}\cdot 2\cdot\pi\frac{\text{rad}}{\text{sec}}} = \mathbf{477.7\ N\cdot m}$$

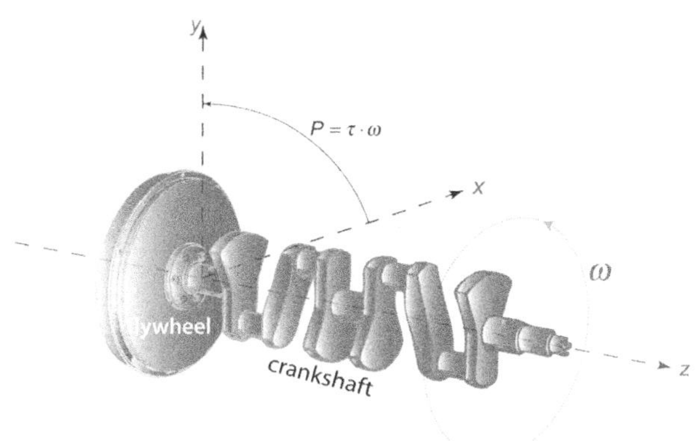

We find out that under this condition the torque delivered by the crankshaft is roughly 478 N·m. Thus, the power and torque output in a car engine are inextricably linked to each other by the above expression with the only adjustable factor being how fast the engine is revving.

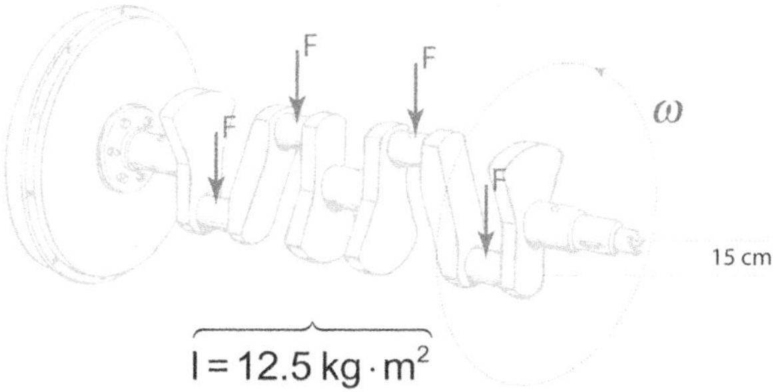

Now, if the moment of inertia of the flywheel and crankshaft assembly is $12.5\ \text{kg}\cdot\text{m}^2$ and the distance that the piston can travel (called the stroke) is 15 cm (0.15 m), then we can calculate the force 'F' required by each of the pistons to supply the desired power as,

$$\text{Work} = F\cdot d = P\cdot\Delta t = \frac{1}{2}I\cdot\omega^2$$

Substituting the parameters for our fictitious four-cylinder gasoline engine gives,

$$F\cdot d = \frac{1}{2}I\cdot\omega^2$$

$$F_{total} = \frac{1}{2\cdot d}I\cdot\omega^2 = \frac{1}{2\cdot 0.15\ \text{m}}(12.5\ \text{kg}\cdot\text{m}^2)\cdot(\frac{2000}{60}\cdot 2\cdot\pi\frac{\text{rad}}{\text{sec}})^2$$

$$= (\frac{3.33}{\cancel{m}})\cdot(12.5\ \text{kg}\cdot\text{m}^{\cancel{2}})\cdot(209.33\frac{\text{rad}}{\text{sec}})^2 = 1{,}823{,}967.9\ \frac{\text{kg}\cdot\text{m}}{\text{sec}^2} = 1.824x10^6\ \text{N}$$

$$F_{per\ cylinder} = \frac{1.824x10^6\ \text{N}}{4} \approx \mathbf{456\ kN}$$

So each piston needs to produce slightly less than a half mega-Newton of downward force per piston stroke. Considering that the generated power under this condition would easily power several large houses with everything turned on, is it any wonder that engines get hot.

Conservation of Angular Momentum and Precession

Previously we discussed that the total angular momentum of a rotating object is equal to $I \cdot \omega$, where $I = m \cdot r^2$. If we get clever and take a look at how fast the angular momentum of a system changes as a function of time 'Δt' we would have the following,

$$\frac{\Delta L}{\Delta t} = \frac{I \cdot \Delta\omega}{\Delta t} = I \cdot \frac{\Delta\omega}{\Delta t} = I \cdot \alpha = \tau$$

So the time rate of change in angular momentum is equal to the applied torque to the system. If we now set the torque 'τ' equal to zero then we have the following,

$$I \cdot \frac{\Delta\omega}{\Delta t} = 0$$

Problem Solving Tip...

In mathematics we call the first answer discussed here the trivial solution and the second the non-trivial solution. We'll choose the non-trivial solution since otherwise we've proven the existence of the number 'zero', which isn't all that exciting.

We have two conditions whereby this equation is satisfied; namely, the case where the moment of inertia 'I' is equal to zero (which is a non realistic solution since in order for the object to have any mass in the universe, it must also poses a moment of inertia), or, the case where the angular speed of the object is constant. If the later case were true, then the rate of change of a constant angular speed (the angular acceleration, α) would need to be zero.

So, if we have a constant angular speed with no angular acceleration, the angular momentum of the system is a constant and does not change. This brings us to a very important concept in classical physics known as the *conservation of angular momentum*. This law states that, *when the external resultant torque acting on an object is zero, the angular momentum of the system remains a constant and is conserved.*

...and it's a good thing too that the non-trivial solution governs the universe; otherwise the planets would have stopped rotating long, long ago.

The Guru Says...

A good example of how angular momentum is conserved (i.e., $I \cdot \omega = \text{constant}$) is to watch an ice-skater spin in a circle. If she starts the spin with her arms outstretched (large moment of inertia), she spins at a certain rate, ω. If she then pulls her arms towards her body (reducing her moment of inertia) while still spinning, this does not change her conserved angular momentum, thus the spin rate must adjust to compensate and she will spin faster.

You can try this at home using a chair or stool that spins. For a really pronounced effect, put an equal mass weight in each of your hands.

Whirling Objects That Defy Gravity?

The mathematical explanation of what goes on when a top is spinning, and how it slows down and ultimately stops, is a long and arduous adventure through the world of vector algebra. We'll skip that and focus on a more conceptual explanation.

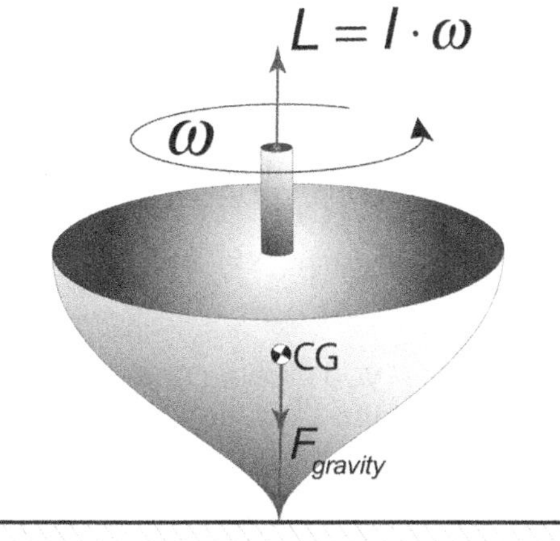

Figure 8.8: The Basic Kinematic Motion of a Toy Top

Once you start a toy top spinning, like the one in Figure 8.8, by applying a rotational torque to the stem at the top, its momentum is conserved. Thus, the angular speed ω is constant and we would expect to see motion like what is shown to the right. Both the angular momentum vector '*L*', as well as the gravity vector '*F*', point straight up and down from the center of gravity of the top, and these two vectors lie along the same line of action. Life is good, and the top, absent any frictional forces such as aerodynamic drag and surface friction, will spin away until the universe ends. You can get very close to this with one of those magnetic levitation tops from a scientific supply company which supports the top in a near-frictionless magnetic field, allowing a really well made one to spin for days on end without human intervention.

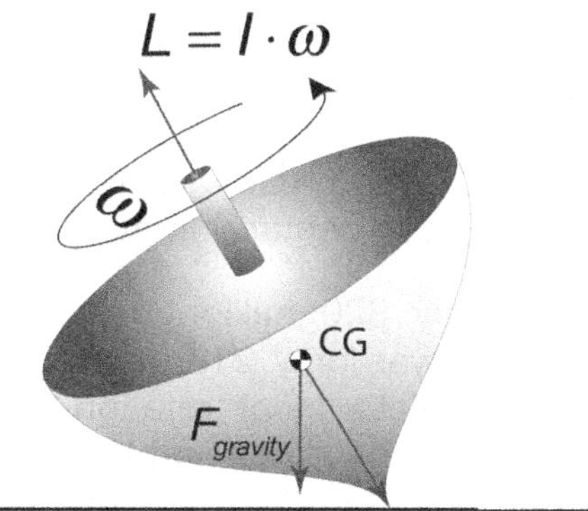

Figure 8.9: The Motion of a Top as it Slows Down and 'Wobbles'

Tops are designed with a pointed end to give them a definite spin axis location. Unfortunately though, things ultimately start to slow down due to friction, and the angular momentum is in reality not conserved. The frictional torque produces a rotational deceleration causing the gravity and angular momentum vectors to misalign as shown in Figure 8.9 (because the gravity vector starts to do what it does best, lower the center of gravity to its lowest possible point).

This off center tipping moment causes the top to lean to the side as it spins, all the while the angular momentum vector '*L*' is desperately trying to keeps things at a status quo. Unfortunately this motion only gets worse as the misalignment of the vectors causes the top to slow down even more since the forces are no longer acting along similar lines of action.

The kinematic effect of this is that the top will begin to *precess* in a circular motion, as shown in Figure 8.10, opposite of the angular speed direction. This motion will cause the angular momentum vector to revolve about a precession axis forming a conical shape in space. The rate of the precession is proportional to the remaining torque left in the top.

As the angular momentum continues to decrease, the precessive motion will grow greater until the gravity vector '*F*' pulling downward is greater than the angular momentum vector '*L*'. At this point the top falls over and stops spinning.

So you can see that analyzing the motion of a spinning object is not too terribly difficult once you understand the basic underlying physics and know which way the vectors are pointing in space.

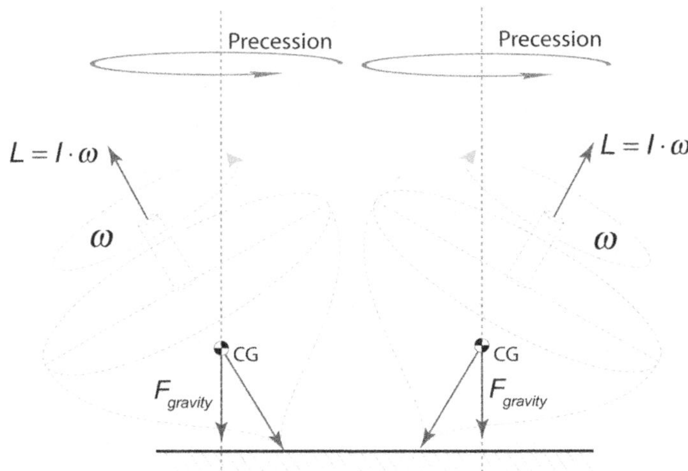

Figure 8.10: The Physical Phenomenon Known as Precession

The phenomenon of precession exists for all freely rotating objects in space. Earth for example, 'wobbles' on its axis due to slight changes in rotational speed associated with tidal friction. One complete precessional wobble of Earth takes a period of 26,000 years and produces variations in the polar ice caps due to differing areas being exposed to more or less solar radiation during the day (i.e., some areas on Earth get warmer over long periods of time, and some areas get colder).

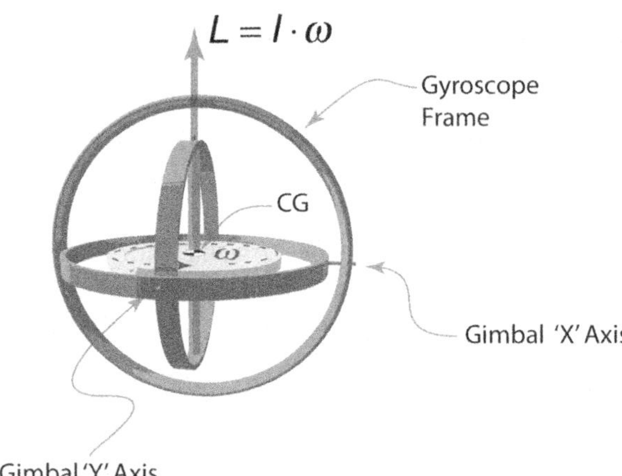

Figure 8.11: A Spinning Top within a Movable Frame – *The Gyroscope*

The effect of planetary precession and tidal friction also has been slowing Earth's rotation and increasing the length of a day by 1.7 milliseconds each century on average.

Roughly 620 million years ago, an Earth day was only 21.9 hours long, today a complete axial rotation takes 23.9 hours. On the other hand, moment of inertia changes due to the 2004 Indian Ocean earthquake caused the rotational speed of Earth to increase, shortening a day by three microseconds, so a plot of rotational speed versus long time periods is not a straight line.

On a different note, a similar rotating device, shown in Figure 8.11, that contains substantially more angular momentum is known as a *gyroscope*. Gyroscopes are principally used for maintaining the orientation of an object such as an airplane, missile, or spacecraft, regardless of its inertial position since the high rate of spin ensures a sufficiently large angular momentum vector, which always points the gyroscope in a set direction of interest.

You can make a simple gyroscope by rapidly spinning a bicycle tire (disconnected from the bicycle of course) and holding on to it by the axle. Once spinning, the bicycle tire will fight any motion you try to impart to it, since the angular momentum vector, which points along the axis of rotation, will want to keep it pointed in the same direction (i.e., momentum is conserved and it takes a torque by you to point the spin axis in another direction).

Try it sometime (being mindful of your fingers around a spinning bicycle wheel) and you'll find out that the wheel puts up quite a fight. You can even balance the bicycle wheel on one finger at the end of the axle and it will sit there without falling to the floor – *appearing to defy gravity*. We know from our previous examination of a spinning top that gravity is still there; it's just that the angular momentum vector is producing sufficient torque to counteract it.

Mechanically, a gyroscope is identical to our spinning bicycle tire in that a spinning disk of sufficient mass is free to 'spin up' in any predefined initial orientation. Once the gyroscope is calibrated in this way, the central axis (which is also the angular momentum vector line of action) will always point towards the calibration direction regardless of how the gyroscope frame tilts or rolls. The remaining mounting hardware (consisting of two pinned hoops, *called gimbals*, connecting to the gyroscope frame) exists only to allow unrestricted motion of the frame with respect the central spinning axis (which does not want to change direction).

The Guru Says...

Despite some nonsense you may have seen on television, *gyroscopes do not defy gravity.*

The Guru has seen some individuals on television demonstrating gyroscopic action and proclaiming that the gravitational force is somehow being 'defied' based upon the fact that spinning gyroscopes appear lighter when in motion, and refuse to fall over.

You of course know, based upon what we've learned about vectors, lines of action, and particularly angular momentum that this is nonsense, and that there are real forces in play. Just because the angular momentum vector acts against gravity, this does not negate the force of gravity.

Since we know (via our right hand) which two positive vectors are required to produce a vector in the direction of the unchanging 'L' vector, we now have a nice inertial coordinate system that will always provide us with the correct orientation in space regardless of whether or not we can see any recognizable reference points. The benefits of this device in a spacecraft or airplane flying at night should be obvious.

Phantom Forces and the Inertial Frame

Finally, things that rotate sometimes have the propensity to produce forces that are not necessarily real, but sure look that way depending on what reference frame you are standing in. Let's start out by looking at a 'force' that almost everyone thinks is real, but is nothing more than an illusion – the best science fiction you'll find out there. That force – *centrifugal force*.

Centrifugal (Latin for the term *center-fleeing*) force is an artifact of how we observe a motion, and the best testament on why we should never, ever, observe anything from a vantage point that is accelerating with respect to something else.

Figure 8.12: The Guru's Coffee Cup Moving to the Right and Out of the Car Window.

To prove our point, let's show the Guru in Figure 8.12 driving an automobile while drinking a cup of coffee. Because the Guru is a clever fellow, he puts his coffee on the dashboard of the car. He then makes a left turn into the parking lot of a donut shop and his coffee cup magically flies off the dashboard to the right and out the side window.

What just happened?

Well, the Guru turned in a circular path to the left and his coffee cup *fled away* from the center of the circle to the right. Certainly a centrifugal force acted on it, what else could it be?

Well not really. Take a look at a slightly different 'birds-eye' view of the same event as shown in Figure 8.13.

As the Guru takes the turn into the parking lot of the donut shop, he is forcing the car to turn to the left by having the front tires exert a frictional torque on the road in that direction. He's seat-belted into the car so he will also travel in this direction as well as everything else that it bolted to the automobile.

The Guru Says...

Most phantom forces arise as a consequence of you being in a different inertial reference frame, or through Newton's third law of motion since objects tend to experience the d'Alembert reaction forces and not the direct forces themselves.

This is the 'trick' in physics, to figure what things in the universe are real, and what is an error in perception. This is why we rely on mathematics to clarify the fuzziness of something we have not seen before.

The coffee cup, on the other hand, is sitting freely on the dashboard and as the car turns, the cup's natural tendency is to continue to move in a straight line (the first law of motion) unless acted upon by an external force. So, looking at Figure 8.13 we can see that in reality, as observed from our inertial coordinate system above the car, the car turns and the coffee cup merely keeps moving in the original straight-line direction. There is no mysterious sideways force acting on the coffee cup.

You should have had a clue that something was amiss with the notion of a centrifugal force, since as you make a turn, you yourself are pulled to the outside of the turn. Recalling d'Alembert forces we know that we always feel the reaction forces and not the actual force vector, thus if we are being pulled to the outside of a curve, the actual force must actually be in the other direction – *which it is.*

The real force pulling you into a turn is directed towards the center of the circle forming the turn and not away from it. This real force is called the *centripetal force* (i.e., accelerating towards the center of the turning circle).

Centrifugal force is just an artifact that the inertia of the object would rather make it travel in a straight line unless it is acted upon by the centripetal force. You can prove this to yourself by tying a mass to a string and spinning it around. The mass moves in a circle pulled centripetally by the string.

The mass might feel like it is being pulled to the outside, but the real force vector is directed along the line of action of the string. Now let go of the string (making sure no one is near you) and what happens? The mass moves in a straight line tangentially to the point of release, just as Newton's first law would predict. Thus, phantom forces do not exist in a non-accelerating inertial reference frame.

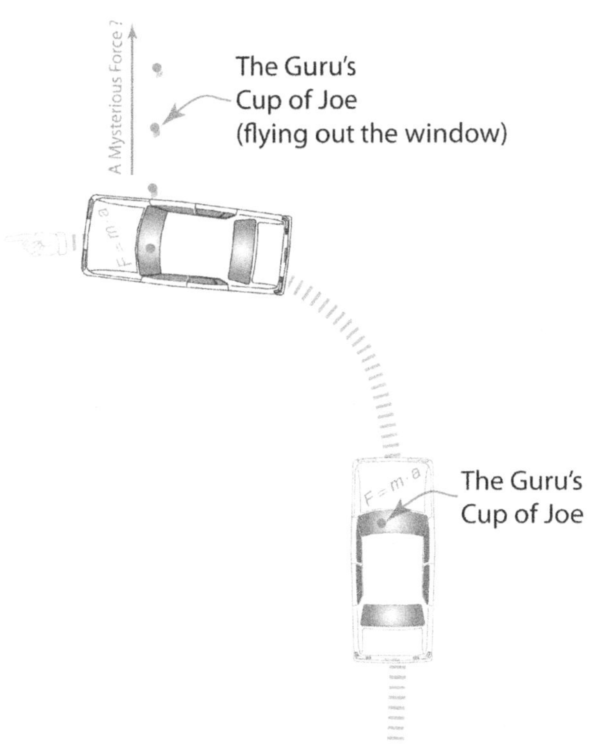

Figure 8.13: Newton's Coffee Being Acted Upon by a Mysterious Physical Force – *Or Is It ?*

Another phantom force is known as the *Coriolis force* or *Coriolis effect* named after French scientist Gaspard-Gustave Coriolis. Like the centrifugal force, the non-existent *Coriolis force* is also an artifact of using Newton's laws of motion in a rotating {non-inertial} reference frame.

The effect of the Coriolis force is to make objects, which are actually traveling in a straight line, appear to move in a curved path when viewed from the rotating frame. This effect can be demonstrated in Figure 8.14, where we have two spinning disks (like carousels at an amusement park), with one experimenter standing at the center of each disk.

In the first disk, an observer is standing in an inertial reference frame and watching the experimenter throw a ball. In this first example, the observer correctly witnesses the ball moving in a straight line. Newton's laws correctly predict the motion.

In the second disk, the observer is standing on the rotating disk while the experimenter throws the ball. In this case, the observer observes the ball traveling in a curved path opposite the direction of rotation of the disk. The ball 'appears' to have a different mysterious force acting on it – the Coriolis force, which is an artifact of the fact that we are not correctly observing the motion from an inertial reference frame and rather are seeing it from a rotating (and therefore accelerating) coordinate system.

The Coriolis effect shows up everywhere on Earth (albeit in small but measurable quantities, because it takes one entire day for a complete rotation of the 'disk' and the effect is only noticeable for motions occurring over large distances and long periods of time).

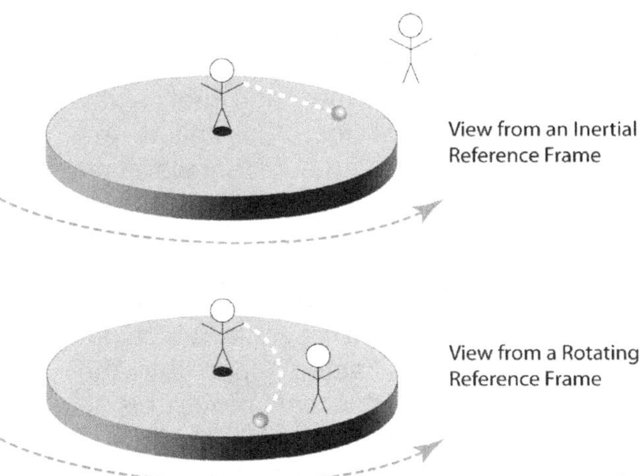

Figure 8.14: The Coriolis 'Force' as Viewed from Two Different Reference Frames – One Inertial, One Spinning (and hence accelerating).

Thus, straight horizontal motion is slightly biased in a *clockwise direction in the northern hemisphere* and in a *counterclockwise direction in the southern hemisphere*. No better example of this can be seen than in looking at a satellite photo of cloud cover on Earth where cyclones and hurricanes are biased in a definite direction depending upon the hemisphere in which they are located. The oceans and flowing water in large bays and rivers can also exhibit some Coriolis effects, but only to a very small extent.

Despite popular belief, Earth's Coriolis effect does not make bathtubs or sinks tend to drain in one particular direction or another. You can make a bathtub drain either in a clockwise or counterclockwise direction just by swirling the water (applying a torque) in a particular direction prior to pulling the plug.

Problem Solving Tip...

Why do we say that the system is still accelerating even though the rotational speed is constant? Simple, all rotational motion is vector motion, and unlike a scalar, which only has magnitude, a vector has magnitude as well as direction.

Change one or the other (i.e., the direction or the magnitude) and you alter the vector. Newton's law of inertia says that objects want to go in a straight line unless acted upon by an external force, while the second law states that the only way to get that force is to apply acceleration.

Thus, objects that are freely rotating are always accelerating towards the center of the rotation.

Section III – What Keeps the Universe Moving

Chapter 9: Fluid Mechanics and Wave Motion

The physics of liquids and gasses and sound

Since we now have a really good understanding of how things interact with each other in the universe, we're going to shift gears away from classical mechanics, and start looking at what powers the universe and keeps things moving right along. To do this, we need to establish some ground work on how non-solid objects interact in space and move energy from one point to another – the topics of *fluid interaction* and *wave motion*.

A *fluid* is any substance in which its constituent particles can move freely and relative to each other. A fluid can be either a *liquid* or a *gas*, and has the unique property of *continuous deformation*, in that it can continuously change shape. Whereas solids can transfer energy through elastic collisions and direct forces within their rigid internal framework, fluids transfer energy through pressure and wave motion. All of the elements in Chapter 3 that can behave as either a liquid or a gas (regardless of the temperature) are fluids.

The study of fluids is typically divided into three general categories called: *hydrostatics*, *hydrodynamics*, and *gas dynamics*.

Hydrostatics (or fluid statics) analyzes the forces on objects submersed in non-moving, or static, fluids like the ocean and the atmosphere and is the easiest to analyze from a mathematical standpoint.

Hydrodynamics, or fluid dynamics, deals with fluids that move when subjected to forces (i.e., changes in pressure), but do not change in physical properties (such as changing density, chemical composition, temperature, or mass). This classification of fluids requires a little more mathematical effort to get to an answer, but you'll find out that this topic is where all the good stuff happens, since it allows us to develop some physical theories along the same lines of what we have done previously in our study of classical mechanics.

Looking at hydrodynamics the other way around, instead of examining fluids moving by a solid object, if we now look at a solid object moving through a fluid, we have a study area known as *aerodynamics*. Aerodynamics is still what the Guru would classify as hydrodynamics, since aerodynamic analysis extensively worries about the forces produced by objects moving through the air. We'll take a cursory look at aerodynamics, since we all want to know what keeps one of those *heavier than air flying machines* aloft.

Finally, gas dynamics deals with the motion of fluids (gasses), which can be subject to changes in temperature, mass, density and chemical composition. This category of analysis involves looking at such things as supersonic shock waves from fighter jets, and gas plasma cutting torches. Although this area is of great interest to the Guru himself, we'll avoid problems in which the fluid changes property mid-stream, since mathematically it has sent many a college engineering undergraduate student screaming into the night.

Given this, we'll concentrate our discussion on two important fluids that we encounter in everyday life: water and air. So, let's start out looking at fluid statics and the mechanisms that make hot air balloons fly and submarines sink.

Fluid Statics – Affecting Objects through Inertia Alone

Fluid statics affects objects through inertial forces alone. Thus, we can think of the effect of a static fluid mass like a swimming pool full of water, as being equal to the total mass of the liquid times its attraction under gravity. This makes sense intuitively, since without a gravitational acceleration acting on a fluid, the concept of a static force produced by a liquid is pretty much a moot point.

Following this logic, it therefore makes sense that the magnitude of the force exerted by a fluid is equal to how much mass is present in any given volume, times the gravitational acceleration. We can therefore define the *specific weight* (γ) of a liquid to be equal to how much a fluid weighs per unit volume. In the SI system we would use the volume of one-cubic-meter, or m^3, as our reference volume.

Table 9.1 shows a sampling of the specific weights of various liquid fluids at room temperature. The bigger the number, the greater the inertial pressure effects produced on an immersed object, since we effectively have a more massive fluid object pushing on it.

Table 9.1: Specific Weights of Different Liquid Fluids

Fluid of Interest	Specific Weight (γ) in $\frac{kN}{m^3}$
Gasoline	6.7
Ethyl Alcohol	7.7
SAE 20 Weight Oil	9.0
Plain Old Fresh Water	9.8
Seawater	10.1
Glycerin	12.4
Mercury	133.7

The specific weight of a fluid also provides a direct indication of whether or not one liquid will float atop another, based solely upon the numerical value of γ. From this table, it should be clear which liquids are lighter than *'plain old fresh water',* and therefore would float in water if placed in it.

Since a fluid can completely surround an object, it's not practical to describe its force effects in terms of a single vector. The net effect of a fluid's weight produces an average force across the surface of the object in question. We'll call this average force over an applied area the *pressure* on an object. Knowing this, we can state the following fact about static fluids:

The amount of 'force' (i.e., pressure) that an object is subjected to when immersed in a fluid is proportional to the depth of the object in the fluid.

This finding is originally attributed to a fellow named Blaise Pascal in 1646, due to his observation that liquids are *incompressible,* in that that they do not change in density due to changes in applied pressure.

Pascal's Principle states that, "*... any change in pressure applied at any given point in a fluid is transmitted undiminished throughout the fluid.*"

Thus, Pascal's Principle states that if you apply a pressure to a confined liquid in one location, the pressure is uniform *everywhere*. Unwittingly, Pascal paved the way for the invention of the modern field of engineering known as *hydraulics*.

In a hydraulic device, the input force is applied over a sufficiently small area through a piston to create a large pressure change somewhere else in the system. Mathematically then, we have the following relationship between the input and output forces F_1 and F_2, and the associated piston areas A_1 and A_2, as shown in Figure 9.1.

$$P_1 = \frac{F_1}{A_1} = P_2 = \frac{F_2}{A_2}$$

$$F_2 = F_1 \cdot \frac{A_2}{A_1}$$

Thus, output force is equal to the input force times the magnification factor (mechanical advantage) of the larger piston versus the smaller one (A_2/A_1).

At first glance this might seem like some type of magic (in that we are getting something from nothing), but that is just the physics of the problem – *reality is sometimes stranger than science fiction.*

Mechanically, hydraulics is no different than applying a long lever arm to move a heavy object. Instead of a physical lever in this case, we are exploiting the fact that liquids are incompressible and using pressure to do the work.

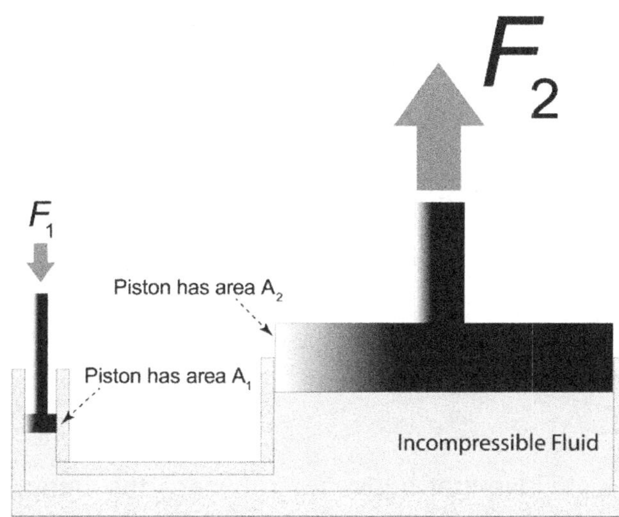

Figure 9.1: The Pressure in a Hydraulic System is the Same Everywhere.

Let's demonstrate some of these new fluid concepts by looking at a couple of examples.

Example 9.1: The wreckage of the RMS Titanic is located at a depth of 3,798 m. What is the pressure on the hull at that depth?

Looking up the specific weight of seawater (it's a little heavier than plain old water due to the dissolved salt content), we see that it weighs 10.1 kN/m^3. Thus the pressure on the hull at a depth 'h' is,

$$\text{Pressure}_{\text{Titanic}} = \gamma_{\text{seawater}} \cdot h_{\text{Titanic}}$$

$$= 10.1\frac{\text{kN}}{\text{m}^3} \cdot 3798\text{ m} = 38359.8\frac{\text{kN}}{\text{m}^2} = \mathbf{38.4\ MPa}$$

This is a pressure greater than the bursting strength of most modern pressure vessels. This pressure would easily crush most submarines and can only be withstood by advanced bathyspheres.

A variant of Pascal's Principle has been around for a very long time, and is known as *Archimedes' Principle*. Archimedes' Principle, named after its discoverer Archimedes of Syracuse (287 BC – 212 BC), is what keeps boats and submarines afloat and allows hot air balloons to rise.

Simply stated, Archimedes' Principle says, "*Any floating object displaces its own volume of fluid.*" Thus, the *buoyant force* produced by an object placed in a fluid is equal to the weight of the volume of the fluid it displaces. Let's see if we can do anything with this little tidbit of knowledge.

Example 9.2: A submarine has the ability to sink or float using a device known as a 'ballast tank'. These tanks are located internal to the hull of the submarine and allow for the vessel to adjust its buoyancy by regulating the amount of water contained within them.

Assuming a submarine has a mass of 6,500 metric tons, how much ballast tank volume is required to maintain neutral buoyancy?

Neutral buoyancy is where the buoyant force balances the gravitational force wanting to sink the object. Thus, we first need to calculate the weight of the submarine and then equate this force to a volume via Archimedes' Principle.

$$\text{Weight}_{\text{sub}} = 6500\ \cancel{\text{MT}} \cdot \frac{1000\text{ kg}}{1\ \cancel{\text{MT}}} \cdot 9.81\frac{\text{m}}{\text{s}^2} = 63765000\frac{\text{kg}\cdot\text{m}}{\text{s}^2}$$

$$= 63765\text{ kN}$$

So we need to produce $6.38x10^7$ N (63,765 kN) of buoyant force to prevent the submarine from sinking. We can find the required volume of the ballast tanks by substitution and solving.

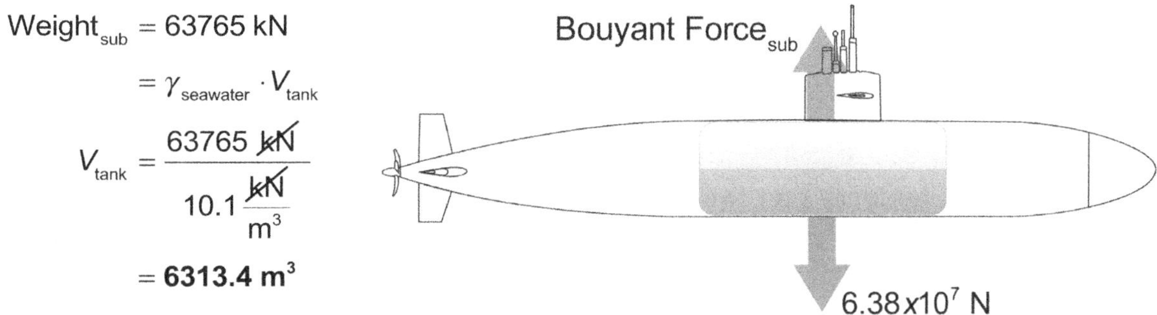

The required volume of the ballast tanks would therefore be 6,313.4 m^3 in order to keep the submarine barely afloat. This example of course assumes that the rest of the sub provides no buoyant force at all.

The function of the main ballast tanks is to slightly maintain a positive buoyancy of the sub. Additional tanks, called trim tanks, which are located in the front (bow) and rear (stern) of the submarine are constantly adjusted to actually complete the task of maintaining neutral buoyancy. When in the neutral condition, a sub is then essentially weightless and can rise or dive using external control surfaces called *dive planes.* This process is called *dynamic diving.*

It is not desirable to flood a submarine and let it continue to sink since, despite what is shown in the movies, Archimedes principle says it will continue to sink – *all the way down to the bottom.* Negative buoyancy is by definition 'sinking', and unless the commander knows the depth of the ocean he is in, the concept of *'putting the boat on the bottom'* can have disastrous consequences.

Example 9.3: A hydraulic lift at a service station is used to raise a 2000 kg automobile a distance of two-meters off the ground. If the input piston has an area of 2.5 m^2 and the output piston has an area of 10 m^2, how much input force is required? How much work is performed? How far do we have to move the input piston?

We know that the mechanical advantage of a hydraulic system is equal to the ratios of the input and output piston areas. Thus, the mechanical advantage of our hydraulic system would be,

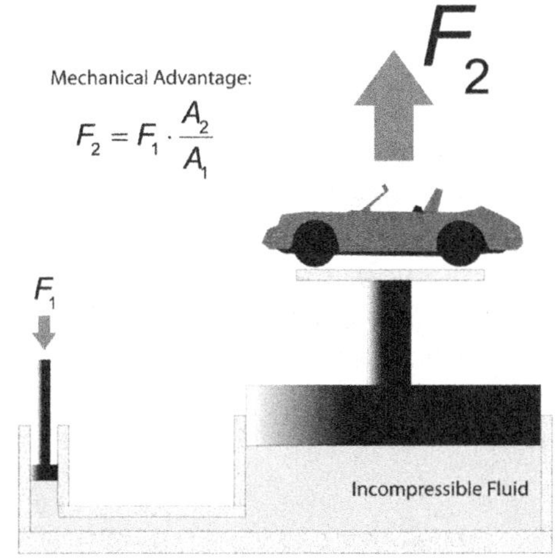

$$F_2 = F_1 \cdot \frac{A_2}{A_1} = F_1 \cdot \frac{10\ \text{m}^2}{2.5\ \text{m}^2} = 4 \cdot F_1$$

$$F_2 = 4 \cdot F_1; \quad F_1 = \frac{F_2}{4}$$

So the amount of required input force to the hydraulic piston would therefore be equal to,

$$F_1 = \frac{2000\ \text{kg}}{4} \cdot 9.81\frac{\text{m}}{\text{s}^2} = \mathbf{4905\ N}$$

Now for the tricky part, we are told that our hydraulic system raises the car to a height of 2 m above the ground. The work is going to be equal to the applied force on the car times the distance that it moves. Since our car weights 2,000 N we have,

$$Work = 2000\ \text{kg} \cdot 9.81\frac{\text{m}}{\text{s}^2} \cdot 2\ \text{m} = 39240\ \text{N} \cdot \text{m}$$

We also note that the input force was only 4,905 N, but the total work has to be conserved; thus the distance the input piston must travel to meet our conservation of work requirement is,

$$Work = 39240\ \text{N} \cdot \text{m} = F_{input} \cdot d_{input}$$

$$d_{input} = \frac{Work}{F_{input}} = \frac{39240\ \text{N} \cdot \text{m}}{4905\ \text{N}} = \mathbf{8\ m}$$

> ***The Guru Says...***
>
> Now you know the dirty little secret of hydraulic systems: the input work is equal to the output work so we compensate for the additional force through distance – *you really do not get anything for free.*
>
> A hydraulic system is nothing more than a force modification device (and a really neat one at that). Think of it as a *'liquid lever'* gizmo.

While similar inertial forces can be demonstrated for gasses, gasses also exhibit forces by virtue of their molecules bouncing about. Take a look at the closed flask in Figure 9.2, that we have filled with oversized nitrogen atoms. Assuming that the nitrogen atoms have some kinetic energy in the form of heat, they will produce numerous elastic collisions with themselves as well as the inside of the container. These repeated collisions produce a pressure within the closed container.

Although both inertial and kinetic energy effects are present in both liquids and gasses, kinetic energy effects are most noticeable in gasses. Additionally, gasses are compressible in nature, and therefore a good way to store energy, but make a terrible working fluid for a hydraulic system, for the same exact reason.

Table 9.2 lists the specific weights for various gasses in a manner similar to what was previously shown for liquids. Note though, that while liquids have specific weights equal to several kilo-Newton's per cubic meter, gasses only range in the mere Newton's per cubic meter.

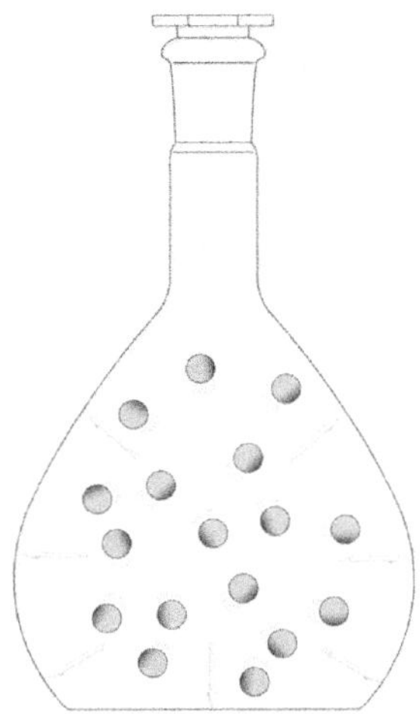

Figure 9.2: Pressure is the Reaction Force of Fluid Acting on an Object.

Thus, the inertial pressure effects due to gasses and liquids vary by a factor of over 1,000. Things crush to 'smithereens' at great depths in a liquid; the same is not true for gasses. To test this hypothesis, let's look at a barometer and the concept we call atmospheric pressure.

If we assumed a uniform density for Earth's atmosphere (which is not wholly accurate, since we know that gasses are compressible – but it's a good place to start), let's see what the average 'uniform' depth of the atmosphere would be under this circumstance.

A barometer at sea level would read a static pressure of approximately 101,325 Pascals. We can see that dividing the total atmospheric pressure by the specific weight of air, we can effectively calculate the 'depth' of the air mass required to produce the measured pressure.

Performing this operation we can guess that the average 'uniform' depth of the atmosphere is,

Table 9.2: Specific Weights of Different Gasses

FLUID OF INTEREST	SPECIFIC WEIGHT (γ) IN $\frac{N}{m^3}$
Hydrogen	0.88
Helium	1.63
Ammonia	7.03
Water Vapor (steam)	7.89
Neon	8.83
Carbon monoxide	11.43
Nitrogen	11.43
Air	11.82
Oxygen	13.06
Argon	16.30
Carbon dioxide	18.07
Chlorine	29.37
Krypton	36.71
Xenon	57.51

$$Depth_{average} = \frac{P_A}{\gamma_{Air}} = \frac{101325\ \frac{N}{\cancel{m^2}}}{11.82\ \frac{N}{m^{\cancel{3}}}} = \frac{101325}{11.82}\ m = 8{,}572.34\ m \approx 8.6\ km$$

Which puts us at the bottom of an ocean of air. The effective force works out to roughly 10 N per every square centimeter on your body. Without that downward pressure force, there would be no liquid oceans on Earth, for they would all simply boil away into space. Given that the atmospheric pressure on Earth is considerable, engineers have developed two different measurements for static fluid pressure that you should be aware of: *absolute pressure (P_A)* and *gauge pressure (P_G).* Absolute pressure is the total pressure exerted on an object. Gauge pressure is simply the pressure exerted on an object independent of atmospheric pressure (i.e., it's the pressure that a 'gauge' reads).

Incidentally, due to the compressibility of air, Earth's atmosphere does not strictly follow the uniform model we developed above. In fact, roughly half of Earth's atmosphere resides below 5.6 km, and 90-percent of it is below 16 km – so our atmospheric ocean is actually quite deep. By definition, 'outer space' begins at 100 km above Earth's surface. At this point 99.99997-percent of the atmosphere is below you. Thus, there really isn't a clear-cut line defining where a planet's atmosphere ends and space begins.

Example 9.4: A hot air balloon is nothing more than a 'submarine for the air'. If a standard hot air balloon has a volume of 2,800 m^3, what is the buoyant force exerted by the hot gas? If the weight of the balloon plus its occupants is 5,000 N, what is the net upward force? How long will it take for the balloon to rise 1,000 m? (Assume the specific weight of heated air is 9.28 N/m^3)

Our hot air balloon has a volume of 2,800 m^3, which gives it a buoyant force equal to the difference between cold (outside balloon, $\gamma = 11.82\ \mathrm{N/m^3}$) air and the heated (inside balloon, $\gamma = 9.28\ \mathrm{N/m^3}$) air. We therefore write,

$$\text{Bouyant Force}_{\text{balloon}} = \left(\gamma_{\text{cold air}} - \gamma_{\text{hot air}}\right)\cdot V_{\text{balloon}}$$

$$= (11.82 - 9.28)\,\frac{\mathrm{N}}{\cancel{\mathrm{m}^3}}\cdot 2800\ \cancel{\mathrm{m}^3}$$

$$= \mathbf{7112\ N}$$

The net upward force is simply equal to the difference between the lifting force and the gravitational weight, thus,

$$\text{Net Force}_{\text{balloon}} = \text{Force}_{\text{lift}} - \text{Weight}_{\text{balloon}}$$

$$= 7112\ \mathrm{N} - 5000\ \mathrm{N} = \mathbf{2112\ N}$$

Now for the 'tricky' part. We know that the weight of the balloon plus its occupants is 5,000 N. If we divide the weight by the acceleration due to gravity we will have the mass of the balloon.

$$\text{Mass}_{\text{balloon}} = \frac{5000\ \mathrm{N}}{9.81\,\frac{\mathrm{m}}{\mathrm{s}^2}} = \frac{5000\,\frac{\mathrm{kg}\cdot\cancel{\mathrm{m}}}{\cancel{\mathrm{s}^2}}}{9.81\,\frac{\cancel{\mathrm{m}}}{\cancel{\mathrm{s}^2}}} = 509.7\ \mathrm{kg}$$

The upward net force is 2112 N. Newton's second law provides a way for us to solve for the magnitude of the required acceleration producing this force on a 509.7 kg mass as,

$$F = m\cdot a$$

$$2112\ \mathrm{N} = 509.7\ \mathrm{kg}\cdot a;\quad a = 4.14\,\frac{\mathrm{m}}{\mathrm{s}^2}$$

Now we just need to invoke our kinematic expression for distance as a function of acceleration, noting that the initial velocity of the balloon is zero when we start the problem.

From Chapter 4 we can write the kinematics of the problem as,

$$y = \frac{1}{2}a \cdot t^2$$

$$1000\,\cancel{m} = \frac{1}{2}(4.14\,\frac{\cancel{m}}{s^2}) \cdot t^2$$

$$t^2 = \frac{2 \cdot 1000}{4.14} s^2 = 483.1\,s^2$$

$$t = \sqrt{483.1\,s^2} \approx \mathbf{22\ s}$$

A very fast trip indeed!! It should be noted that hot air balloons develop their lift over a finite time interval (i.e., from neutral buoyancy to full lift) and thus the actual ascent is a little tamer than what we analyzed.

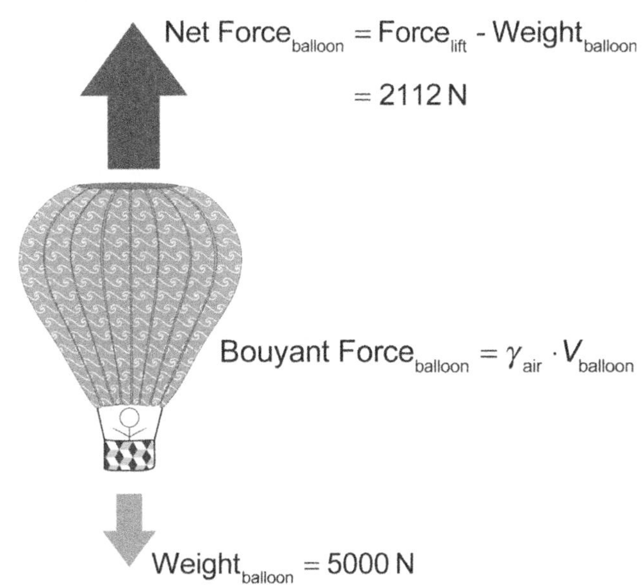

Net Force$_{balloon}$ = Force$_{lift}$ - Weight$_{balloon}$
= 2112 N
Bouyant Force$_{balloon}$ = $\gamma_{air} \cdot V_{balloon}$
Weight$_{balloon}$ = 5000 N

Continuity, Conservation, and Bernoulli's Equation – Fluid Dynamics

A flowing fluid not encountering any external forces will flow in a nice smooth 'glasslike' manner with negligible loss in energy. We call this type of fluid motion a *laminar flow*. An example of laminar flow would be cold honey flowing out of a container and into the Guru's cup of tea. On the other hand, *turbulent flow* is characterized by swirling fluid motion, which cause the fluid to twist and swirl. An example of this would be Cirrus and Stratus cloud patterns. Turbulent flow patterns lose energy primarily due to frictional forces.

To figure out how fast a fluid is moving, we need to define something called the *volume rate of flow*, or 'Q'. The volume rate of flow is simply the rate at which a volume of fluid passes through a certain point. Think water flowing through a pipe – the volume of water past a certain point in the pipe in a certain period of time is the volume rate of flow. This is what is measured by a typical household water meter (or a household gas meter for that matter).

We define the volume rate of flow as shown in Figure 9.3 as being equal to,

$$Q = v \cdot A$$

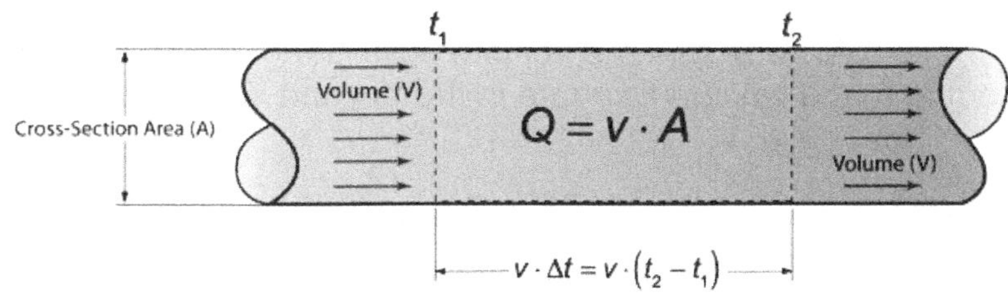

Figure 9.3: Defining the Volume Flow Rate of a Fluid.

The quantity 'Q' can be a vector, but we'll treat it as a scalar to simplify the math. This expression is known as *the continuity equation*. The concept of continuity expresses the notion that fluids are 'continuous' in nature. Continuity is a big deal in fluid mechanics from a mathematical standpoint since it implies that you cannot 'fold' or 'tear' a fluid. Given this, we can now define the velocity of any flow of fluid as the volume rate that passes through a certain cross-sectional area, and can write,

$$v = \frac{Q}{A}$$

The 'cross-sectional area' of an object is also very important in the study of *fluid mechanics*, since this is the effective area that the fluid 'sees' as it passes by a certain point in space – an important parameter when determining forces. Looking at Figure 9.3 we can see that the amount of fluid into a certain volume is equal to the fluid out of the same volume.

Thus,

$$Q_{in} = Q_{out}$$

Which states that mass is conserved. This makes sense since we know that even for a 'leaky' pipe, the water has to go somewhere and is not magically lost. Now, if it follows that mass is conserved, then without friction (or negligible friction), we can also write,

$$(\rho \cdot v)_{in} = (\rho \cdot v)_{out}$$

And the momentum of the fluid is also conserved. Thus, we now know that fluids exhibit the same properties as solid matter, namely, a conservation of mass and a conservation of momentum. Fluids also conserve energy as well, but we haven't gotten to that quite yet.

Example 9.5: Water is flowing through a pipe with a cross-sectional area of 1 m^2 at a rate of 2.0 m/s. If we reduce the cross sectional area of the next length of pipe by one-half, what happens to the velocity of the flow?

We know that the volume flow rate going through the first section of pipe must be equal to the volume flow rate in the second section and we can write,

$$Q_{\text{Section 1}} = Q_{\text{Section 2}}$$
$$(v \cdot A)_{\text{Section 1}} = (v \cdot A)_{\text{Section 2}}$$

We can write the answer as,

$$v_{\text{Section 2}} = v_{\text{Section 1}} \cdot \left(\frac{A_{\text{Section 1}}}{A_{\text{Section 2}}} \right) = 2\,\frac{\text{m}}{\text{s}} \cdot \left(\frac{1}{0.5} \right) = 2\,\frac{\text{m}}{\text{s}} \cdot (2) = \mathbf{4\,\frac{m}{s}}$$

So when we decrease the cross-sectional area (or diameter, or volume, or whatever) of a fluid's flow, we increase the velocity in order to maintain continuity. The opposite is also true, if we increase the cross-sectional area (or diameter, or volume, or whatever), then the velocity drops. Since the 'push' on the fluid (i.e., the pressure) is what makes this all happen, then something else must be changing here as well – but how do we figure this out? Enter mathematician Daniel Bernoulli, who found the solution to this problem in 1738.

Bernoulli's Equation, named after its inventor, was derived from the Work-Energy expression we examined in Chapter 6. Recalling it, we stated that kinetic energy equals mass 'times' acceleration 'times' distance, which also equals our expression for work.

$$KE = Work = \text{m} \cdot \left(\frac{\Delta v}{\cancel{\Delta t}} \right) \cdot \left(\frac{\Delta v \cdot \cancel{\Delta t}}{2} \right) = \frac{1}{2} m \Delta v^2$$

$$= KE_1 - KE_2 = \Delta KE$$

So that work was nothing more than the change in kinetic energy. To model this mathematically for a fluid, the Guru is going to use a bucket and a faucet fixture. We'll place the faucet fixture at a height of 'h' to demonstrate pressure pushing the water up to the spigot, which then free falls back into the bucket as shown in Figure 9.4.

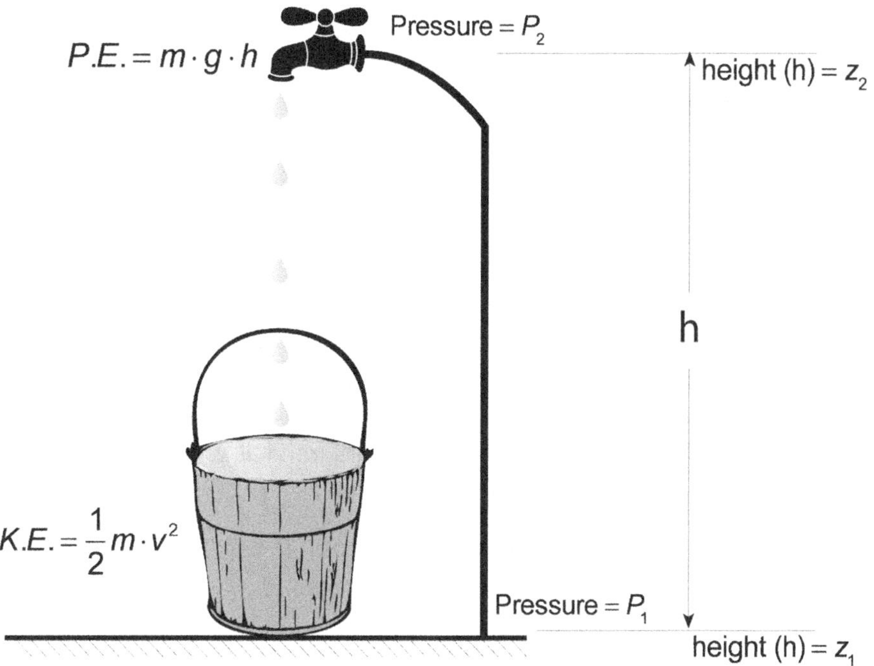

Figure 9.4: Using the Guru's 'Bucket of Water Analogy' to Derive Bernoulli's Equation.

Let's try to relate the kinetic and potential energies of the free falling water into the bucket, to the amount of work required to pump it up to the faucet in the first place. If we can do that, then we will have derived Bernoulli's equation.

Using fluid statics, we know that the pressure in the pipe leading to the faucet is greater near the bottom of the pipe than it is near the spigot due to gravity. Thus, we'll measure the pressure in the bottom of the pipe as P_1, and near the spigot as P_2. The pressure change ΔP is therefore going to be $P_1 - P_2$ so that we have a positive pressure difference.

Since we know that the pressure in the pipe is equal to the force divided by the area of the pipe, we can write the work required to get the water to the top of the spigot as,

$$\begin{aligned} Work_{\text{Due to Pressure}} &= \Delta P \cdot A_{\text{Pipe}} \cdot h = (P_1 - P_2) \cdot A \cdot h \\ &= (P_1 - P_2) \cdot V_{\text{Pipe}} \end{aligned}$$

In the above expression, the big 'V' represents a volume of fluid flow and not a velocity. Velocities are shown in our derivation with a little 'v' so as not to confuse things.

Also, since our pipe does not leak, we have conserved the mass of the system between the bottom of the pipe and the spigot. We can re-express the volume of water in the pipe in terms of its mass 'm'. This will make for some nice simplifications in our final equation in a moment.

Additionally, we know that the density (ρ) of any object is equal to its mass (m) divided by its volume (V). Using this, our work expression due to pushing water up in the pipe becomes,

$$\rho = \frac{m}{V}; \quad V = \frac{m}{\rho}$$

$$Work = (P_1 - P_2) \cdot V = (P_1 - P_2) \cdot \frac{m}{\rho}$$

So far, so good – we now have the 'work' contribution of the 'work-energy' theorem taken care of – *now for the 'energy' part.*

The potential energy stored by a mass of water 'm' at a height 'h' is simply equal to $-m \cdot g \cdot h$ (it's negative because we're assuming that +z is pointing straight up). We'll add this term to the *work side* of the equation since this stored energy has the ability to move the water, plus the work-energy theorem tells us that we have to write it this way – *work equals the change in kinetic energy*.

The change in kinetic energy of the falling water drops is equal to,

$$\Delta K.E. = \frac{1}{2} m \cdot v_2^2 - \frac{1}{2} m \cdot v_1^2$$

Where 'v' is now a velocity. In our derivation we're assuming that the mass 'm' is the same both before and after the droplet falls. Combining this into the work-energy theorem we get,

$$W = \Delta K.E.$$

$$(P_1 - P_2)\cdot\frac{m}{\rho} - m\cdot g\cdot h = \frac{1}{2}m\cdot v_2^2 - \frac{1}{2}m\cdot v_1^2$$

Which is effectively Bernoulli's equation for fluids (albeit in a slightly untidy fashion). Performing some cleanup and taking advantage that we were clever and wrote everything in terms of the mass 'm', we get,

$$(P_1 - P_2)\cdot\frac{m}{\rho} - m\cdot g\cdot h = \frac{1}{2}m\cdot v_2^2 - \frac{1}{2}m\cdot v_1^2$$

$$P_1\cdot\frac{\cancel{m}}{\rho} - P_2\cdot\frac{\cancel{m}}{\rho} - \cancel{m}\cdot g\cdot h = \frac{1}{2}\cancel{m}\cdot v_2^2 - \frac{1}{2}\cancel{m}\cdot v_1^2$$

$$\frac{P_1}{\rho} + \frac{1}{2}v_1^2 - g\cdot h = \frac{1}{2}v_2^2 + \frac{P_2}{\rho}$$

We also know that the height 'h' is nothing more than the difference between when the fluid is at ground level (some point z_1) and when it is at the spigot (point z_2), thus $h = z_2 - z_1$. Performing this final substitution we get,

$$\frac{P_1}{\rho} + \frac{1}{2}v_1^2 + g\cdot z_1 = \frac{P_2}{\rho} + \frac{1}{2}v_2^2 + g\cdot z_2$$

This final answer is formally known as *Bernoulli's equation* and is a very helpful tool relating fluid pressure, velocity, density, and elevation height.

You can also take Bernoulli's equation and divide everything by $\rho\cdot g$ leaving each term having units of length. Engineers refer to these length terms as *hydraulic head*. The *hydraulic head* of Bernoulli's equation is a useful form to see how far a fluid can be moved under varying conditions. In this form, terms on the left side of the equation generate fluid motion, and terms on the right side take energy away from the fluid.

The Guru Says...

Bernoulli's equation has everything in it: pressure, velocity, cross-sectional area and force (hidden in the pressure term), the fluid density (or specific weight if you want to use that), gravity, and elevation potential.

With this one single expression, you can solve pretty much any fluids problem that happens your way since you now know that all of these parameters are related to each other. Change one, and the others must follow.

Thus, the first term of Bernoulli's equation would be termed the 'pressure head', the second would be the 'velocity head' or 'dynamic pressure', and the third term would be the elevation head. We can also add artificial fluid movement devices (such as mechanical pumps) to the left side of the equation and call it *pump head* h_P, which would be the maximum elevation that a pump could lift a given liquid. Alternatively, we could take energy out of the moving fluid using a device known as a *turbine* and add the term h_T for this. Energy losses (due to friction, turbulent flow, etc.) are referred to as, you guessed it, *'head losses'* or h_L. The important thing to remember here is that devices that put energy into a fluid go on the left side of Bernoulli's equation, and things that consume energy go on the right side.

Given this, we can write our final form of Bernoulli's equation as found in any engineering fluid mechanics book as,

$$\frac{P_1}{\gamma}+\frac{1}{2g}v_1^2+z_1+h_P=\frac{P_2}{\gamma}+\frac{1}{2g}v_2^2+z_2+h_L+h_T$$

This one equation will solve pretty much any problem you may ever encounter in life that contains a moving fluid. Knowing this, let's see what kind of trouble we can get into using it.

Example 9.6: Pumping water up into a large skyscraper is actually quite an engineering challenge. Suppose you're on the observation deck of the Empire State Building at the 102nd floor and decide that a restroom break is in order. Assuming no losses in the pipes, how much head does a pump have to supply to get water up that high at a flow rate of 1.0 cubic-meters-per-second (m^3/s)?

(Assume the pipe cross-sectional area is 0.5 m^2. One skyscraper story is approximately 3 m.)

We are told that there are no friction losses in the pipe, and clearly we are not generating any power from this activity; thus h_L and h_T equal zero. We don't know the flow rate at the pump, but we are told that it has to be 1.0 m^3/s at the exit. From our continuity equation we find that the required water velocity at the specified pipe cross-sectional area is,

$$Q=v\cdot A$$

$$v=\frac{Q}{A}=\frac{1.0\,\frac{m^3}{s}}{0.5\,m^2}=2.0\,\frac{m}{s}$$

Also, we are not told anything about what the final pressure would be, so we'll skip that part for now. Our form of Bernoulli's equation would therefore look like,

$$\cancel{\frac{P_1}{\gamma}}+\cancel{\frac{1}{2g}v_1^2}+0+h_P=\cancel{\frac{P_2}{\gamma}}+\frac{1}{2g}\left(2\,\frac{m}{s}\right)^2+102\cdot 3\,m+0+\cancel{h_T}$$

Where we can directly solve for the head required by the pump, h_P, to move the water as,

$$h_P = \frac{1}{2 \cdot 9.81 \frac{\text{m}}{\text{s}^2}} \cdot 4 \frac{\text{m}}{\text{s}^2}^2 + 306 \text{ m} = (.204 + 306) \text{ m} = \mathbf{306.204 \text{ m}}$$

So the answer is just slightly larger than the total elevation head that we knew we needed. So, what did we just model? How about a pump pumping water through a 0.5 m^2 pipe to a height of 306 meters with a little additional 'nudge' in hydraulic head to keep the 1 m^3/s flow rate trickling out of the end. This will never work in a bathroom setting as we need to also specify the pressure of the flow rate (you see, you can indeed have water flowing without any measurable pressure).

The interesting part is that you can also have electrical current (our corollary to fluid flow) without voltage (the electrical corollary to pressure) – that type of device is called a *super-conductor.*

Example 9.7: The previous engineer designing the plumbing system for the Empire State Building has been fired and a new engineer now wants the same flow rate but at a pressure of 100 kPa. How much hydraulic head must the pump supply?

Going back to our Bernoulli expression from the previous problem with the addition of some pressure head and using our specific weight from Table 9.1, we have,

$$h_P = \frac{P_2}{\gamma} + 306.204 \text{ m} = \frac{100000 \frac{\text{N}}{\text{m}^2}}{9800 \frac{\text{N}}{\text{m}^3}} + 306.204 \text{ m} = 10.2 \text{ m} + 306.204 \text{ m}$$

$$= \mathbf{316.404 \text{ m}}$$

Which will give us the proper flow rate at the proper pressure. Note though, that the biggest portion of the work required by the pump is still in the shear lifting of the water to the required height. In larger buildings, such as the one modeled here, this becomes prohibitive from a pump and electrical consumption standpoint fairly fast. This is why most large skyscrapers pump water in stages to various floors within the building to store it there in temporary tanks.

Example 9.8: The middle term in Bernoulli's equation (the term with the 'v' in it) is known as the *dynamic pressure* term since it has the ability to create a pressure from the movement of a fluid. If you put your hand out a car window when you are traveling at 100 km/hr, what is the dynamic pressure on your hand?

The beauty of Bernoulli's equation is that we can rearrange terms to get into the desired units we need, since every term is specified with identical units. In past examples we've been working in units of hydraulic head (m), so let's now adjust that form of the equation to give us the 'pressure form' where 'P' is all by itself.

$$\frac{P_1}{\gamma}+\frac{1}{2g}v_1^2+z_1+h_P=\frac{P_2}{\gamma}+\frac{1}{2g}v_2^2+z_2+h_L+h_T$$

$$P_1+\frac{\gamma}{2g}v_1^2+\gamma\cdot z_1+\gamma\cdot h_P=P_2+\frac{\gamma}{2g}v_2^2+\gamma\cdot z_2+\gamma\cdot h_L+\gamma\cdot h_T$$

In the problem given, there are no pumps, turbines, frictional losses or elevation or velocity changes, so we can set those terms equal to zero. Additionally, we are told that we are not applying any initial air pressure (P_1) to your hand, thus that term is equal to zero as well. Doing this we get the following simplified expression.

$$\cancel{P_1}+\frac{\gamma}{2g}v_1^2+\cancel{\gamma\cdot z_1}+\cancel{\gamma\cdot h_P}=P_2+\cancel{\frac{\gamma}{2g}v_2^2}+\cancel{\gamma\cdot z_2}+\cancel{\gamma\cdot h_L}+\cancel{\gamma\cdot h_T}$$

$$\frac{\gamma}{2g}v_1^2=P_2$$

Plugging in some numbers and looking-up the specific weight of air from Table 9.2 gives,

$$P_2=-\frac{11.82\,\frac{\text{N}}{\text{m}^3}}{2\cdot 9.81\,\frac{\text{m}}{\text{s}^2}}\left(100\,\frac{\cancel{\text{km}}}{\cancel{\text{hr}}}\cdot\frac{1000\text{ m}}{1\,\cancel{\text{km}}}\cdot\frac{1\,\cancel{\text{hr}}}{3600\text{ s}}\right)^2=\mathbf{462.3\ Pa}$$

This is quite a bit of pressure due to the movement of the air against your hand. Dynamic pressure is a very important component of aerodynamics (well actually it's a big component since this is what makes the plane fly) and we'll noodle-around with it a little before we leave our fluids chapter.

Problem Solving Tip...

At this point you should be able to close your eyes and visualize which way the forces are working in a problem. If you can't then by all means draw a sketch. The mathematics will tell you if your assumption was correct.

Generating Power From Fluid Motion

Power is equal to the time rate of change of either work or energy (or some mixture of the two). Therefore, it would be a pretty good way to wrap up Bernoulli's equation by showing how much power we can get out of a moving fluid. This sets the stage for our discussion of thermodynamics in the next chapter, in that we will see that given a certain set of physical circumstances, there is a definite 'upper end' to the amount of work or energy (or power) you can get out of any physical process.

In order for a fluid to generate power it has to produce a net force at a certain flow rate. Starting with our fundamental expression for power we have,

$$P = F \cdot v$$

The force produced by the fluid is equal to its inertia, which we've already defined by the quantity γ. Substituting this into our expression we have the approximate relationship,

$$P \approx \gamma \cdot v$$

Where we are using the approximate sign ($\approx$) since this is not an exact expression because the units on both sides of the equation are slightly different. In fact, the right side of the equation is 'off' by a factor of m^3, or some type of volume. We can therefore rewrite this expression as,

$$\begin{aligned} P &= \gamma \cdot v \cdot (volume) \\ &= \gamma \cdot v \cdot (length) \cdot (area) = \gamma \cdot v \cdot (length) \cdot A \end{aligned}$$

This basically says that the power produced by a liquid is equal to the specific weight of the liquid, times its velocity, times the volume of fluid that pass by at any given time.

The volume of the liquid in a section of pipe is equal to the cross-sectional area of the pipe times the length of the section of pipe. If the length of pipe works against gravity then it works against the hydraulic head and we would have to put work into the system to move the fluid. *This type of device is known as a pump.*

Knowing this, we can immediately write the expression for the power required by a pump to lift a liquid through a hydraulic head h_P as,

$$\begin{aligned} P_{pump} &= \gamma \cdot v \cdot A \cdot h_P \\ &= Q \cdot \gamma \cdot h_P \end{aligned}$$

If we assume that the hydraulic head works with gravity, then we get work out of the system *and we have a device known as a turbine.*

The expression for the power produced by a turbine is therefore equal to,

$$P_{turbine} = Q \cdot \gamma \cdot h_T$$

Example 9.9: Revisiting our Empire State Building plumbing problem in Example 9.7, what is the minimum power required by the pump to raise the water to the desired elevation?

In this previous example, the head required by the pump was 316.404 m. Thus we easily find the required power as,

$$P_{pump} = Q \cdot \gamma \cdot h_P$$

$$= (1.0 \frac{\not{m}^3}{s}) \cdot (9800 \frac{N}{\not{m}^3}) \cdot (316.32\text{ m}) = 3099936 \frac{N \cdot m}{s} = 3099.94\text{ kW} = \mathbf{3.1\,MW}$$

This is a ridiculous amount of power equal to the total power consumption of roughly 130 residential households. There are two morals to be learned here. First, it becomes extremely prohibitive both from a cost and an engineering standpoint to pump water straight up against gravity meaning that something has got to give, namely either the water pressure or the flow rate (hence the need to perform this work incrementally and pump the water in stages).

Secondly, this example should give you a pretty good idea as to the vast amounts of power a hydroelectric plant can produce since the opposite of a pumping water is generating power in a turbine. Hey, why don't we look at that as well.

Example 9.10: Hoover Dam takes in water to its hydroelectric turbines through sixteen intake tunnels called 'penstocks'. The combined flow rate for all the penstocks is 14,000 m^3/s with a maximum hydraulic head of 180 m at the turbines. Under this condition, how much potential power can the dam generate?

The problem states that water can fall a maximum of 180 vertical meters before it hits the turbine blades. Using our expression for the power produced by a turbine, we can calculate the amount of available power from Hoover Dam as,

$$P_{turbine} = Q \cdot \gamma \cdot h_T = (14000 \frac{\not{m}^3}{s}) \cdot (9800 \frac{N}{\not{m}^3}) \cdot (180\text{ m})$$

$$= 2.4696x10^{10} \frac{N \cdot m}{s} = 2.4696x10^{7}\text{ kW} = \mathbf{24696\ MW}$$

This is the theoretical upper-limit of power production given the current dam design. It should be obvious that installing larger generators at the dam would never exceed this power threshold, since it is determined solely by the amount of water flowing (the continuity equation) and the hydraulic head (Bernoulli's equation). Incidentally, this amount of power can also be expressed as 24.696 Gigawatts (GW).

The 'Ups' and 'Downs' of Fluid Motion

Previously, we looked at how objects moved up and down by adjusting their buoyancy within a static fluid. For many centuries humans knew that lighter things floated and heavier things sank in a static fluid – *but there were still those pesky 'heavier than air' birds flying about.*

Since we're talking about fluids, we might as well digress slightly into the world of aerodynamics and figure out why things can fly at all.

Bernoulli's equation in 1738 actually predicted the existence of airplanes, since objects are capable of generating a net-upward force, known as 'lift', by virtue of the dynamic pressure term in Bernoulli's equation. With that in mind, take a look at Figure 9.5, and notice that a cross-section of a typical aircraft wing, called an *airfoil*, has distinctively more curvature along the top surface than at the bottom. *Why is that?*

Well, one really good reason is that by curving the top of an airfoil and not the bottom, you force the air to move faster over the top of the airfoil than over the bottom. Recalling some basic kinematics from an earlier chapter, we know that velocity is equal to the distance traveled divided by the time it takes to travel said distance. Since laminar air particles, like all fluid particles, like to hang-out near each other (that whole continuity thing again), the only way that the upper path of air particles can meet up with their neighbors taking the lower path is to speed up.

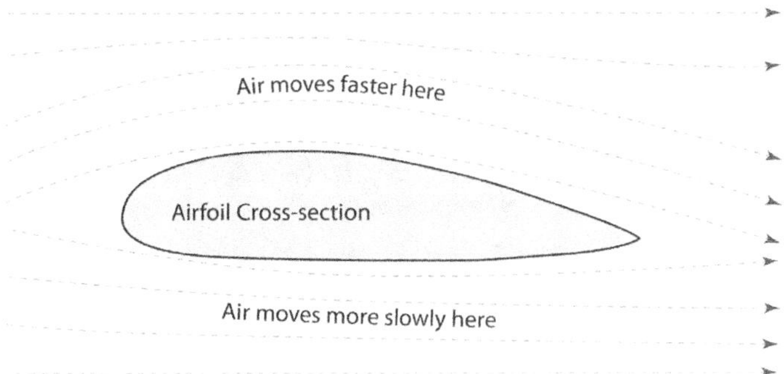

Figure 9.5: Cross-Section of an Airplane Wing Called an 'Airfoil'.

Hence, the laminar airflow increases velocity by virtue of traveling that extra distance. This doesn't only apply for an airfoil, any geometric shape (like a cylinder or sphere) will exhibit increased velocity as a fluid moves around it. Grabbing the relevant pieces of Bernoulli's equation, we can write the following simplified expression for the total pressure (P_T) valid for either the top or bottom of the airfoil as,

$$P_T = P_A + \frac{\gamma}{2g}v^2 = P_A + \frac{1}{2}\rho \cdot v^2$$

Where P_A is the atmospheric (static) air pressure and v is the velocity of the airflow at any point. The scalar γ is the specific weight of air and ρ is the density of air, which is nothing more than γ/g.

Thus, the total pressure on the airfoil at any point is simply equal to the ambient pressure plus the dynamic pressure at any point. Let's now look at the net effect on the upper and lower surfaces of the airfoil by adding a small Δv term to the upper flow to account for the curvature.

Given this, our new equation would be,

$$P_T = P_{lower} + \frac{1}{2}\rho \cdot v^2 = P_{upper} + \frac{1}{2}\rho \cdot (v + \Delta v)^2$$

$$P_{lower} + \frac{1}{2}\rho \cdot v^2 = P_{upper} + \frac{1}{2}\rho \cdot (v + \Delta v)^2$$

$$P_{upper} = P_{lower} + \frac{1}{2}\rho \cdot v^2 - \frac{1}{2}\rho \cdot (v + \Delta v)^2$$

Where the combination of the lower static and dynamic pressure is called the *stagnation pressure* in aerodynamics. We can already see that *the upper pressure will always be less than the lower pressure* by the finite amount of velocity increase (Δv) attributable to the upper surface of the airfoil being slightly more curved than the lower surface. Let's fool around with the Δv term a little more, since the Guru believes that we can still inflict additional damage on this poor equation.

Expanding the last term we get,

$$P_{upper} = P_{lower} + \frac{1}{2}\rho \cdot v^2 - \frac{1}{2}\rho \cdot (v + \Delta v)^2$$

$$= P_{lower} + \cancel{\frac{1}{2}\rho \cdot v^2} - \cancel{\frac{1}{2}\rho \cdot v^2} - \frac{1}{2}\rho \cdot (2 \cdot v \cdot \Delta v + \Delta v^2) = P_{lower} - \frac{1}{2}\rho \cdot (\underbrace{2 \cdot v \cdot \Delta v + \Delta v^2}_{C_L \cdot v^2})$$

$$\boxed{P_{upper} = P_{lower} - C_L \cdot \frac{1}{2}\rho \cdot v^2}$$

Since the additional increase in velocity (Δv) due to the curvature of the wing compared to the velocity of the fluid (v) is typically small, and the Δv effects are produced solely by the geometry of the wing, aeronautical engineers typically lump this into a single term called the *coefficient of lift* (or C_L).

C_L is an experimentally determined value dependent on the geometric shape of the wing being analyzed. Multiply this net-upward pressure difference times the surface area of the wing (S), and you have the total amount of lift-force generated by the wing as,

$$\boxed{\textit{Lift Force} = C_L \cdot S \cdot \frac{1}{2}\rho \cdot v^2}$$

All things being equal, the total lift generated by the wings in level flight had better equal the weight of the aircraft with everyone on board.

Example 9.11: A small private aircraft has a maximum takeoff weight of 1000 kg and cruises at 100 km/hr. If the coefficient of lift of the wings in level flight is 0.8, what must the surface area of the wings be to maintain level flight?

The maximum weight of the aircraft is equal to its 1,000 kg mass times the acceleration due to gravity. This value has to equal the lifting force of the wings if the plane is to stay aloft.

Thus, using the values from Table 9.2 we get,

$$C_L \cdot S \cdot \frac{1}{2}\rho \cdot v^2 = 1000\,\text{kg} \cdot 9.81\frac{\text{m}}{\text{s}^2};\quad \rho = \frac{11.82\,\frac{\text{N}}{\text{m}^3}}{9.81\,\frac{\text{m}}{\text{s}^2}} = 1.204\,\frac{\text{kg}}{\text{m}^3}$$

$$0.8 \cdot S \cdot \frac{1}{2} \cdot \left(1.204\,\frac{\text{kg}}{\text{m}^3}\right) \cdot \left(\frac{100\,\cancel{\text{km}}}{\cancel{\text{hr}}} \cdot \frac{1\,\cancel{\text{hr}}}{3600\,\text{s}} \cdot \frac{1000\,\text{m}}{\cancel{\text{km}}}\right)^2 = 9810\,\text{N}$$

$$0.8 \cdot S \cdot \frac{1}{2} \cdot \left(1.204\,\frac{\text{kg}}{\text{m}^3}\right) \cdot \left(27.8\frac{\text{m}}{\text{s}}\right)^2 = 9810\,\text{N}$$

$$372.2 \cdot S = 9810$$

$$\mathbf{S = 26.4\ m^2}$$

Therefore, the wings should have a minimum effective upper surface area of roughly 26.4 square-meters to stay aloft while cruising in a straight line. Now you know what actually keeps planes up in the air – *not a whole lot, huh*.

We've discovered that the increase in dynamic pressure due to the air unequally hitting one side of the wing more than the other will produce lift and make the airplane fly. This lift, however, comes at the expense of another fluid force known as frictional *drag*.

Drag is literally – *a drag*. It slows objects moving through a fluid by producing frictional forces and taking energy away. The analysis of drag, like that of lift, is quite a complex subject relegated to entire textbooks. Luckily, aeronautical engineers have also given us another unitless 'fudge factor', similar to C_L, known as the *coefficient of drag*, or C_D.

The drag coefficient is also an experimentally determined value based in this case on how smooth the surface is and how laminar it keeps the flow moving by it (remember, turbulent flow equals lots of drag - *always*). For the case of a drag force, however, the only area that is important is the cross-sectional area that the fluid sees as it hits it head-on.

Thus, we can write a similar expression for the total drag force on an object as,

$$\textit{Drag Force} = C_D \cdot A_P \cdot \frac{1}{2}\rho \cdot v^2$$

Where the variable A_P is called the *area of projection,* and is the cross-sectional area of the object that comes into head-on contact with the fluid. In most cases, drag is something to be avoided, like in the design of racecars, airplanes, and rockets. In the case of a parachute, however, lots of drag force is a good thing.

A drag force on a falling object also implies that for a given freefall speed there will be an upward force on an object. Thus, a falling object should only fall so fast, since there will be a finite point where the object's gravitational force is equal to the upward frictional force and equilibrium is achieved.

The Guru Says...

Lift coefficients can range between slightly less than 1.0, to slightly greater than 1.0, depending on the geometry of the wing. Typical values are around 0.8 to 1.2. The higher the number the better.

Drag on the other hand is always designed to be as close to zero as possible although values as high as 1.28 are possible for a 'flat plate' flying perpendicularly through the air.

This point is called the *terminal velocity* of an object, and it applies equally to gasses as well as liquids. Despite what you might have previously heard, this is not a fixed value, and is dependent on many parameters, as can be seen in the following example.

Example 9.12: A 90 kg person with an effective cross-sectional area of 1.0 m^2 and a drag coefficient of 0.9 is freefalling from an unspecified height. What is this person's terminal velocity?

Plugging in the numbers in a similar fashion to what we did before in our airplane example we get,

$$C_D \cdot A_P \cdot \frac{1}{2}\rho \cdot v^2 = 90\text{ kg} \cdot 9.81\frac{\text{m}}{\text{s}^2}$$

$$0.9 \cdot 1.0\text{ m}^2 \cdot \frac{1}{2} \cdot \left(1.204\frac{\text{kg}}{\text{m}^3}\right) \cdot v^2 = 882.9\text{ N}$$

$$v^2 = 1629.6\frac{\text{m}^2}{\text{s}^2}; \quad \mathbf{v = 40.4\frac{m}{s}}$$

This is the equivalent speed of a drag racing car in a 100 m sprint. Clearly the need for a parachute is evident. This same concept applies to drag forces in any fluid. A sinking ship will only hit the ocean bottom so fast (run that calculation as practice sometime using the Titanic example and see if you can figure out how long it took the ship to hit the ocean bottom). Similarly, skydiving on Jupiter should be no problem to calculate, just as long as you adjust the above equation for the gravitational attraction of that planet, and change the density of air to that of methane.

Whooshing Through a Whooshing Fluid – Wave Motion and Sound

To *'whoosh'* means to, *'move quickly or suddenly with a rushing sound'*. Thus, it's time to take a look at things that move quickly through a fluid – *very quickly*.

There are actually two different types of motion that occur in a fluid – *particle and wave motion*. The first type we have already looked at. We've seen that the movement (i.e., velocity) of fluid particles produces dynamic pressure that can be used to do everything from moving water in a pipe to allowing an airplane to fly. In addition to this physical energy transfer through the shear movement of matter, energy is also conveyed in a fluid through a mechanism known as *wave motion*.

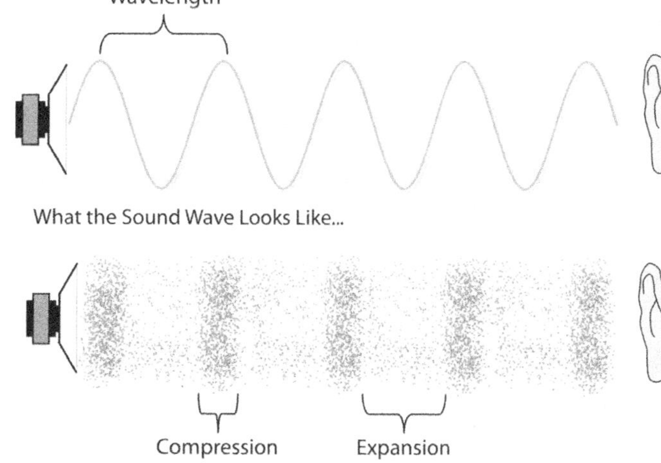

Figure 9.6: Example of Sound Propagating From a Loud Speaker Through the Air.

Although wave motion is not restricted to fluids (indeed waves can also travel in solids, as well as though empty space itself in the form of *electromagnetic waves*), motion in a fluid is the easiest to understand, since we experience it everyday under another name – *sound*.

Sound waves (also known as acoustic pressure waves) are linear mechanical waves. The medium transmitting this wave motion oscillates in the direction of propagation of the wave as shown in Figure 9.6. All sound waves originate from some sort of vibrating surface. Whether this surface is the vibrating string of a violin or a person's vocal cords, a vibrating column of air from an organ or clarinet, or a vibrating panel from a loudspeaker, drum, or aircraft, the sound waves generated are all similar. All wave motion of this type alternatively compresses the surrounding air on a forward movement, and expands it on a backward movement. This compression and expansion moves energy without having to move the entire fluid

The Guru Says...

Humans can hear sound waves from between 50 cycles per second (also called Hertz, Hz) to about 20,000 cycles per second. Sound waves above that are referred to as being 'ultrasonic waves'; those below it are called 'subsonic waves'. This is more of a matter of definition rather than of any hard physical scientific fact.

mass. Since fluids can only be compressed or expanded at a finite rate, it therefore follows why sound waves have a tendency to fade away faster at higher frequencies, since each compression and expansion consumes energy which is ultimately converted to heat. This is a very important aspect of most wave motion in general.

Sound waves, if unimpeded, will spread out in all directions from a source. Upon entering the ear, these waves produce the sensation we call sound. The faster the movement of the air, the higher the frequency of the sound wave; the larger the amplitude of the wave, the louder the sound. In addition to having the neat effect of communicating information, sound waves serve a very important purpose in the physical universe. They tell fluids to 'get out of the way' of an approaching object.

Table 9.3: Speeds of Sound in Different Liquids

FLUID OF INTEREST	SPEED OF SOUND (C) IN $\frac{m}{s}$
Methyl Alcohol	1103
Ethyl Alcohol	1207
Mercury	1450
Plain Old Water	1497
Sea Water	1531

Pressure disturbances in a fluid (i.e., sound waves) travel forward at a speed much faster than any individual particle could ever reach. This speed is called the *speed of sound* or, 'c' for the fluid. The speed of sound is different for each material one encounters.

Table 9.3 shows values for the maximum pressure wave speed (i.e., the speed of sound) in various liquid fluids. As can be seen, the speed of sound in water is almost 1,500 meters per second, faster than the speed of a bullet. Table 9.4 shows a similar table for the speed of sound in gaseous fluids, which is much less than in liquids. The speed of sound in air is shown as being 343 meters per second at sea level.

Table 9.4: Speeds of Sound in Different Gasses

GAS OF INTEREST	SPEED OF SOUND (C) IN $\frac{m}{s}$
Carbon Dioxide	259
Oxygen (O_2)	316
Argon	319
Nitrogen	334
Air	343
Neon	435
Steam	494
Helium	965
Hydrogen (H_2)	1284

Let's look at an automobile traveling down the highway at 100 km/hr (27.7 m/s). There is an acoustic pressure wave traveling in front of, and forwardly-away from, the car at a speed of 343 m/s alerting the air particles far ahead that there is an object approaching. Without this mechanism, the automobile would literally 'smash' into the oncoming motionless air, which as thin as it might seem, does have a fair amount of inertia.

As one travels faster towards the speed of sound, there is less and less time for the fluid (air) to make this adjustment and the resulting pressure changes become greater and greater. At the speed of sound, air literally forms a solid wall in front of an object with the ability to cause significant damage. This was a problem encountered by aeronautical engineers in the late 1940's and early 1950's when it was believed that the speed of sound might be a fastest speed limit achievable by humans. It turned out that with some serious rethinking of the geometries of aircraft airfoils and fuselages, this problem could be tackled.

This is why subsonic airplanes (like biplanes and commercial jets) look entirely different from supersonic airplanes (such as the SR-71 Blackbird and the SST Concorde). This is also the study area of fluid mechanics that we previously termed as gas dynamics.

The Guru Says...

A non-dimensional number known as the 'Mach number' is simply the ratio of the speed something is moving relative to the speed of sound in the fluid in which it is moving. Mostly attributed to aircraft flying in air, the Mach number tells us how close something is traveling to the speed of sound.

Mach numbers below 0.6 are typically called 'subsonic', while those between 0.6 to 0.9 are called 'transonic'. A Mach number of 1.0 is called 'sonic'. Mach numbers above 1.0 are called 'supersonic' until we get to about Mach 4 or 5, at which point things are typically termed as being 'hypersonic'.

As you can imagine with all that air moving by, there is bound to be a whole lot of friction. Thus, supersonic and hypersonic flight is currently limited by heat generation (and how to get rid of it).

On a different topic, we should also note that the relationship between the frequency (i.e., how fast a wave passes by), and the wavelength (the distance between any two equal points in a wave), is not an independent relationship as shown in Figure 9.7. In fact, they are related to each other by the speed of sound of the wave. We can write this simple relationship as,

$$\lambda \cdot f = c$$

Or, wavelength λ in meters, 'times' frequency f in Hertz, equals the speed of sound in meters per second. Referring back to Table 1.6, we can see that the units of this expression are consistent with our SI derived units. This expression implies that as the frequency of a pressure wave is increased, the wavelength must get smaller and vice versa.

A fellow by the name of Christian Doppler in 1842 discovered that the frequency and wavelength of a sound wave could also be dependent on your particular reference frame.

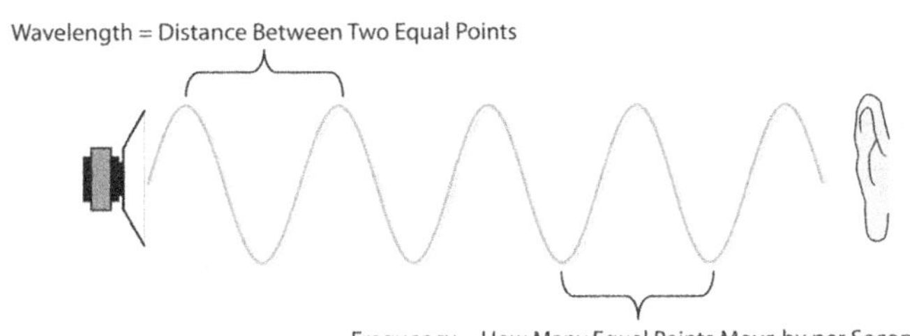

Figure 9.7: The Relationship between Wavelength and Frequency.

As an example, lets redraw Figure 9.7 as Figure 9.8 on the following page where first we have the speaker moving towards the receiving 'ear' (all the time leaving our reference markers indicating the wavelength and frequency unchanged). What we would hear would be a slightly higher frequency sound level indicating that the wavelength must be getting smaller. Thus, a sound source that is approaching a receiver is *frequency shifted* upwards in the auditory spectrum.

Similarly, if we now allow the sound source to retreat from the receiver, we would notice that the frequency of the sound is decidedly lower, and we could therefore conclude that the wavelength of the sound is getting longer.

This concept of 'stretching' or compressing the wavelengths of the sound should not cause you any heartburn, since it is a common sense conclusion based upon what we know about the kinematics of moving objects. Why should wave propagation in the universe be any different?

Mathematically, we can describe this apparent shift in frequency and wavelength with the following simple expression as,

$$f_{observed} = \alpha \cdot f_{source}$$

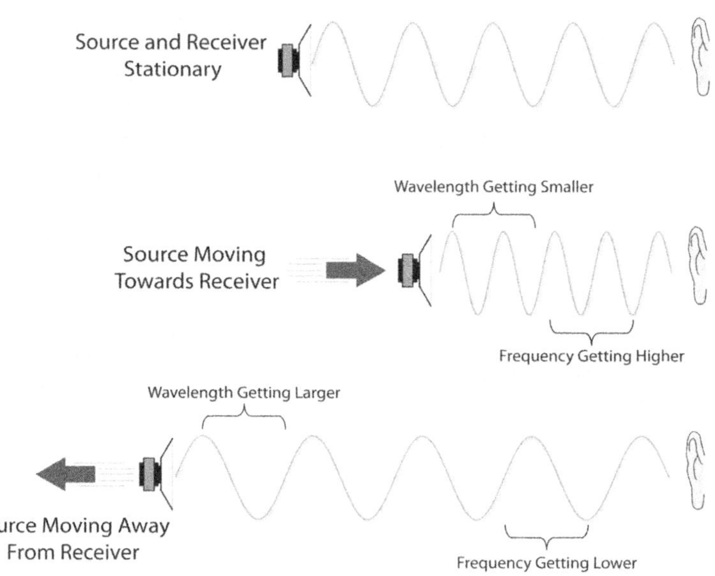

Figure 9.8: How the Doppler Effect Changes the Frequency and Wavelength of a Sound From a Loudspeaker.

Where our 'fudge factor' α allows us to perform a 'shift' to the frequency depending on whether or not the source is approaching the receiver or the other way around. We already know that when the source and the receiver are stationary, the observed frequency is the same as the source, therefore α must equal '1' in this case.

Now, when we have a sound source that is moving towards us, we know that the observed frequency will be higher, and thus α must be greater than '1'. Similarly, when the source is moving away, α must be some value less than '1'.

We also know that a stationary source produces its sound at the local speed of sound in the fluid; thus any additional velocity contribution by virtue of the source moving through the fluid would be merely equal to the ratio of that additional velocity, to the speed of sound. We can therefore write our expression for α as,

$$\alpha = \frac{c \pm v_s}{c} = 1 \pm \frac{v_s}{c}$$

Where the '$\pm$' expression is used to signify that the source is either traveling towards '+' the receiver, or away from '-' the receiver. Given this, our final expression for the Doppler effect in air can be written as,

$$f_{observed} = \left(1 \pm \frac{v_s}{c}\right) \cdot f_{source}$$

Welcome to the method by which a sonar system works. Let's wrap this chapter up with a couple of examples to test out what we have learned.

Example 9.13: An aircraft is flying at 300 km/hr. What is its Mach number?

The speed of sound in air is approximately 343 m/s, thus we can calculate the Mach number as follows,

$$Mach = \frac{Speed_{object}}{Speed_{sound}} = \frac{300\,\frac{\cancel{km}}{\cancel{hr}}\cdot\frac{1\,\cancel{hr}}{3600\text{ s}}\cdot\frac{1000\text{ m}}{1\,\cancel{km}}}{343\,\frac{\text{m}}{\text{s}}}$$

$$= \frac{83.3\,\frac{\text{m}}{\text{s}}}{343\,\frac{\text{m}}{\text{s}}} = \mathbf{0.24}$$

Example 9.14: A police siren is audible at 1,000 Hz. If the police car is traveling straight towards you at 100 km/hr, what is the observed frequency of the siren?

Using our expression for determining the Doppler shift of the incoming frequency wave we can write,

$$f_{observed} = \left(1 \pm \frac{v_s}{c}\right)\cdot f_{source}$$

Since the vehicle is approaching us, the frequency would be expected to increase since the wavelengths of each sound wave reaching our ear would be getting shorter. Substituting in the problem variables we can write,

$$f_{observed} = \left(1 + \frac{100\,\frac{\cancel{km}}{\cancel{hr}}\cdot\frac{1\,\cancel{hr}}{3600\text{ s}}\cdot\frac{1000\text{ m}}{1\,\cancel{km}}}{343\,\frac{\text{m}}{\text{s}}}\right)\cdot 1000\text{ Hz}$$

The final frequency would therefore be,

$$f_{observed} = \left(1 + \frac{27.8}{343}\right)\cdot 1000\text{ Hz} = 1.08\cdot 1000\text{ Hz}$$

$$= \mathbf{1080\ Hz}$$

With all this sound bouncing about, how come we don't hear things echo forever? Well, fluids move about when sounds pass by, and the rubbing of molecules generates friction, which is converted to heat. So, the simple answer would be – all sound is ultimately converted to heat.

Everything that was ever said has been converted into so much heat in the universe (the subject of our next chapter). Imagine the untapped thermal energy present in politicians worldwide. *Wait a minute, the Guru may just have invented a new renewable power source…*

The Guru Says…

Doppler originally performed his work with light waves, thus we historically call sources, which are frequency shifted 'up' as being *'blue shifted',* since when looking at light the tendency is to shift the light source towards the blue {higher} part of the spectrum. Similarly, light sources that are moving away and shifted towards the red part of the spectrum, are labeled as *'red shifted'*.

Astronomers love the Doppler effect since it allows them to determine the direction a star is moving based upon its light frequencies – and also how fast it is moving towards or away from us.

Chapter 10: Heat, Temperature and Thermodynamics

The 'fuel' that propels the universe

The universe is a moving object, albeit a super-amazingly-large moving object. Galaxies, light, comets, asteroids and planets of one type or another have all been whirling about for at least 20 billion years and currently show no signs of stopping anytime in the distant future.

We learned so far that to move something about requires work, and work requires stored energy. The universe is not a power stingy mechanical system and thus the batteries required to keep the lights on are enormous indeed. Those universal batteries are simply known as *heat*. How good a quantity of heat is at doing work is called *temperature*, and the rules that make it all work is called *thermodynamics* (the dynamics of hot stuff).

Thermodynamics is that part of physics that deals with the relationships and conversions between heat and the other forms of energy we've already covered in the Guru's guide.

College students typically fear the topic of thermodynamics (or *'thermo'* as it's called), because it looks scary. In actuality, the physics of thermodynamics is quite simple, straightforward, and based upon common sense. Moreover, they are probably the most important laws in all of physics. Sure, the math involved in a particular problem might be tricky, but the actual 'thermo' is not. The Guru will cover the basics in a couple of pages – so keep reading and be amazed.

What is Temperature? The Venerable Zeroth Law

Initially there were only two fundamental laws of thermodynamics. They dealt with how to relate work and energy to heat (the first law), and which way heat liked to travel in the universe (the second law). Later, as an implication of the second law, someone added a trivial third law without much fanfare. Things were good, and early physicists thought themselves to be clever people indeed.

Then there was a problem because someone caught an omission regarding the most fundamental property of thermodynamics – *temperature*. It was easy to overlook. Everyone knows what temperature is, but after the ink way dry on the first two fundamental laws I'm sure someone shouted in a university lecture hall, *"...hey, what about temperature? Isn't the basic definition of that more fundamental than the first two fundamental laws?"*

Sure enough it was, and to avoid upsetting a lot of people by saying 'oops' and renumbering the laws, scientists merely added this 'fundamental of fundamental laws' before the first two, and called it the zeroth law of thermodynamics. So, in essence, we have a zeroth law of thermodynamics because of a clerical screw-up.

Now this does not mean that we actually have three laws of thermodynamics. Physicists have way too much pride to admit anything as trivial as a numbering scheme error, so they still just refer to there only being two laws of thermodynamics and quickly say with lots of arm waving, *"...oh, by the way, there's this zeroth law regarding temperature that is so basic it's not worthy of any further discussion."*

Indeed... So, without breaking any long standing tradition or taboo, the Guru will refer to thermodynamics as having two main laws (the biggies) and reference everything else around it in an anecdotal way to maintain the cover story.

The Guru Says...

The zeroth law is 'kind of' like Newton's first law in that it implies that without some external force 'temperature' there is no momentum. Unlike the first law though, there is no free motion absent external forces.

If there is no force 'temperature' there is no motion. Thermal inertia (heat), as it were, is an immovable object without temperature to drive it.

The zeroth law of thermodynamics deals with thermal (i.e., temperature) equilibrium. Thermal equilibrium occurs when two objects can no longer transfer heat to each other, and have reached the same temperature. Ultimately the universe will die courtesy of the zeroth law, since all this star-power being radiated throughout the universe will heat up other colder areas until everything reaches the same temperature. This is called the *heat death* model for the universe. When the temperature is the same everywhere, the universal batteries are officially dead.

The zeroth law tells us that heat transfer only occurs where there is a temperature difference. No difference in temperature – no heat transfer, nothing happens. This phenomenon is called *heat conduction,* and would be classified as a property of matter (but not space, since that's a different concept called thermal *radiation*).

In Figure 10.1 we have our old friend the 'hot headed' star. Let's place this irritable character at one end of a very long piece of some type of conductive material and put a block of ice at the other end, all the time measuring the temperature along the material.

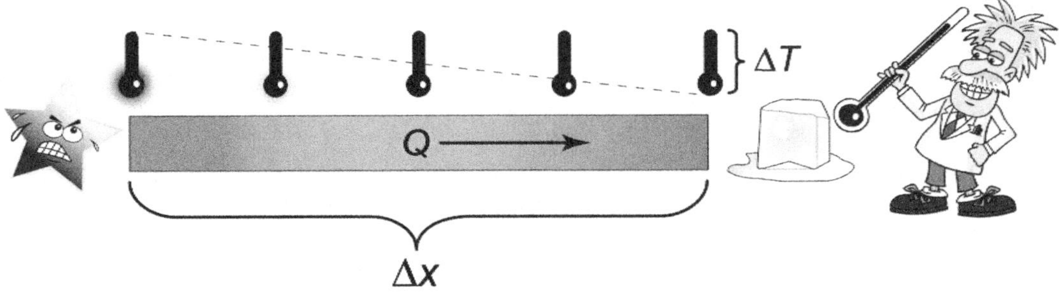

Figure 10.1: Measuring the Heat Transfer Through an Object – *The Guru's Way.*

What we would observe is that as we move away from *old-hot-head* and towards the block of ice, the temperature of the material would drop accordingly. Over time, the star would heat up its side, and the ice would cool off its end, until a 'happy medium' was achieved somewhere along the length of the material.

Let's define the variable 'Q' as the heat that flows through the material, and make some physical observations about what is going on in terms of the heat movement from our series of thermometer measurements. One of the first things that we would observe is that the amount of temperature measurable by the thermometer would vary linearly as a function of distance from the star, and would be proportional to how big of a surface area 'A' is exposed to the heat source. Closest to the star, the temperature would be greatest; further away, the temperature would be the lowest. Thus, we know that the movement of heat is related to the temperature through distance, and we can write,

$$heat_{movement} \approx A \cdot \frac{\Delta T}{\Delta x}$$

The quantity of $\Delta T / \Delta x$ is often referred to as the *thermal gradient* (i.e., the change in temperature as a function of distance). We also know that as time goes by, the hot end of the rod is going to get hotter, and the cold end colder, so there has to be some relationship between the movement of the heat itself and how much time has elapsed. Let's rough-out an expression for this as well.

$$heat_{movement} \approx \frac{\Delta Q}{\Delta t}$$

Since we have identified that both of these physical processes are going on at the same time, it's probably a safe bet that we can equate these two different types of heat movement together using some sort of thermal 'fudge factor'. We'll call this fudge factor the *thermal conductivity* of the material and denote it with the letter 'k'.

A list of the thermal conductivities of various materials is shown in Table 10.1. It should be noted that the units of thermal conductivity are in *power per degree per unit length of the material.* This would somehow imply that heat transfer involves the movement of energy, since 'power' times 'time' equals energy.

Table 10.1: Thermal Conductivities of Various Materials

Material of Interest	**Conductivity (k) in** $\frac{W}{{}^{o}C \cdot m}$
Air (0° C)	0.02
Styrofoam	0.03
Asbestos	0.08
Wood	0.12
Helium (20°C)	0.14
Hydrogen(20°C)	0.17
Water (20° C)	0.60
Ice	1.60
Mercury	8.30
Lead	34.70
Steel	50.20
Iron	79.50
Aluminum	205.00
Gold	314.00
Copper	385.00
Silver	406.00

Let's 'bolt' our two previous expressions for heat movement together, and to keep the direction of our heat flow 'Q' moving in a positive sense (like we've drawn in Figure 10.1), we'll also place a minus sign in front of the thermal conductivity parameter 'k' so that heat flows from hot to cold just as we observe it going on in the universe. Thus, we have the following expression describing the thermal equilibrium process defined by the Zeroth Law of Thermodynamics.

$$\frac{\Delta Q}{\Delta t} \approx A \cdot \frac{\Delta T}{\Delta x}; \quad \frac{\Delta Q}{\Delta t} = -k \cdot A \cdot \frac{\Delta T}{\Delta x}$$

$$\Delta Q = \left(-k \cdot A \cdot \frac{\Delta T}{\Delta x}\right) \cdot \Delta t$$

Thus the '*change in heat is equal to the thermal conductivity of the material, times the area of contact, times the thermal gradient, times how long the heat is applied'*. Thermal equilibrium occurs when there is no additional heat transfer occurring between two objects (i.e., $\Delta Q = 0$). This occurs at two points, the first one is when we first apply the heat ($\Delta t = 0$). Can you guess what is the second one is?

Example 10.1: A block of ice is placed atop a 10 mm thick steel table in a 20 °C room. If the block is 0.5 m^2 on each side, what is the change in heat through the table in one minute?

Looking up the value for the thermal conductivity of steel and substituting into our expression we get,

$$\Delta Q = -k \cdot A \cdot \frac{\Delta T}{\Delta x} \cdot \Delta t$$

$$\Delta Q = -50.20 \frac{\text{W}}{^{\circ}\text{C} \cdot \text{m}} \cdot 0.5\,\text{m}^2 \cdot \frac{(20-0)\,^{\circ}\text{C}}{0.01\,\text{m}} \cdot (60\text{ s}) = -3012000\text{ J} = \mathbf{-3012\ kJ}$$

So there's actually a fair amount of energy moving from the top surface of the table to the bottom as the block of ice melts. This movement of heat is the fundamental mechanism that makes the universe work, since we can reduce all physical processes down to the simple movement of heat from one point in space to another.

The Guru Says...

There's a neat scientific gizmo called a 'Stirling engine' that works entirely due to the movement of heat. This simple heat engine, which was invented in 1816, is an amazing thing to see work, since it requires no 'fuel' other than a source of heat. You can place a toy Stirling engine atop a cup of coffee and it will run for well over a half hour powered by nothing more than the heat from the cup. Check one out sometime.

Now that we know that heat is the 'fuel' that makes the universe 'go', we need to derive some type of equation that relates 'heat energy' to 'work energy'. This is the role of the *First Law of Thermodynamics*. After the first law, it's all downhill from here on out.

The First Law of Thermodynamics – The Ultimate Law of Conservation

The first law of thermodynamics is an easy one to grasp: heat equals work, or work can be turned into heat – *that's it!*

Don't believe it? Then, try the following experiment. While you're reading the next couple of paragraphs, start rubbing your hands together rapidly – and make sure to expend a good deal of work doing it. Your hands start to get hot, don't they?

Because you're a clever physics person, you know that the heat was generated by friction. Let's say now that you decide to slide down a rope attached to a high tree branch, but forgot to wear a pair of gloves. In this case, you are converting your potential energy into kinetic energy, and transferring this into the rope by virtue of trying to slow yourself down. The results are painful first-degree burns on your hands. Clearly we are well aware of the concept of converting energy or work into heat. Now, how about going the other way around?

Let's boil a pot of water on a stovetop. In this case, we are putting heat into the system (i.e., thermal potential energy into the pot of water) and as a result, the water starts to perform a phase change and converts to steam. If we now put a lid on the pot and wait a minute or two, the lid starts to bounce up and down as the steam expands.

The Guru Says...

There are a whole bunch of elegant and clever mathematical ways to derive the First Law of Thermodynamics. These look really great on paper, but believe it or not, the whole idea got started in the 1600's by looking at how steam could be used to produce mechanical motion.

Thus, the Guru's way of conceptually deriving the problem is very much at home in a historical context (plus, most physics problems are derived by first having a conceptual idea of the problem, and then attacking it with mathematics).

Because the lid has mass, it requires a force to move it. Since the lid bounces up and down, there must indeed be force acting over some type of distance (the bounce), and thus work is being generated. Thus, it looks like we can also go from heat to work. We could even get cleverer still and run a pipe from the lid to some sort of turbine or piston and make a rudimentary steam engine, and tap some of the kinetic energy of the steam moving through the pipe.

So, it looks like we can go from heat to work or energy, and back again, with very little difficulty. Knowing this, and our previous understanding that both matter and energy is conserved in the universe, we can write the first law by simply stating that all the conserved energy in a system 'U' has to be equal to any heat we apply 'Q' plus any work 'W' we do, or,

$$U = Q + W$$

Recalling our previous example with the pot of water on the stovetop, we know that any energy stored in the pot is trapped in the water. Thus when the lid bounces up and down, that work is being taken out of the system. The difference in energy in the system, ΔU, is therefore going to be equal to the heat we put into the system minus the work being performed moving the lid.

We can write this new, and slightly modified expression as,

$$\Delta U = Q - W$$

This form of the equation is what is known as the *First Law of Thermodynamics,* and it basically states that the change in energy of a system is equal to any heat you put into it, minus any work you get out.

Either form of the first law is correct depending on whether you are trying to sum up all the energy in a system (the first form), or trying to figure out what you can do with heat generated work (the second form).

The First Law of Thermodynamics is nothing more than a 'super duper conservation equation', which can never, ever be violated. Energy can be changed from one form to another, but it cannot be created or destroyed.

This *most important of all equations* in all of physics is the first of two fundamental operating rules governing how things must behave in the universe.

Problem Solving Tip...

When we start talking about thermodynamics, invariably the topic of 'perpetual motion machines' arises. Perpetual motion machines (or PMM's) are those quirky science-fiction-like devices or concepts that seem to generate something from nothing.

There are two types of perpetual motion machine classifications in the universe – namely, a Type 1 and a Type 2 machine. These classifications arise by virtue of whether or not the system in question violates either the first law (a Type 1 machine), or the second law (a Type 2 machine). Therefore, any device that violates $\Delta U = Q - W$ is a Type 1 perpetual motion machine.

Claims and hoaxes of this type are still commonplace in our modern world. A five-minute search of the internet by the Guru found 'plans' for four such devices being sold. You now know how to tell if something is mechanically too good to be true, since first a law violation is pretty easy to spot.

Example 10.2: A car is speeding down the road at 120 km/hr. How much aerodynamic power is consumed by the vehicle if C_D is 0.2 and the cross-sectional area of the car is 1 m^2? If the heat of combustion of gasoline is 47.0 kJ/g, how much additional fuel must be burned per minute just to overcome the drag force? If the car were a gasoline-electric hybrid, how much 'free energy' would we get to reduce the fuel consumption?

From the last chapter, we can calculate the power required to overcome aerodynamic drag forces by multiplying that force by how fast we are moving. Doing this we have the following relationship describing how much power is required to overcome aerodynamic drag.

$$P = F \cdot v = \left(\frac{1}{2} C_D \cdot A_P \cdot \rho \cdot v^2\right) \cdot v = \frac{C_D \cdot A_P \cdot \rho}{2} \cdot v^3$$

Notice how our final expression is a function of the cube of the velocity. Substituting in the applicable problem values, we determine the power required to be equal to,

$$P = \frac{C_D \cdot A_P \cdot \rho}{2} \cdot v^3$$

$$= \frac{0.2 \cdot (1\ \text{m}^2) \cdot 1.204\ \frac{\text{kg}}{\text{m}^3}}{2} \cdot \left(120 \frac{\cancel{\text{km}}}{\cancel{\text{hr}}} \cdot \frac{1\ \cancel{\text{hr}}}{3600\ \text{s}} \cdot \frac{1000\ \text{m}}{1\ \cancel{\text{km}}} \right)^3$$

$$= 0.1204\ \frac{\text{kg}}{\text{m}} \cdot \left(33.3\ \frac{\text{m}}{\text{s}} \right)^3 = 4445.9\ \frac{\text{kg} \cdot \text{m}^2}{\text{s}^3} = \mathbf{4.45\ kW}$$

Thus we find out that it takes roughly 4.5 kilowatts of power by the engine just to overcome aerodynamic drag alone. For every minute of engine operation, the energy required would therefore be,

$$\Delta U = P \cdot t = 4445.9\ \frac{\text{kg} \cdot \text{m}^2}{\text{s}^3} \cdot 60\ \text{s} = 266754\ \frac{\text{kg} \cdot \text{m}^2}{\text{s}^2} = 266.75\ \text{kJ}$$

If gasoline has a heat of combustion (i.e., how much 'umph' you can get out of it by burning it) of 47.0 kJ/g, then the additional amount of gasoline required per minute would be,

$$\Delta Fuel_{drag} = \frac{266.75\ \text{kJ}}{47.0\ \frac{\text{kJ}}{\text{g}}} = \mathbf{5.67\ g}$$

Thus it takes roughly 5.7 additional grams of fuel per minute just to counteract the air resistance and maintain the indicated speed (and this is in addition to the fuel required to move the vehicle in the first place, and overcome rolling resistance and engine friction). In the case of a gasoline-electric hybrid, we know that,

$$\Delta U = Q + W = 266.75\ \text{kJ}$$

Which states that our system (the car) must have 266.75 kJ no matter what (i.e., an absolute total, which is why we are using the first form of the first law). In the case of a purely gasoline engine, we obtain this amount through heat energy from fuel combustion. If the car is augmented by an electric motor, then the above expression states that some additional work must come in the form of stored electrical energy (i.e., batteries), and that this must be accounted for in the total required energy to overcome aerodynamic drag.

There is no free ride in a hybrid car. You obtain the needed electrical energy by swapping with thermal energy (burning gasoline) or from gravitational potential or kinetic energy recovery through what is called *regenerative braking*. Other than being able to temporarily store some of your energy in batteries for reuse, hybrid cars are no different than any other type of motorized conveyance, from a thermodynamics standpoint.

Example 10.3: For the same car in Example 10.2, assuming that the fuel being burned was a 20% ethanol (alcohol) 'blended' mixture, how much additional fuel would have to be burned to maintain the previously stated thermodynamic equilibrium of the system?

(Ethanol has a heat of combustion of 29.7 kJ/g)

A 20% ethanol mix to gasoline would yield a final heat of combustion 'H' equal to a simple ratio of the heats of combustion of the two individual fuels. Thus, our new value would be,

$$H_{mix} = (0.2 \cdot 29.7)\frac{\text{kJ}}{\text{g}} + (0.8 \cdot 47.0)\frac{\text{kJ}}{\text{g}} = 5.94\frac{\text{kJ}}{\text{g}} + 37.6\frac{\text{kJ}}{\text{g}} = 43.54\frac{\text{kJ}}{\text{g}}$$

Reperforming our previous calculation with the new 'blended' fuel mixture gives the following result,

$$\Delta Fuel_{drag} = \frac{266.75\ \text{kJ}}{43.54\ \frac{\text{kJ}}{\text{g}}} = \mathbf{6.13\ g}$$

So our fuel economy will not be as good using the 'blended' fuel over straight gasoline, since the amount of heat energy that this fuel can develop is not as great. It's even worse economically when you pay more for this fuel, since you're taking a double-hit in terms of vehicle performance and cost at the fuel pump. *...but that's just the Guru ranting for a moment.*

Example 10.4: Professor Flim Flam is selling some little known (and physically small) power generating 'secret' on the internet that will make your house energy independent. Does this claim sound plausible?

This is an actual example of an internet claim made by multiple different authors showing altered photos of electric motors, power inverters, Tesla coils, or even internet data routers. Each author spends large amounts of time telling interested parties how the power companies hate this 'device', and how (depending on the grandioseness of the claims) it will cut your energy bills, or even make your electrical meter run backwards.

Most perpetual motion claims are made without numerical validation. Think for a moment as to the total amount of power every electrical circuit in your house can consume. A typical house requirement would be on the order of 24,000 W with everything turned on.

Using that we could write for each hour,

$$\Delta U = Q + W = 24{,}000\ \text{W} \cdot \text{h}$$

$$Q + W = 24{,}000\ \text{W} \cdot \text{h} = 24\ \text{kW} \cdot \text{h}$$

So somewhere we have to generate 24 kilowatt-hours of energy if we are going to 'keep the lights burning', as it were. The First Law of Thermodynamics states that this required change in energy must be developed through the conversion of potential or kinetic energy to work 'W', or through the combustion of some type of 'fuel' to generate the required amount of heat.

Looking at the first case, we know that in order to obtain this energy through kinetic energy we would need a very large flywheel with an equally large moment of inertia (or some similar type device), to produce this amount of energy, but (as we're going to see in the next section), this spinning device would ultimately come to a stop, and require at a minimum an equal amount of energy to get it going again (which really negates the point of having the gizmo at all).

Even if we stored the energy potentially, say in batteries, they would eventually go dead requiring a recharge, which also defeats the purpose of the device. We might be able to do the trick with a small radioisotope power plant, but you can already see the logistical headaches associated with operating such a device (plus, it is illegal to sell one mail order anyway). Clearly a small, highly energetic device capable of producing the required ΔU is out of the question.

So what about using heat?

Sure, we could build such a device, which burned fuel to generate the required ΔU (think emergency generator), but we would need a large fuel source requiring refilling – so what's the benefit? In all likelihood, the claim is nothing more than a PMM of Type 1.

The Second Law of Thermodynamics – The '*Universal One Way Sign*'

There's a very old saying that, *"The Devil is in the details"*. Nothing can be truer than this statement applied in reference to the Second Law of Thermodynamics – *a law of utter simplicity, but extremely devilish in its ramifications.*

In our example of the hybrid car, we commented that there is a limit to how much energy you can get back into the batteries to reuse, and all of that electrical energy started out as gasoline at one point in time. Nowhere in a hybrid car's design can it ever convert unused potential or kinetic energy back into gasoline in the tank. There's a reason for this – *the second law.*

Also we commented that any storage device would ultimately run down and require 'recharging' at some point. No energy storage device is perfect, and the conversion from gasoline in our hybrid automobile to electricity in the storage batteries is never on a *Joule-for-Joule* basis. There's a reason for that too – *yup that mean old second law again.* In fact, the Second Law of Thermodynamics places limits on just how much conversion from potential to kinetic to heat energy (or any combination thereof) that the first law allows.

The second law also states that you'll never get all of your energy back again through any of the above conversions, no matter how hard you try otherwise – *muahahahahahaha* (sorry, that was the second law laughing at you yet again). I told you it was a 'big meanie'.

The *Second Law of Thermodynamics* states that the *entropy* '*S*' of a closed-system, which is a measure of chaos or equilibrium in a system, is *always increasing*. Only in one special case can it be equal to zero, and we'll talk about that in a second. Mathematically, the second law can be written as,

$$S = \frac{\Delta Q}{\Delta T} \geq 0$$

Thus, from a 'universal heat-moving around-the-place standpoint', the second law states that the entropy is equal to the change in heat of a system divided by its change in temperature, and that this change is *always greater than or equal to zero*. It therefore implies that large changes of heat result in large changes in entropy. This makes sense, since doesn't a large change in heat result in a more uniform and chaotic distribution of heat across space, or our previous definition of thermal equilibrium?

Consequently, the second law also says that a large change in temperature generally results in lower levels of entropy. This is an interesting concept in its own right since this implies that not all 'heat' is created equal, and heat that is generated at high temperatures and transferred to a low temperature has a better 'quality' than heat transferred under lesser temperature extremes. Clearly there is some 'snobbery' going on in the universe between the various classes of stars in terms of whose heat content is better.

The Guru Says...

The term 'entropy' is one of those words tossed about by everyone, but few know what it actually means.

Entropy is a measure of disorder or chaos. It quantifies the level randomness or ultimate equilibrium that all processes in the universe seek to attain. Entropy is a 'cartoonishly destructive' concept, as it is one of the main driving forces that will ultimately destroy the universe.

In the peculiar case of where the entropy is equal to zero we have the following equation describing what is called an *isentropic* (or constant entropy) system,

$$S = \frac{\Delta Q}{\Delta T} = 0$$

An isentropic system is one in which there is no net change in the chaos due to the conversion of heat into energy or work. The system is left in the same state of equilibrium (or order) both before and after some type of event has taken place. From looking at the above equation we can see that this occurs when there is no heat change in the system ($\Delta Q = 0$) or the change in temperature is really, really large (like a star radiating onto the cold vacuum of outer space. Isentropic systems have another interesting property, namely they can operate in one direction, and then switch directions without any thermal or mechanical losses. These types of systems are called *reversible systems*.

Under no circumstances can the entropy of a system be less than zero. Devices or processes that claim this feat are known as perpetual motion machines of Type 2 (or PMM's Type 2). These types of machines are a little more difficult to spot, since the key here is to identify a PMM that actually improves the order in the universe (like a broken dish jumping off the floor and landing on a table completely reassembled). The second law prohibits such things from ever happening, and keeps a certain level of sanity in the universe.

Problem Solving Tip...

So what does this mean in plain terms? Well, the second law states that all energy processes are imperfect or inefficient and that everything in the universe is going to 'run down' since entropy is driving all temperature differentials to an absolute equilibrium. Entropy wants to distribute every atom of matter uniformly in the universe and give each atom the exact same temperature.

At the point in the very distant future that this occurs, all functioning processes within the universe will stop. This is the concept known as the *heat death* of the universe.

Considering that the second law defines *one and only one direction* that processes must move, it is in essence a *universal one-way traffic sign* telling heat and energy the correct direction of motion. So, if the first law keeps all matter and energy in check such that nothing can be magically created or destroyed within the universe, the second law most definitely defines the *direction that time itself must travel.*

The Guru takes a sip of coffee to let that concept register with the reader...

If we take the second law at face value (and there is no reason why we shouldn't), then the direction of time in the universe is not arbitrary and time travel (at least backwards in time) is out of the question, since this would imply a reduction of universal entropy, which is a definite *'no no'*.

Example 10.5: Suppose in the automobile example in 10.2, the Guru drove for an hour. If the car's engine burned the fuel at a temperature of 3000 °C and dumped the unused heat out the radiator at 150 °C, what is the increase in entropy?

So, for an hours worth of travel, the energy requirements to overcome wind drag would be,

$$\Delta Q = P \cdot t = 4445.9 \, \frac{\text{kg} \cdot \text{m}^2}{\text{s}^3} \cdot 3600 \text{ s} = 16 \text{ MJ}$$

The increase in entropy for the car engine would therefore be equal to this amount of heat generated divided by the starting and ending temperatures.

Thus we have,

$$S = \frac{\Delta Q}{\Delta T} = \frac{16\text{ MJ}}{(3000 - 150)\ ^\circ\text{C}} = \mathbf{5615.8\ \frac{J}{^\circ C}}$$

Which is greater than zero, so we know that the chaos of the system has increased and we can never recover this energy fully again. Notice that it does not matter in the context of this problem whether or not the car is a hybrid, we will always lose energy in the deal.

Example 10.6: Professor Flim Flam is at it again... This time he's selling a gizmo for your car's engine, which looks a lot like an air conditioner compressor with a blinking light. He states that it can actually increase the efficiency of your engine by further reducing the temperature of the exhaust gas. His device sure enough gets cold to the touch, and he says that it takes a starting exhaust gas at 300 °C and cools it to 50 °C – *guaranteed* (assuming you can find him after you give him your money).

Does this device work as claimed? What is it really?

Being a shrewd consumer, you reach for the Guru's Guide and our handy entropy equation. We know that the device bolts to the engine and consumes power, thus it has to consume a positive 'ΔQ'. We are told that as a result of this positive 'ΔQ', the exhaust temperature drops from 300 °C to 50 °C (for a net drop of -250 °C). Plugging in the values we find,

$$S = \frac{\Delta Q}{\Delta T} = \frac{\text{Positive } \Delta Q}{-250\ ^\circ\text{C}} = -\text{ Some Entropy } \frac{\text{J}}{^\circ\text{C}} < 0$$

Thus the device, as advertised, is a PMM of Type 2. In actuality *the change in temperature observed is due to a commensurate removal of heat*. Professor Flim Flam likes to try to confuse these two quantities all the time, and sometimes even tries to say that they are the same. Thus, the entropy is again positive and we conclude that the device is indeed some type of air conditioning compressor with a cool light on it. We offer to trade the professor several acres of prime oceanfront property in Arizona for the patents to the device.

The Third Law of Thermodynamics (Well, it's *kind-of, sort-of* a law)

The Third Law of Thermodynamics is a natural consequence of the second law, so it's not really a law in its own right – it's more like an axiom. By definition, entropy is a measure of energy or heat that has 'changed hands' into some other form. In the limit where all the heat in a system has been transferred, the system is by definition at its lowest energy state and has approached zero entropy.

Mathematically this is written in the following form as,

$$\lim_{\Delta T \to 0} S = \lim_{\Delta T \to 0} \frac{\Delta Q}{\Delta T} = 0$$

Where the mathematical term *'lim'* is pre-calculus 'lingo' for the *limit as a variable approaches a specific value*. In this case, we are looking at the limit of the entropy equation as the change in temperature ΔT approaches zero.

At first glance one would think that dividing by ΔT would cause the entropy equation to explode, but it can be shown using Calculus that the entropy indeed hits zero at some finite temperature point that is always the same. At this point, no additional energy can be removed from the process.

The Guru Says...

There is an old joke which sums up the laws of thermodynamics nicely. Imagine that the universe is just one big table game, like a Roulette table.

If this is the case, then these are the *Official Rules of Play* for the 'Universal Game'...

- Zeroth Rule: You must play the game.
- First Rule: You can't win the game.
- Second Rule: You can't even break even at the game.
- Third Rule: You can't quit the game - *ever*.

This temperature, where at all atomic motion stops, is called *absolute zero*. Absolute zero is defined as zero-degrees Kelvin (0 °K) in the SI system, which is roughly equal to -273.15 °C. *Nothing is colder than absolute zero, and to get there is an impossibility.*

This concludes our whirlwind discussion of thermodynamics. The goal of this chapter was to outline the basics of what powers the universe, and that the First and Second Laws of Thermodynamics pretty much run the whole show – *universally speaking*. The most important concept to remember is that all physical laws must obey the laws of thermodynamics, without exception!

At Absolute Zero – Not a Whole Lot Happens

Chapter 11: Fundamental Forces in Nature

The glue that holds the universe together

After all of our discussion on the different types of mechanisms that move things around in the universe, it turns out that from a force standpoint, everything boils down to just four (*read that – just four*, fundamental forces that control everything). These forces are collectively called: *the strong nuclear force, the weak nuclear force, electromagnetic forces, and the gravitational force*. Whereas the laws of thermodynamics provide the framework wherein the universe operates, these four forces provide the necessary push and pull to get things done.

Let's start out with the fundamental force that holds the very foundation of matter itself together – the strong nuclear force.

The Strong Nuclear Force – The Glue that Holds Matter Together

The strong nuclear force is the attractive force that holds protons and neutrons together inside an elemental nucleus. Just as the name implies, it is the strongest of all the forces in the universe. If we normalize all forces in the universe to be relative to that of gravity (i.e., we scale everything so that gravity equals '1'), the strong nuclear force would be somewhere in the range of $1x10^{38}$ times that of gravity. This should give you a pretty good idea as to just how strong this force is – *it's huge!*

So in essence, the strong nuclear force holds matter itself together, like what is shown in Figure 11.1. The upside of having a force this great holding matter together is that it's pretty unlikely that the atoms in your right arm will spontaneously fly apart due to entropy, since the raw matter is 'glued' together with the strongest force in the universe. *Whew* – that is a reassuring thought.

On the other hand, with a force having a magnitude equal to $1x10^{38}$ times that of gravity it's amazing that the whole universe itself just doesn't 'crush' under the tremendous pressure. Well, the other 'upside' of the strong nuclear force is that it is only strong over a *very small distance* – a distance like a couple of femtometers ($1x10^{-15}$ meters). At distances beyond roughly 10 femtometers, the strong nuclear force is essentially non-measurable, and definitely has no great effect on the motion of the universe as a whole. In this regard, the strongest force in the universe is very much self-limiting, and is only interested in hanging onto matter within its own little atomic 'universe'.

The best that modern physics can figure out about the strong nuclear force is that it appears to be constant over distance and uniform throughout. Although definitely an attractive force (and we should comment here that even though it might be a great force to look at, *attractive* in this context means that it likes to pull objects closer to it), the strong nuclear force does have the ability to repulse other objects. Thus even though matter within a nucleus is close together, it never actually touches.

The Weak Nuclear Force

In terms of ranking the four fundamental forces by order to relative strength, the weak nuclear force is actually third on the list compared to gravity. Sandwiched between the two nuclear forces is the electromagnetic force. Weak nuclear forces are, however, about 1×10^{25} times stronger than gravity.

The weak nuclear forces are also forces that only work within the nucleus of an atom, but as the name implies, they are, well, 'weak'. This force is responsible for nuclear reactions within the atom, such as electrons being 'kicked off' of neutrons - a process called *beta decay*, and one that produces the observable phenomenon called *radiation* that we can see in Figure 11.1.

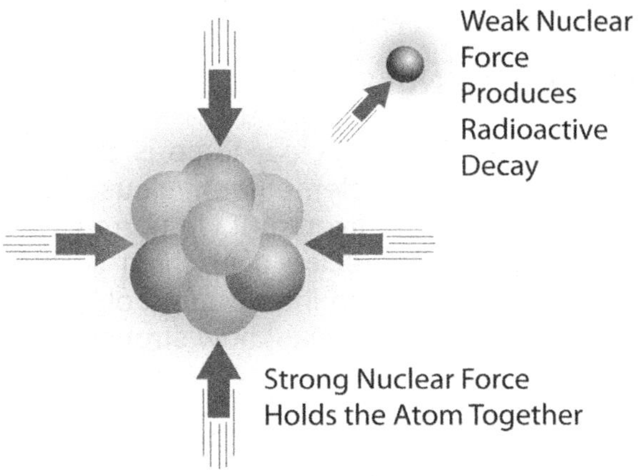

Figure 11.1: The Strong Nuclear Force Holds Atomic Nuclei Together, Weak Ones Split Them Apart

Thus, without this 'weak' force within the universe, radioactive decay of an atom could never occur. This might not seem like a big deal, but remember from Chapter 3 that without radioactive decay, fusion reactions could not occur within stars, and thus there would be no stars to speak of. So the weak force serves a very important purpose, and without it the universe would be an uneventful dead collection of matter.

Weak nuclear forces also act over extremely small distances on the order of one attometer (i.e., 1×10^{-18} meters). Instead of applying a constant force over that distance, the weak nuclear force drops-off at a rate inversely-proportional to the distance 'r' from the center of the nucleus (or, at a rate of 1/r). Physical laws that behave this sort of rule are called *inverse laws*. The weak nuclear force is repulsive by nature (no, it's not disgusting to look at, it just likes to push away, or *repulse*, things).

The Electromagnetic Force

Now that we know what holds matter together and causes radiation (and stars to function), we should note that we only have two fundamental forces left. Since the first two forces (the strong and weak nuclear ones) only really care about what goes on in the neighborhood of the atomic nucleus, they are quite oblivious that there is even such a thing as a universe. That must mean that all observable interactive forces in the larger universal scheme of things must be due to the remaining two other forces. Let's look at the first one, the forces of electricity and magnetism or *electromagnetic forces*.

The electromagnetic forces are responsible for all electrical and magnetic effects in the universe. Forces such as electrical charge, electrical charge potential (voltage), electron movement (current), and the ability to exhibit any form of magnetism is all due to this fundamental force. The electromagnetic force is responsible for electrons moving from atom to atom in a chemical reaction (and thus responsible for the totality of chemistry), as well as being responsible for that motion killing force known as friction. Yes indeed, without the electromagnetic force the universe would come to a grinding, but frictionless, halt.

Electromagnetic forces are about $1x10^{36}$ times stronger than gravity and extend their influence out to infinity '∞'. This makes them a formidable force in the universe on a cosmological scale. Try placing a powerful magnet near a metallic object. It will literally lift the object up easily overcoming the gravitational force. So why isn't gravity perpetually 'drowned-out' by this overwhelming electromagnetic force? It's quite simple, electromagnetic forces always show up in pairs.

Take a look at Figure 11.2 to see what the Guru is talking about. For every positively charged (+) particle out there, there is a negatively charged one lurking about. The same applies with magnetic fields. For every north magnetic pole (N), there has to be a south magnetic pole (S) connected to the same object. Different electromagnetic charges or poles attract each other; like ones repel each other. Think of it like two sides of the same coin.

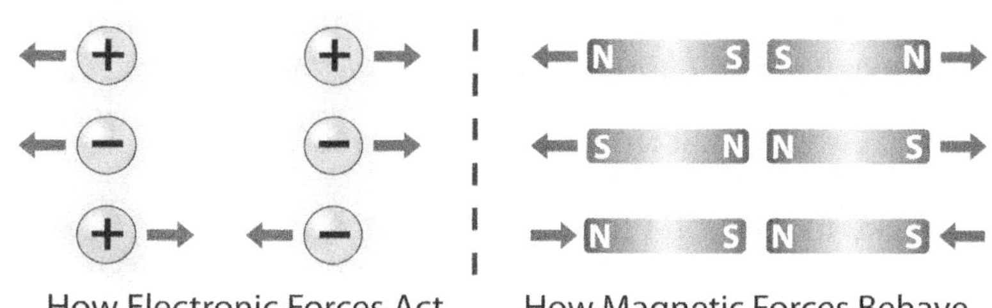

Figure 11.2: Examples of Electronic and Magnetic Attraction and Repulsion

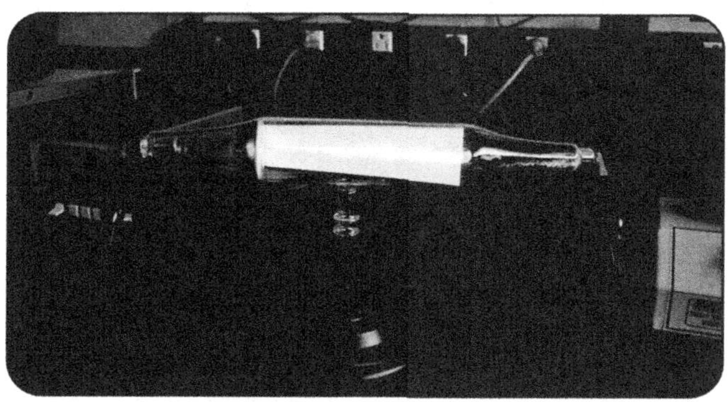

Figure 11.3: Example of Electromagnetic Radiation Emission Using the Guru's Cathode Ray Tube

When electromagnetic forces act, like those in the cathode ray tube shown in Figure 11.3, they exactly cancel each other out in the process of doing whatever they were doing. Electricity has to flow from an electron source to a ground. The net result is that even though the electromagnetic force is far greater than gravity, the annihilation of that force to accomplish something still allows gravity to impose its effects on matter.

Electromagnetic forces drop-off at a rate inversely-proportional to the square of the distance 'r' from the center of the charged object (or, at a rate of $1/r^2$). Physical laws that behave this sort of rule are called *inverse square laws,* and there are a great many of them in the universe. Electromagnetic forces are so important that we're going to dedicate an entire chapter to them.

The Gravitational Force – The Real Universal Weakling (or is it?)

So now we've come full circle, and we're back to gravity again, the principal force that we have examined in one way or another in the Guru's Guide since the very first chapter. Considering that gravity is indeed a universal weakling from a shear force magnitude standpoint, it's amazing that it produces any net effect on anything.

If we were to build a *'fundamental force-o-meter',* we could easily verify the claim that indeed gravity is a very, very, very small force which drops-off at $1/r^2$ just like electromagnetic forces, and can theoretically extend its influence out to infinity '∞'. The fundamental difference with a gravitational force is that it has three extremely unique things going for it, namely, 1) it's <u>always</u> an attractive force, 2) it only acts on matter, and, 3) it is immutable.

Unlike electromagnetic forces, which can be conducted and blocked using various elements, like the metals and metalloids we've examined earlier, gravity is *'all knowing'* and *'all seeing'*. You cannot block or shield gravity. Gravity cannot be absorbed, stored, deflected or transformed in any way. Gravity is its own animal – *and it is everywhere and knows where you are at all times.*

Gravity likes mass. It likes mass so much that the two are inseparable. It's picky too; it won't touch anything that does not have mass (like energy). However, if you have mass, then you also have a gravitational attraction which is a function of how much mass you have. Gravity, like matter itself, is a weird entity that physicists really don't know how or why it works in the manner that it does.

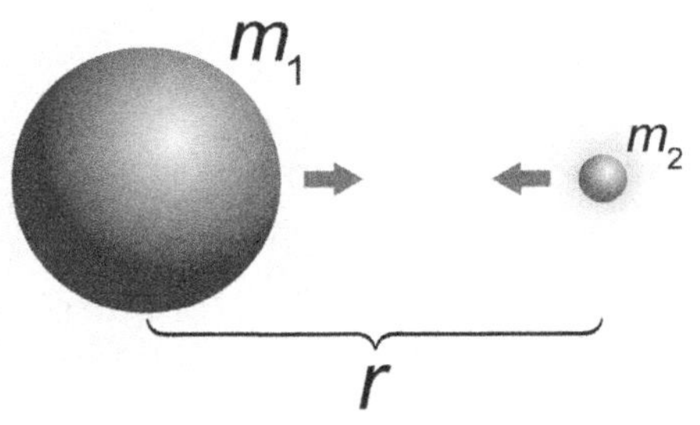

m_1 m_2 r

$$F = G \cdot \frac{m_1 \cdot m_2}{r^2}$$

Figure 11.4: The Simplicity of Newton's Law of Universal Gravitation

Isaac Newton obsessed about gravitational forces and derived a very simple equation relating the mass of two objects to how much gravitational force there is. This equation, known as *Newton's Law of Universal Gravitation*, can be stated simply in the following manner as,

$$F = G \cdot \frac{m_1 \cdot m_2}{r^2}$$

Where, as shown in Figure 11.4, m_1 and m_2 are the masses of two different objects, and 'r' is the distance between them. We can see that Newton's universal gravitation law is also an inverse square law, consistent with what we know about gravity. The term denoted as 'G' is called the *gravitational constant* and is equal to $6.673\text{x}10^{-11}\frac{\text{m}^3}{\text{kg}\cdot\text{s}^2}$. Thus 'G' is the *mother of all fudge factors*, and works surprisingly well to calculate gravitational forces to a high degree of accuracy in a Newtonian reference frame. Let's do the obligatory example problem to demonstrate what Newton was thinking.

Example 11.1: Suppose we have a person who weighs 80 kg. What is the gravitational attraction of this person to Earth?

(The mass of Earth is 5.976×10^{24} kg. The radius of Earth is 6,378.1 km.)

We can directly calculate the gravitational force by substituting into our equation as follows.

$$F = G\cdot\frac{m_1\cdot m_2}{r^2}$$

$$= 6.673\text{x}10^{-11}\frac{\text{m}^3}{\text{kg}\cdot\text{s}^2}\cdot\frac{5.97x10^{24}\ \cancel{\text{kg}}\cdot 80\ \text{kg}}{\left(6378.1\ \cancel{\text{km}}\cdot\frac{1000\ \text{m}}{\cancel{\text{km}}}\right)^2} = 783.43\,\frac{\text{kg}\cdot\text{m}}{\text{s}^2}$$

$$= \mathbf{783.43\ N}$$

If we divide this answer by the mass of the person (80 kg) using Newton's second law we can determine what the acceleration of the person is. Doing the math we find out that it is 9.793 $\frac{\text{m}}{\text{s}^2}$, which is pretty close to the average gravitational acceleration on Earth that we have been using throughout this book.

The unique ability of gravity to be everywhere and anywhere at anytime in the universe makes it responsible for such things as weight, planetary orbits, the structure of galaxies and bizarre gravitational singularities known as *black holes*. Gravitational forces, as weak as they are, are what produce the immense heat within stars required to start fusion reactions, since even a small force multiplied by the mass of a star produces a very respectable response.

The Guru Says...

Note though that we always talk about the acceleration due to gravity that an object feels. What about the acceleration due gravity that Earth feels?

In our previous example, if we divide the gravitational force by the mass of Earth we find that Earth is also accelerating towards the 80 kg person at a rate of $1.3123x10^{-22}\,\frac{\text{m}}{\text{s}^2}$. You can't have one acceleration without the other.

So, despite gravity being the weakest force in the universe, it is this weakest force that actually holds the visible universe together. The Guru told you that physics is stranger than any science fiction.

The GUTs of the Universe

We would be remiss in any physics book without at least mentioning that one of the goals of modern physics is to find a *'master equation'* that links the four fundamental forces together. It is believed that during the early moments of the universe (i.e., shortly after the so-called 'big bang'), all four of the fundamental forces were combined into one single *super-duper* force.

Using some pretty impressive theoretical physics and mathematics, physicists have postulated that gravity was the first defector at about $1x10^{-43}$ seconds after the big bang. The strong force took off shortly after that at about $1x10^{-35}$ seconds. At $1x10^{-11}$ seconds after the big bang the weak force said goodbye to the electromagnetic forces. The universe is currently about $6.3x10^{17}$ seconds old and none of these fundamental forces have talked to each other since.

In an effort of universal goodwill (and good scientific advancement) physicists are currently working out the details of a single force theory, which could be used to describe how the universe fundamentally works. This theory is called the *Grand Unified Theories* (or GUTs).

Theories that try to *'solve for everything'* wind up deep in the world of quantum mechanics and general relativity. So far, the weak nuclear and electromagnetic forces seem to play nicely with each other, remembering that they did talk to one another at sometime in the distant past. These forces can be combined into a single theoretical model called the *electro-weak unification theory*. Some progress has been made to get the strong nuclear force to cooperate as well under what is called the *gauge coupling unification theory*, but the strong force it still somewhat reluctant to share its power with any of the other forces.

For all of this, though, there is still a really big problem child – gravity. Gravity is a loner, a rebel, and it plays by its own set of rules.

Currently there are no consistent theories to combine gravity to any other force or even fully describe what this solitary force really is. Although we know that gravity likes matter, there currently appears to be not enough matter in the universe to cover off on all the observed gravity inside objects like galaxies. Clearly something very strange is at work behind the scenes in the universe, so maybe gravity isn't a single force, maybe there is a fifth fundamental force required to describe current theories involving stuff like *dark matter* and *dark energy*? Maybe gravity isn't even a force, but an observable effect of something else like matter.

The Guru will save that topic for another book, someday, maybe, *gravity permitting…*

Section IV – The Lighter Side of the Universe

Chapter 12: Electricity and Magnetism

Some rather shocking findings the Guru won't insulate you from

There isn't a day that goes by in our modern world when we are not exposed to the effects of electricity and magnetism. In fact, with the exception of gravity, electricity and magnetism is responsible for <u>all phenomena</u> we encounter in everyday life – *including things like Newton's third law since any type of collision or reactive force is just electronic forces pushing back.*

Just think about it, household electricity, radio waves, lightning storms, electric lights, telephones, friction, static electricity, magnets, generators, motors, computers, any chemical reaction, *and on and on and on, are all a product of only one of the four fundamental forces* in the universe. If you think about it, gravity might be what holds the universe together, but it's the electromagnetic forces that get things done.

Given this, the topic of electricity and magnetism could, in itself, cover several textbooks, as well as multiple college level physics courses. Since the Guru is an impatient person, this chapter is going to address the more significant findings dealing with electricity and magnetic fields, as well as provide some real-world examples of their application.

Electric Charge and the Electric Field

Normal 'uncharged matter' has, by definition, an equal number of positively and negatively charged particles, as shown in Figure 12.1.

Imbalances occur as atoms interact with one another and electrons are loosely exchanged between dissimilar materials. The result is that one object will become more positively charged (i.e., it loses some electrons), and the other will become more negatively charged (i.e., it gains some electrons). Thus, electric charge is all about the movement of electrons, and which object has the highest number of them determines the electrical charge.

The Guru Says...

As a side note, a phenomenon called the *Hall Effect* shows that indeed only negatively charged particles (i.e., electrons) can move around to affect the charge on an object. Positively charged particles remain fixed to their respective objects.

This makes sense since electric charge only trades electrons between objects and not the whole atom itself.

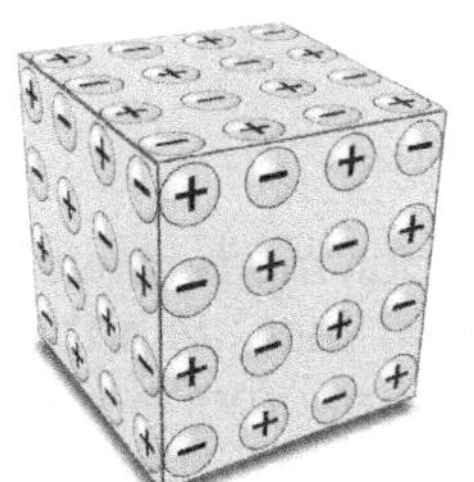

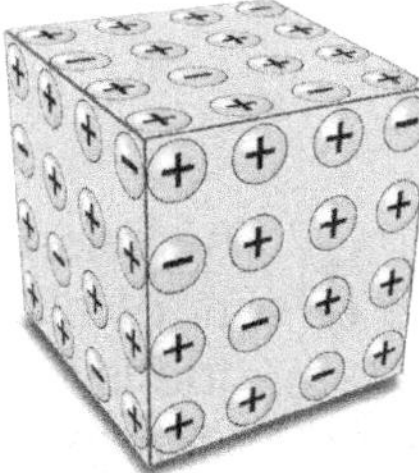

Equal (+) and (-) Charges *Object has No Charge* | More Negative (-) Charges *Object is Negatively Charged* | Less Negative (-) Charges *Object is Positively Charged*

Figure 12.1: A Simplified Example of What the Electrical Charge on an Object Looks Like

Some materials are more likely to develop a charge than others. For example, elements that have 'loose' or 'free' electrons lying around, such as metals, are not inclined to develop a charge since they can easily grab or give away excess electrons to a neighboring element, and therefore are more inclined to maintain electrical neutrality. Materials of this type are called *conductors* and transfer electrical charge, but do not typically store it. Objects that contain a lot of water, such as humans, are also great conductors of electric charge, as is Earth itself.

On the other hand materials that don't readily want to give up an electron, like glass made from silicon, fused quartz materials, ceramics, rubber and plastics, tend to buildup excess charge, which accumulates on the surface of the material. Materials of this type are called *insulators* or *dielectrics*.

A physicist by the name of Charles Augustin Coulomb, after having been zapped by static electricity enough times in his life, noticed that the magnitude of the 'zap' was proportional to the distance between the objects. The closer the objects were placed with respect to each other, the greater the 'zap'. He also noticed that the greater the charge differences between the objects, the greater the 'zap'. Coulomb wrote the initial form of his observations as,

$$F \approx \frac{q_1 \cdot q_2}{r^2}$$

Where 'F' is the force of attraction (or repulsion) between two charges 'q_1' and 'q_2', and 'r' is the distance between them.

Now Coulomb was a 'smart cookie' and noticed that the form of his equation looked a lot like Newton's law of universal gravitation, which at this time had been around for about 100 years. Believing that there had to be some form of proportionality constant (i.e., fudge factor), which made his equation an exact relationship, he then proceeded to painstakingly measure the force between two differently charged spheres in an attempt to quantify this value.

This result became formally known as *Coulomb's Equation* or *Coulomb's Law* as shown below.

$$F = \frac{1}{4\pi\varepsilon_o} \cdot \frac{q_1 \cdot q_2}{r^2}$$

Where 'ε_o' is the *permittivity constant of space* (i.e., a vacuum). This value is given as $\varepsilon_o = 8.9x10^{-12}\ C^2 \cdot N^{-1} \cdot m^{-2}$. Thus, the quantity $1/4\pi\varepsilon_o$ works out to be roughly equal to $9.0x10^9\ N \cdot m^2 \cdot C^{-2}$.

The unit designated as 'C' is an SI unit called the *Coulomb* (C), and is a measure of how many electrons are present in a given sample of charge. Similar in form to Avogadro's number, the Coulomb contains $6.241x10^{18}$ electrons. As a side note, the unit of electrical current, the *Ampere* (A), is defined as one Coulomb passing through a wire per second, so this should give you a pretty good idea as to the amount of electrical 'jolt' available from a Coulomb.

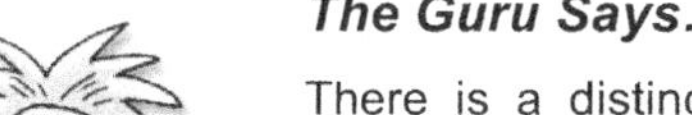

The Guru Says...

There is a distinct difference between 'static' electricity and 'current' electricity, which has to do with whether or not the electrons are moving (i.e., is the electricity at rest or not).

Static electricity is just a build-up of electrons on the surface of an object, and while the voltage and the 'zap' might be spectacular, the electrons only slightly move to ground when the charge force breaks down the dielectric of air. There is no force other than the static discharge itself moving the electrons. Current electricity, on the otherhand, has a voltage continuously pushing electrons along a specific conductive path (like along a wire), and will continue to do so regardless of any discharge to ground.

Static electricity is relatively safe (and fun) to play with. Current electricity (like that in your house) should be treated with extreme care.

The permittivity constant changes for different types of materials as shown in Table 12.1. To adjust the permittivity constant for different materials (defined as ε, and not ε_0), we merely multiply the permittivity constant of space times the material corrections shown. Air is almost the same as a vacuum in this context, but the Guru has provided the exact value for the purists out there.

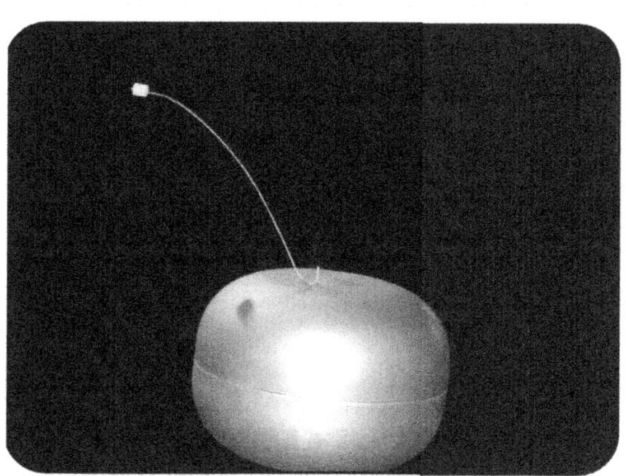

Figure 12.2: Example of Static Electric Force Produced by a Van De Graff Generator

An interesting mechanical device known as a Van de Graff generator creates a large static electric charge by moving a rubber belt inside an aluminum sphere. The motion of the belt near a small metal rake inside the sphere strips electrons off the belt and transfers them to the sphere. This produces a net charge on the surface of the sphere.

Figure 12.2 shows the Guru's Van de Graff machine with a small wooden ball attached to it using a piece of string. As you can see, the generator charges the wooden ball causing it to repel from the sphere in accordance with Coulomb's law. This repulsive force easily overwhelms gravity, causing the wooden ball to extend far away from the aluminum sphere.

Discharging the sphere causes the wooden ball to drop under the action of gravity demonstrating that electromagnetic forces can easily overwhelm this smallest of the fundamental forces.

Table 12.1: Permittivity Constants of Materials Relative to a Vacuum ε_o

MATERIAL OF INTEREST	PERMITTIVITY CONSTANT ε
Vacuum	1.00
Air	1.00058986
Teflon	2.10
Polyethylene	2.25
Polypropylene	2.36
Polystyrene	2.70
Paper	3.85
Concrete	4.50
Glass	4.70
Rubber	7.00
Graphite	10.00
Silicon	11.68
Water	80.10
Plastic	100.00

Another simple hand-cranked device invented in 1883 called a Wimshurst machine (named after British inventor James Wimshurst) uses two insulated discs with attached metal strips that rotate in opposite directions. As the disks spin, two pairs of metal brushes touching the opposite disks produce an imbalance of charge. This charge is accumulated in small dielectric containers, called *Leyden Jars.*

Static electric charges can actually be stored between alternating layers of conducting and dielectric material. This type of storage device is known as a *capacitor*. Leyden Jars are nothing more than antique capacitors.

The Guru's Wimshurst machine is shown in Figure 12.3, and after about 15 spins of the counter-rotating disks it produces a static electric spark of roughly 100,000 volts. A Wimshurst machine is a great way to generate, and perform scientific experiments using static electricity – *and they're relatively inexpensive to get your hands on too.*

Let's look at a couple of examples of static electricity and things that we can do with it.

Example 12.1: The Wimshurst machine in Figure 12.3 develops a net charge of $5x10^{-12}$ C just before the dielectric strength of air breaks down and the unit generates a spark from one rod to the other. If the distance between the rod ends is 20 mm, what force is developed? What happens if we further insulate the rods by placing a 20 mm thick piece of polystyrene (Styrofoam) between them?

Coulomb's equation states that the net force produced is equal to the permittivity 'fudge factor' times the scaled distance between the charges. Since the Wimshurst machine develops equal but opposite charges on the two rods, the net charge difference between the two points is what produces the electronic force and we can modify the expression slightly as shown below.

$$F = \frac{1}{4\pi\varepsilon_o} \cdot \frac{q_1 \cdot q_2}{r^2} = 9.0x10^9 \, \frac{\mathrm{N} \cdot \mathrm{m}^2}{\mathrm{C}^2} \cdot \frac{\Delta q}{r^2}$$

The numerical value of the net force produced by the charge difference can therefore be determined as follows.

$$F = \frac{1}{4\pi\varepsilon_o} \cdot \frac{\Delta q}{r^2} = \frac{1}{4\pi\varepsilon_o} \cdot \frac{\Delta q}{r^2}$$

$$= 9.0x10^9 \frac{N \cdot \cancel{m}^2}{\cancel{C}^2} \cdot \frac{5x10^{-12} \cancel{C}^2}{(0.02 \cancel{m})^2} = \mathbf{112.5\ N}$$

So, the force developed by a very small amount of static electric charge is actually quite significant.

The force produced by the spark shown in the figure is enough to noticeably shake the end rods and make the entire machine vibrate each time it discharges to the grounding rod.

Figure 12.3: Example of an Approximate 100,000 Volt Static Electric Spark Produced Using a Wimshurst Machine.

Now, looking at the second part of this example problem, the dielectric correction for polystyrene is 2.70. Thus, the permittivity 'ε' would be calculated as follows,

$$\varepsilon = 2.70 \cdot \varepsilon_0 = 2.70 \cdot 8.9x10^{-12}\ \mathrm{C^2/N \cdot m^2} = 2.4030x10^{-11}\ \mathrm{C^2/N \cdot m^2}$$

Then, the value of $1/4\pi\varepsilon_o$ for polystyrene works out to be $3.312x10^9\ \mathrm{N \cdot m^2/C^2}$. Our final value for the force produced would therefore be,

$$F = \frac{1}{4\pi\varepsilon} \cdot \frac{\Delta q}{r^2} = 3.312x10^9 \frac{N \cdot \cancel{m}^2}{\cancel{C}^2} \cdot \frac{5x10^{-12} \cancel{C}^2}{(0.02 \cancel{m})^2} = \mathbf{41.1\ N}$$

So adding a dielectric material other than a vacuum (or air) increases the effective resistance between the two charges, thereby requiring a larger *electromotive force* to breakdown the material and ultimately produce a spark. We'll see shortly that electromotive force is another word for a term we all know – *voltage*.

Now that we know that two charges located in space can produce either attractive or repulsive forces on each other, and we know how to figure out what the magnitude of that force is, the question remains – *how does a charged particle know where another one is located?* This is the job of – *the Electric Field*.

The electric field was originally observed by physicist Michael Faraday, and was found to extend out to infinity. Thus, it is theoretically possible that every charged particle in the universe knows about every other charged particle floating around. In this regard, the electric field and the gravitational field are quite similar. The interaction between two charged particles is indicated using what are called *field lines*, as shown in Figure 12.4.

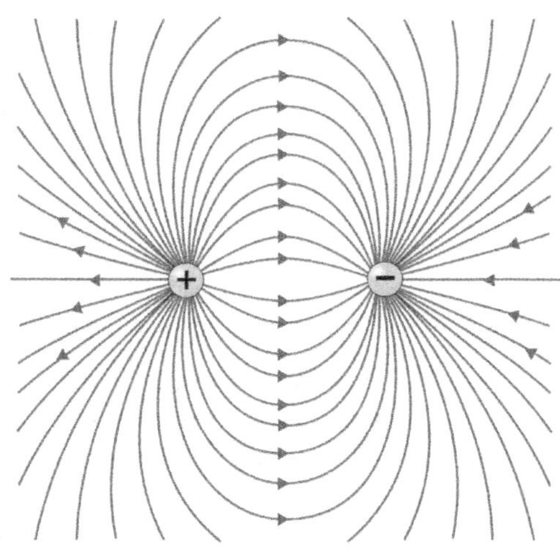

Figure 12.4: Positive Sense of How Two Charged Particles Communicate Using the Electric Field.

Field lines are vectors (as you can imagine from the arrows), and there are technically an infinite number of them radiating from any given particle, although one could get a wicked case of writers-cramp drawing an infinite number of lines; so we typically settle on showing just a few of them to make the point.

Since charge likes to travel from negative (-) to positive (+), a movement we call *electron current*, it must follow from Newton's third law that the positive particles are 'calling' negative ones and we draw the positive sense of the field lines as shown.

For most practical analysis, it is acceptable to treat the electric field as a scalar quantity, since we are usually only interested in knowing the magnitude of the field's effect between the two particles, and we already know that the vector representation is going to look like Figure 12.4.

The Guru Says...

The American patriot and statesman Benjamin Franklin was also America's first experimental physicist. He had a fascination with electricity and 'zapped' himself, or his assistant, into unconsciousness a couple of times performing experiments in his lab.

Franklin correctly observed the field lines shown in Figure 12.4, but being unable to see how the electrons moved (or even know what an electron was for that matter), had a 50-50 chance on guessing which way electricity flowed – *and he guessed wrong.* He concluded that electricity flowed in the same direction as the field – a direction we call *'conventional current'*. Some electrical circuit components, like diodes and transistors, still draw the flow in a conventional manner.

Incidentally, Franklin did indeed fly a kite in a thunderstorm to study lightning (a study that led to the invention of the lightning rod), but he never knew that lightning actually starts from the ground, and travels upward towards the sky – *but you do.*

So, in an ideal universe, a single charged particle would radiate an infinite number of straight field lines outward in all directions. Once another particle moves through this field, it produces an electric disturbance, which bends the field lines, and lets the other particle know that something else with a charge is out there.

If these charges are dissimilar they will tend to accelerate towards each other. If they are alike, they will accelerate away from each other. Charged particles behave like any other type of matter in the universe. Knowing this, we can mathematically define the electric field using a form identical to Newton's second law.

This gives us the following expression,

$$F = q \cdot E; \quad E = \frac{F}{q}$$

Where effectively 'q' is the charged object having mass, and 'E' is the accelerating force producing the motion. This is similar to gravity, except that gravity is always attractive, and the electric field can be either attractive or repulsive.

Example 12.2: In Example 12.1, what is the magnitude of the electric field in a vacuum?

The electrostatic force was found to be equal to 112.5 N on the rods having an effective net charge of $5x10^{-12}$ C. Thus the magnitude of the electric field generated by the Wimshurst machine would be,

$$E = \frac{F}{q} = \frac{112.5\,\mathrm{N}}{5x10^{-12}\,\mathrm{C}} = \mathbf{2.25x10^{13}}\ \frac{\mathbf{N}}{\mathbf{C}}$$

This is a sizeable electric field produced by a simple hand-cranked machine designed 250 years ago. Finally, we could combine Coulomb's law with Faraday's observations and write the expression for the electric field as,

$$E = \frac{F}{q} = \frac{\frac{1}{4\pi\varepsilon_o} \cdot \frac{q \cdot \cancel{q}}{r^2}}{\cancel{q}} = \frac{q}{4\pi r^2 \varepsilon_o}\ \frac{\mathrm{N}}{\mathrm{C}}$$

This expression is extremely close to our basic definition of voltage, in that we have the 'force' part of the voltage expression correct. Let's massage this expression a little more and see how we arrive at our SI definition of voltage.

Electric Potential and Work

The Guru defines voltage as, *"An electromotive force, or potential difference in an electric field, expressed in the units of volts."* Thus, the concept of electromotive force implies an electric field pushing an electron through space (or a conductor like a copper wire). Voltage is expressed in the unit of the Volt, which is named in honor of the Italian physicist Alessandro Volta, inventor of the first practical chemical battery.

Voltage in a wire is in many ways like water pressure in a pipe. The higher the pressure, the more water is able to flow over a specific distance, and consequently more work is obtained. Likewise, the greater the voltage, the more pressure is pushing on electrons to move past a certain point in space – *and again more work can be obtained.* So in essence, voltage is nothing more than a measure of the work able to be performed by an electric field on a charged particle. Good thing we know a lot about the concept of work at this point.

Going back to our definitions, we know that work 'W' equals force times the distance 'd' traveled, which in this case equals the electric field force 'E' times the distance we want to move an electron 'd'. Given this, we can directly write the following.

$$W = F \cdot d = E \cdot r$$

$$\boxed{V = E \cdot r}$$

Thus, the voltage or *electromotive potential* can be derived from the electric charge as follows,

$$V = E \cdot r = \frac{q}{4\pi r^{\not 2} \varepsilon_o} \frac{\text{N}}{\text{C}} \cdot \not r = \frac{q}{4\pi r \varepsilon_o} \frac{\text{N} \cdot \text{m}}{\text{C}} = \frac{q}{4\pi r \varepsilon_o} \frac{\text{J}}{\text{C}}$$

So we find out that one volt is equal to one Joule per Coulomb, making voltage indeed a measure of the amount of work applied to a group of electrons.

Example 12.3: What is the voltage potential between two charged metallic spheres separated by a distance of 1 m and with a net charge of 0.5 C?

The electrostatic voltage between the two spheres can be calculated using the expression we just derived. Thus,

$$\Delta V = \frac{q}{4\pi r \varepsilon_o} \frac{\text{J}}{\text{C}} = 9.0x10^9 \cdot \frac{q}{r} \frac{\text{J}}{\text{C}} = 9.0x10^9 \cdot \frac{0.5}{1} \frac{\text{J}}{\text{C}}$$

$$= \mathbf{4.5x10^9 \text{ Volts}}$$

Which is quite sizeable indeed. The *dielectric strength* of air is approximately 3×10^6 V/m, which means that for every meter of air, it takes roughly 3,000,000 volts of potential to convert it from an insulator to a conductor.

Clearly our example is roughly 1,000 times more voltage than is required, so expect a loud 'zap'. This example would not be a safe one to conduct without large amounts of insulation since, as we're going to find out next, the movement of this many electrons comes with a serious consequence – *electrical shock.*

Problem Solving Tip...

Determining the dielectric strength of a particular substance is as simple as a table look-up. Physicists over the years have calculated the dielectric strength of practically every substance in the known universe for applications just like the one shown in this example.

Example 12.4: In the previous example, what is the minimum distance the Guru has to place the metallic spheres to avoid an electric spark?

To solve this problem, we need to equate the dielectric strength of air to ΔV and solve for 'r'. Thus we have,

$$\Delta V = 3\times10^{6}\ \cdot r\frac{\text{V}}{\text{m}} = \frac{q}{4\pi r\varepsilon_o}\frac{\text{V}}{\text{m}}$$

$$3\times10^{6}\ \cdot r = \frac{4.5x10^{9}}{r}$$

$$r^2 = \frac{4.5x10^{9}}{3\times10^{6}} = 1.5\times10^{3}\ \text{m}^2$$

$$r = \sqrt{1.5\times10^{3}\ \text{m}^2} = \mathbf{38.73\ m}$$

The metallic spheres would have to be placed a minimum of roughly 38.7 meters away from each other to stop an electric spark from jumping through the air – *and this is only for 0.5 C of charge*. Imagine what a full Coulomb would do.

Current, Power, Resistance and Simple Circuits

Pretty much everyone has heard the unit of the Ampere (A) tossed around. It is a unit of electrical current indicating how many electrons flow by a certain point in space over a given period of time. Household appliances might be rated for 0.5 A for a light bulb to roughly 12.5 A for a hair dryer. Most household electrical outlets are rated for a maximum of 15 A. If this value is exceeded, the odds are you would 'trip' the circuit breaker in the electrical panel.

An Ampere of current (denoted by the symbol 'I') is one of the basic units in the SI system, and is defined as being equal to one Coulomb of charge passing by a point in space per second. Thus,

$$Current = I = 1\,\text{A} = 1\frac{\text{C}}{\text{s}}$$

So an Ampere carries a large amount of electrons behind it and should always be respected.

The Guru Says…

In our previous example we saw what charge potential a mere 0.5 C produced and the large separation distance required to prevent a spark in air.

Now imagine that the Guru was neglegent and got his hand in the way of the spark. Considering that a spark can jump an air gap in a milllisecond or less, the resulting current would be close to 500 A – the Guru's hand could be burned from the electrical shock.

Are you starting to appreciate just how much 'punch' one Coulomb of charge carries?

Typically when we talk about electrons moving through space, we work in the unit of the Ampere, since an Ampere defines the rate at which electrons flow, which is an indication of the amount of work that can be performed. An Ampere is also a nicer unit, since it quantifies a large 'chunk' of electrons doing work, instead of focusing on individual charges, which don't by themselves do all that much work.

If we were to take the unit of the Ampere and multiply it times the unit of the Volt, we would have a familiar unit representing the amount of work performed over a fixed time period – *namely a unit of power.* Thus we can state that the *electrical power* 'P' produced by any device or process is equal to the voltage times the current, as follows,

$$\text{Power} = P = V \cdot I$$

Electrical power, being a scalar, is like any other form of power we have discussed previously, and is fully interchangeable with anything we've done so far.

Example 12.5: An electrical cable supplies 10 A at a voltage of 240 V. What is the available power?

We simply substitute the example values into the above equation to determine the available power.

$$P = V \cdot I = 240\ \text{V} \cdot 10\ \text{A} = 2400\ \text{VA} = \mathbf{2400\ W}$$

Example 12.6: If we run the electrical power in the previous problem for one-minute, how high could we theoretically raise a 100 kg mass?

If we run a power source for a period of time, we get a certain quantity of energy. Thus for one minute (60 seconds), we supply the following available energy,

$$\begin{aligned} \text{Energy} &= P \cdot t = V \cdot I \cdot t \\ &= 240\ \text{V} \cdot 10\ \text{A} \cdot 60\ \text{s} = 240\ \frac{\text{J}}{\cancel{\text{C}}} \cdot 10\ \frac{\cancel{\text{C}}}{\cancel{\text{s}}} \cdot 60\ \cancel{\text{s}} \\ &= \mathbf{144000\ J} \end{aligned}$$

Recalling from Chapter 6, the work required to lift an object against gravity is equal to,

$$W = m \cdot g \cdot h$$

Thus, substituting and solving for 'h' we can determine the theoretical lifting height we can obtain from our quantity of electrical energy.

$$W = m \cdot g \cdot h$$

$$144000\ \text{J} = 100\ \text{kg} \cdot 9.81\frac{\text{m}}{\text{s}^2} \cdot h$$

$$h = \frac{144000\ \text{N} \cdot \text{m}}{100\ \text{kg} \cdot 9.81\frac{\text{m}}{\text{s}^2}} = \frac{144000\ \cancel{\text{N}} \cdot \text{m}}{981\ \cancel{\text{N}}} = \mathbf{146.8\ m}$$

It is of course understood that the above example is a hypothetical one in which there is a complete conversion of energy from electrical energy to potential energy, with no losses. In this context, it is informative to know that 10 A at 240 V switched on for one minute is equivalent to the potential energy of a heavy mass falling from a height of almost 150 m – *or a 45 story building*.

The German physicist Georg Simon Ohm noted in 1827 that current and voltage are also related to an 'electrical friction' force called *resistance*. Resistance, expressed in the SI unit of the Ohm (after its discoverer), prevents or restricts the flow of electrons by forcing a dielectric change. Resistance is everywhere. It's present in every electrical device and every object. It's even present in you. What Ohm discovered was the following linear relationship between voltage, resistance and current,

$$V = I \cdot R$$

Where 'V' is the voltage, 'I' is the current, and 'R' is the resistance in Ohms. This equation is known as *Ohm's Law,* and is one of the fundamental laws that all electrical circuits must obey as it relates all three fundamental electrical parameters in the universe: voltage, current, and resistance.

Problem Solving Tip...

Electricity follows what is called the *'path of least resistance'*.

More directly said, electrical energy will <u>always</u> seek out a path which has the minimum electrical resistance. Normally this is the wire or conduit that conducts the electricity, but sometimes it could be 'you', in which case that's called an electric shock. Always keep this in mind when solving electrical problems or working with electricity.

Using Ohm's law you can always determine any missing parameter if you know the other two. The best way to understand Ohm's law is to work through a couple of examples.

Example 12.7: The Guru creates a simple electrical circuit containing a light bulb and a battery, as shown in Figure 12.5. If he measures the resistance of the light bulb as 10 Ω, and the current using an ammeter as 2 A, what must the voltage be?

To solve this problem, we write Ohm's law in terms of the unknown voltage quantity 'V'. Thus we have,

$$V = I \cdot R$$

$$= (2\ \text{A}) \cdot (10\ \Omega) = \mathbf{20\ Volts}$$

So we can see from this simple circuit that our choice of voltage is not arbitrary. If we are going to have a light bulb with a resistance of 10 Ohms and force a current of two Amperes through it, the required 'electrical pressure' must be 20 Volts.

Similarly, if we set the voltage to 50 V, and leave in the same light bulb, the current will proportionately increase to 5 A, which would damage our light bulb if it were only rated for 2 A.

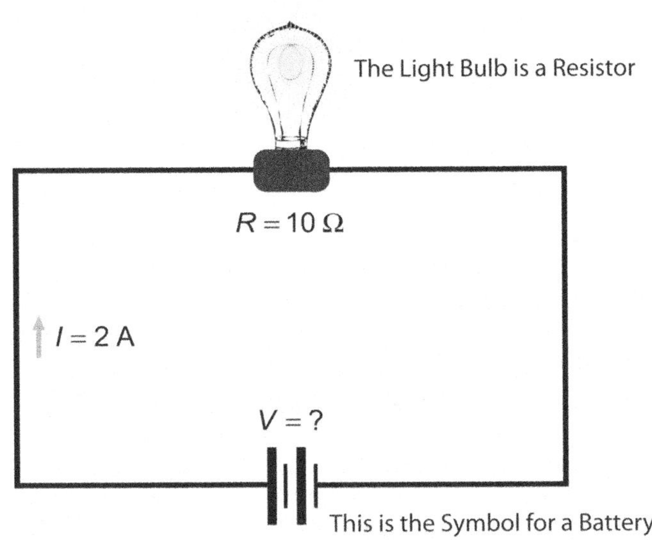

Figure 12.5: A Simple Electrical Circuit Consisting of a Battery and a Light Bulb

We call Figure 12.5 a simple *series circuit*. It is called a series circuit because all the electrical current passes through every component in the circuit. A series circuit is like a strand of Christmas lights where if one bulb goes out, they all go out. Let's look at another example of our Christmas bulb circuit in Figure 12.6.

In a series circuit, any break in any component causes current to stop flowing in the circuit. Likewise, the total *voltage drop* required to maintain the current shown in Figure 12.6 is going to be a function of the total resistance of the circuit.

Thus in a series circuit, the total resistance is equal to the arithmetic summation of the resistive loads. Mathematically, the total series resistance 'R_T' is calculated as,

$$R_T = R_1 + R_2 + R_3 + \ldots + R_n$$

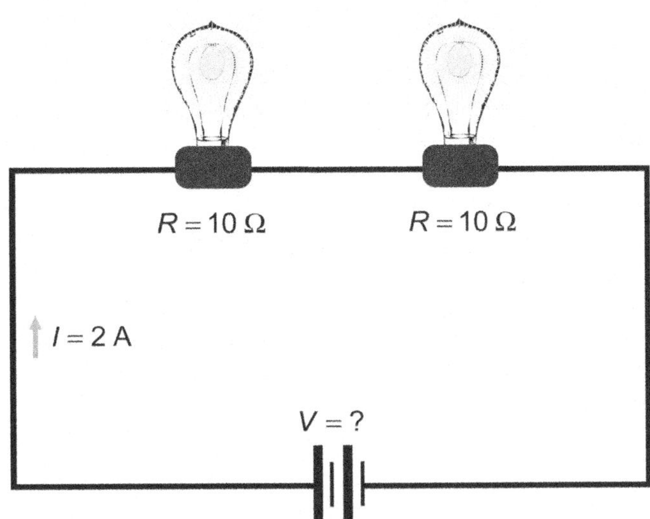

Figure 12.6: A Simple Series Circuit – If One Bulb Goes Out, The Other One Does Too

In Figure 12.6, this total resistance would be equal to 20 Ohms. Indeed, for any number of series resistors, we can just add them together to arrive at a single equivalent resistive load making the circuit 'electrically look like' the one shown in Figure 12.5. Once we make this mathematical adjustment to the circuit, we can always apply Ohm's law with the ease that we did in the previous example.

Example 12.8: What is the voltage required by the battery to operate the series circuit as shown in Figure 12.6?

The total resistance of the circuit is 20 Ω; thus, the required voltage to maintain a 2 A current flow through the circuit is,

$$V = I \cdot R$$
$$= (2\text{ A}) \cdot (20\ \Omega) = \mathbf{40\ Volts}$$

Problem Solving Tip...

There's an inexpensive and handy device called a 'multimeter' which can measure voltage, current and resistance of electrical components allowing you to apply Ohm's law to real world objects. The Guru recommends getting one and exploring electricity and magnetism yourself.

So as an application of Ohm's law we can see that if one of the three fundamental parameters in an electrical circuit change, one of the others (or both) must adjust accordingly.

There is also another circuit configuration known as a *parallel circuit*. In a parallel circuit, not all of the current flows through a single path (also called a branch). In this case, there can be multiple branches for the current to flow.

Take a look at Figure 12.7 to see what we're taking about. In a parallel circuit, the current is split among the branches in a manner inversely proportional to their relative resistive loads. Also, if one branch of the circuit fails, the other part will still function, as the current will still have a complete circuit path in which to travel.

So looking at Figure 12.7, the initial current out of the battery is 2 A. Since we know that an Ampere is nothing more than a Coulomb of charge per second, we also know that this Coulomb must split itself up in the branches in a manner that does not violate the conservation of mass (after all, we are still taking about electrons here).

For our example, the current is split equally, since the resistive loads are the same. If we reduce (or increase) either of the resistances in the light bulbs, the electrical current will flow more along the branch having the *least resistance*.

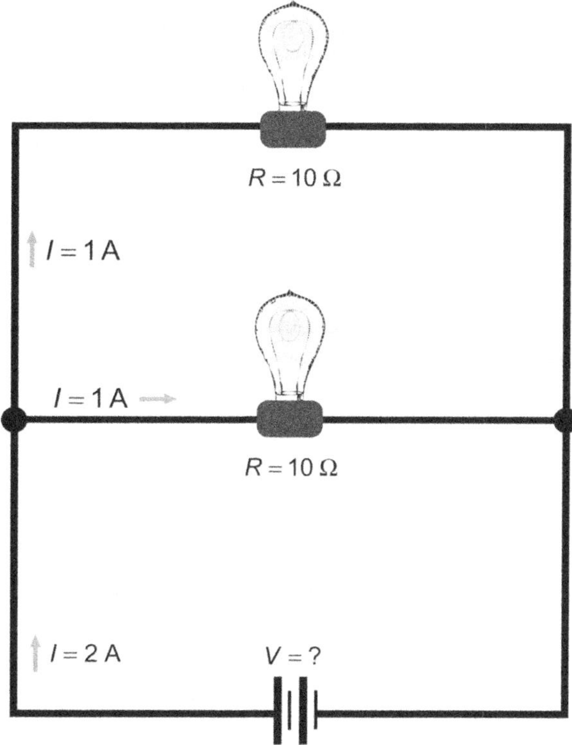

Figure 12.7: A Simple Series Circuit – If One Bulb Goes Out, The Other One Stays On

If we reduce one light bulb to say 5 Ω, then the current in that branch would increase to 1.33 A, since 2/3 of the total two amperes would flow through it, while in the other branch it would drop to 0.67 A, since with its greater relative resistance only 1/3 of the circuit current would flow.

Just like what we did with series circuits, we can reduce a parallel circuit to an equivalent single resistive load, like the circuit shown in Figure 12.5. Mathematically, the total parallel resistance of a circuit, 'R_T' is calculated as,

$$\frac{1}{R_T}=\frac{1}{R_1}+\frac{1}{R_2}+\frac{1}{R_3}+\ldots+\frac{1}{R_n}$$

The computation is just a little trickier algebraically in that we are adding the fractions of the individual resistors, and then flipping the result over to get the final answer. If we only have two resistive loads like that shown in Figure 12.7, then the equation for the total parallel resistance reduces to,

$$R_T=\frac{R_1\cdot R_2}{R_1+R_2}$$

So in a pinch, you can always reduce a parallel circuit two branches at a time by combining results using the above equation until you get to the final answer.

Example 12.9: What is the equivalent resistance of the light bulb circuit shown in Figure 12.7? What is the voltage drop required by the battery?

Since there are only two resistive branches, we can calculate the equivalent resistance of the circuit as,

$$R_T=\frac{10\ \Omega\cdot 10\ \Omega}{10\ \Omega+10\ \Omega}=\frac{100\ \Omega}{20\ \Omega}=\mathbf{5\ \Omega}$$

It should be noted that in a parallel circuit, the total equivalent resistance of all the branches together will always be smaller than the smallest individual branch resistance. This is a good thing to know when designing a parallel circuit.

We can now calculate the required voltage drop by the battery to maintain the current flow shown using Ohm's law as,

$$\begin{aligned}V&=I\cdot R\\&=(2\text{ A})\cdot(5\ \Omega)=\mathbf{10\ Volts}\end{aligned}$$

Magnets and the Magnetic Field

In 1820 the Danish physicist Hans Christian Orsted, while preparing to give a classroom demonstration on electricity, made the remarkable, and truly accidental, discovery that an electric field that changes with time (such as being switched on and off repeatedly) deflected the needle of a compass sitting nearby on his desk. This was the first experimental proof that electricity and magnetism were inextricably related to each other.

So, just as we drew electric field lines, we can also show magnetic field lines in a similar manner. Figure 12.8 demonstrates what the magnetic field lines from a simple bar magnet would look like as observed by Orsted.

Unlike the electric field 'E', the magnetic field 'B' does not radiate from a single point *or pole*. Instead, magnetic fields must always originate from a 'north' magnetic pole, and end at a 'south' magnetic pole. Thus, if we were to take the bar magnet shown in Figure 12.8, and break it in half, we would wind up with two smaller magnets, each with a distinct north and south pole and a different magnetic 'B' field pattern than what is shown.

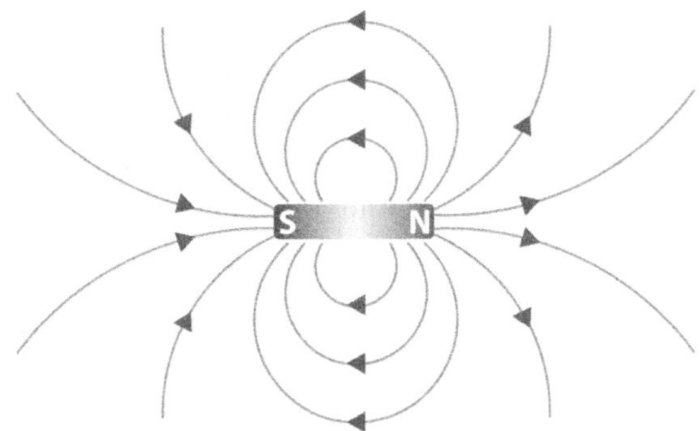

Figure 12.8: How Magnetic Field Lines are Developed Around a Simple Bar Magnet.

Thus in our universe, there is no such thing as a *magnetic monopole*, although mathematically there is no restriction placed on the existence of one within the laws of physics.

The magnetic field, like the electric field, is a vector quantity, and it's typically a good idea to have your right hand handy when working with it, since the orientation of the 'B' field is now important. Following the same logic that we used to derive the electric field, we know that the resultant force on a charged particle 'F' is equal to a charged mass term we'll call 'q', times an acceleration due to the presence of the magnetic field.

This was an easy operation for the electric field since we merely multiplied the scalar charge 'q' times the vector electric field 'E' to get the vector force on the charge 'F'. The magnetic field, however, presents an additional wrinkle to the problem.

The Guru Says...

Incidentially, Earth's magnetic north pole is actually physically the south magnetic pole on a magnet.

High energy particles are magnetically pulled in from space, and ionize the upper atmosphere and cause the production of light – something known as the *Aurora Borealis*.

Unlike the electric field, *the magnetic field is only observed acting on moving charges*. A stationary charge is invisible to the magnetic field. Since 'q' is always a scalar, we therefore need some type of velocity vector 'v', that we can multiply times 'q', to give it a direction in space.

Let's define this direction-oriented *charge vector* as follows,

$$\text{Charge Vector} = q \cdot \bar{v}$$

When we multiply this charge vector times the magnetic field, which is actually the acceleration part of Newton's equation, we should get the force on the charged particle produced by this field.

We can write this mathematically as,

$$\bar{F} = q \cdot \bar{v} \otimes \bar{B}$$

Our choice of using cross-product terminology is necessitated by the fact that we are multiplying two vectors together to create yet another vector (hence the Guru's statement that you need a right hand to work with magnetic fields). The fact that our final expression is in this form gives us some valuable insight into how a magnetic force behaves.

Recall that the cross product of two vectors yields a new vector 90-degrees away from either original vector; thus, the magnetic force must always act at a 90-degree angle with respect to the motion of the particle (remember, think electron moving in space) and the magnetic field. The scalar of our magnetic force expression would be given by the following equation,

$$F = \left|q \cdot \bar{v}\right| \cdot \left|\bar{B}\right| \sin\theta = q \cdot v \cdot B \cdot \sin\theta$$

Since the sine of an angle is at a maximum only when the angle is equal to 90-degrees, this would imply that *the magnetic force is maximal on a charged particle when it is traveling at a 90-degree angle with respect to a magnetic field.* This is another good point to know if we start building things that rely of magnetic fields like electric motors or generators.

Following the form of the previous equation, it is therefore implied that the magnetic field must have units of $\text{kg} \cdot A^{-1} \cdot s^{-2}$ in order to make both sides of the equation dimensionally balance. We derive in the SI system the unit of the *Tesla* (after the inventor of alternating current, Nikola Tesla) to be equal to 1.0 $\text{kg} \cdot A^{-1} \cdot s^{-2}$. The Tesla is abbreviated with the letter 'T'.

Example 12.10: Assuming that a charged particle is moving through space at a speed of 1,000 m/s perpendicular to a 10 T field. If the particle has a charge of 0.5 C, what is the force exerted by the magnetic field?

Since the path of the particle and the magnetic field are perpendicular, the sine of the angle is '1' and our equation can be evaluated in the following form.

$$F = 0.5\ \text{C} \cdot 1000\ \frac{\text{m}}{\text{s}} \cdot 10\ \text{T} = 0.5\ \text{C} \cdot 1000\ \frac{\text{m}}{\text{s}} \cdot 10\ \frac{\text{kg}}{\text{A} \cdot \text{s}^2}$$

$$= 0.5\ \not{\text{C}} \cdot 1000\ \frac{\text{m}}{\not{\text{s}}} \cdot 10\ \frac{\text{kg}}{\frac{\not{\text{C}}}{\not{\text{s}}} \cdot \text{s}^2} = (0.5) \cdot (1000) \cdot (10)\ \frac{\text{kg} \cdot \text{m}}{\text{s}^2} = \mathbf{5000\ N}$$

There are actually quite a few scientific devices that take advantage of the principle demonstrated in this example. The most notable of which would be a mass spectrometer, which is a device that measures the relative quantity or concentration of a substance by looking at its atomic composition. It works by firing the material of interest (usually a gas) into a chamber, which first ionizes it to a certain level (i.e., gives it a known electrical charge), and then pushes against it with an adjustable magnetic field. All things being equal, the amount of 'push' required by the magnetic field to deflect the material a specified distance is in direct correlation to the atomic mass of the substance (i.e., more mass, more inertia, more force required to produce an acceleration).

Finally, we should note that since humans are typically interested in moving current through wires instead of space itself, we can specialize our magnetic force equation for the specific case where the charges are moving through a linear conductor. The Guru presents this without the long and painful proof, as being equal to,

$$\bar{F} = I \cdot \bar{L} \otimes \bar{B}$$

With the scalar form,

$$F = \left| I \cdot \bar{L} \right| \cdot \left| \bar{B} \right| \sin\theta$$

Where 'I' is the current in the wire in Amperes, 'L' is the vector part of the current and points in the direction and length of the wire in space, and 'B' is the same old magnetic field from before. As with our more general expression for magnetic force, the above equation also implies that it would be a pretty-good idea to apply the magnetic field at a right angle to the wire carrying the current in order to maximize the force produced (and hence the work). Sure enough, if you tear apart a device that uses this equation, like an electric motor for example, you'll find that the magnetic field in the fixed *stator* is always applied at 90-degrees to the current passing through the spinning *commutator*.

Example 12.11: A one-meter long length of wire with a current of 10 A is held perpendicular to a 10 T electromagnet. What is the force exerted on the wire?

Again, since the direction of the current and the magnetic field are perpendicular to each other, the sine of the angle is '1' and our expression can again be written in a simplified form.

$$F = 10\ \text{A} \cdot 1\text{m} \cdot 10\ \text{T}$$

$$= 10\ \cancel{\text{A}} \cdot 1\text{m} \cdot 10\ \frac{\text{kg}}{\cancel{\text{A}} \cdot \text{s}^2} = (10)\cdot(1)\cdot(10)\frac{\text{kg}\cdot\text{m}}{\text{s}^2} = \mathbf{100\ N}$$

Alternating Versus Direct Current – Edison's Nightmare

An interesting aspect of our magnetic force equation is that it very much adheres to Newton's third law of motion, in that it becomes irrelevant from a force standpoint whether or not the electron is doing the moving ($q \cdot \bar{v}$), or the magnetic field ($\bar{B}$). Thus, if an electron (or electrical current for that matter) is moving in a stationary magnetic field like that produced by a bar magnet, current flows and a force is produced as we have previously seen.

Likewise if a moving magnetic field (think moving bar magnet) is passed by a stationary wire, electrons will also start to move and a force is again produced. This will continue for as long as we keep moving the magnet and changing the position and/or magnitude of the magnetic field lines shown in Figure 12.8. Stop the motion, and the resulting electric field in the wire stops. This phenomenon is known as *induction* and was originally observed by Michael Faraday in 1831. Induction is what makes modern power generation and transmission possible.

The Guru Says...

Up until now, we have only talked about current that flows continuously in one direction or another. This direct current (or DC) travels from a source, like an automobile battery, through a designated resistive load, and back to the other terminal (called the ground).

Since the current is a constant value that does not change with time, there really isn't any magnetic field produced (remember, Oersted only observed the movement of the compass needle when he opened and closed the electric switch). Electrical power in this case is generated through the shear volume movement of electrons through a wire from point 'A' to point 'B'.

As it works out, electricity and magnetism are just two sides of the same coin. Just as you can't have a front without a back, you cannot have an electric field without a magnetic field because a changing electric field automatically generates a magnetic field, and a changing magnetic field automatically generates an electric field. With the addition of the physical concept of induction, it now became possible to develop a new way to generate electrical current – the *Alternating Current* (or AC).

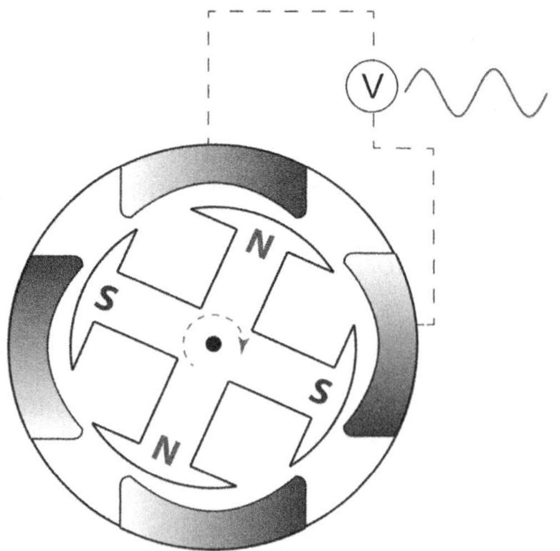

Figure 12.9: How A Simple Synchronous AC Electric Generator Works

Alternating current, as the name implies, is a time varying electrical current in a wire produced by, you guessed it, a corresponding time varying magnetic field similar to what is shown in Figure 12.9. As the magnet spins around, the magnetic field is constantly changing, producing the classical sine-wave voltage pattern shown at the output. In the Unites States, the magnetic rotor spins 3,600 times per minute, which provides electrical power at a frequency of 60 Hz. In Europe and parts of Asia, the magnetic rotor spins 3,000 times per minute, which provides electrical power at 50 Hz.

In the early days of electrical power generation and transmission, *Direct Current* (or DC) was the only show in town, especially if you happened to live in the United States. The noted American inventor Thomas Edison built a small business empire geared towards the generation, storage, transmission, and billing of DC electrical power. His company, Edison General Electric is still in business today – it's called General Electric.

The only problem with Edison's plan for total power domination was that other inventors, notably Nikola Tesla, were working on perfecting an alternating current generation system. Alternating current (AC) had numerous advantages over DC in terms of performance and reliability, but the one big reason houses today are powered by AC, and not DC, again has to do with Faraday's observations about inductance.

Line loss (yes, that frictional pest called resistance) becomes a problem when you are stretching electrical cables any sizeable distance. Since DC power has to move *every electron from one point to another*, the resistive losses in the wires per Ohm's law really start to add up. To combat this problem, Edison had to limit the distance between electrical power generation and its consumption to distances of roughly a mile, install additional power generating systems along the transmission corridor, and use thicker transmission cables to combat resistive losses. *This became a very expensive proposition for Mr. Edison.*

The Guru Says...

Since a time varying magnetic field can automatically induce a current in a length of wire consistent with our magnetic force equation, there really need not be any physical connection between what generates the magnetic field and the point where it is converted to electrical current, the space between the two areas transfers the magnetic wave just fine.

There's actually a really cool use for radiating magnetic fields freely in space by varying the frequency (i.e., the rate) of the magnetic field and picking up the results in a wire – *it's called radio transmission*.

AC power, on the other hand, took advantage of the inductance occurring due to the oscillation of the magnetic field. A few years earlier, various researchers discovered that inductance allowed for the swapping of current with voltage and vice versa using a newly invented electrical device known as a *transformer*.

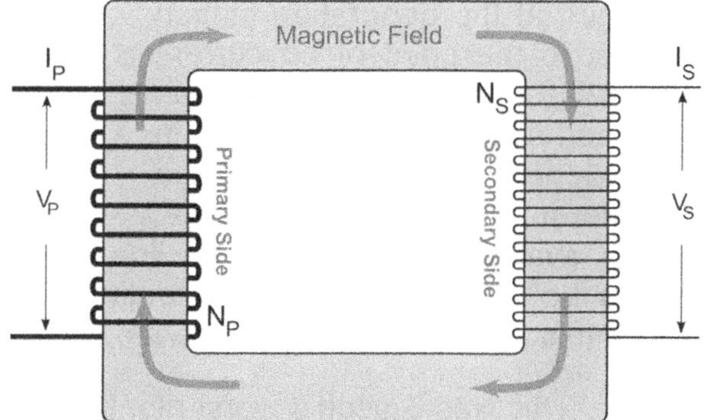

Figure 12.10: The Fundamental Principle Behind a Single-Phase Transformer

A transformer is a mechanical device with no moving parts, consisting of a conducting, but insulated, metal 'core' around which wires are wrapped at either end. Transformers work by generating a magnetic field in the core for an alternating current on one side, known as the 'primary' side of the transformer, and then converting it back to an electric potential in a wire on the other side, known as the 'secondary'. A simple single-phase schematic of one is shown in Figure 12.10. The actual physics behind a transformer is that the power 'in' equals the power 'out' with a little bit of loss in heat during the transfer. Thus, voltage and current can be shifted up or down so that the following condition is satisfied,

$$Power_{in} = Power_{out}; \quad (V \cdot I)_{in} = (V \cdot I)_{out}$$

If both the primary and secondary sides of the transformer have the same number of wire wraps (i.e., a 1:1 ratio), then the conversion of power behaves as though there is no transformer there at all. Such a device does have a practical purpose; it's called an *isolation transformer,* and is used to electrically isolate one side of a circuit from another. All stereo systems use isolation transformers to prevent one audio channel from interfering with the other.

Some clever fellow though discovered early on that if we wrap the primary and secondary sides of the transformer at different ratios, we induce a commensurate ratio difference in the induced electric field. For example, a 2:1 (secondary to primary) wrapping of the transformer, or 'turns ratio', produces twice the electrical voltage on the secondary side with half the amount of current. The power transfer is still the same in either case. Thus, we can relate the voltage and current in the primary part of the transformer (P), to the secondary part (S), via the turn ratio (N) as,

$$\frac{V_S}{V_P} = \frac{N_S}{N_P} = \frac{I_P}{I_S}$$

Where 'V' is the voltage, 'I' is the current, and 'N' is the number of turns of the primary and secondary sections of the transformer. To accomplish this in the DC world would be a difficult task at best, especially over 100 years ago.

With the advent of the transformer, electrical power could be 'stepped up' to very high voltages at the generation station almost to the point of no current and all electric field, and then 'stepped down' to usable levels in a residential neighborhood using a transformer on a power pole, which everyone has seen. The line losses were extremely small since with a small amount of current (electrons) actually being moved, the resistance of the line as viewed by Ohm's law was quite small indeed.

The Guru Says...

Today DC power has applications in small, portable devices such as consumer electronics, automobiles, municipal subways and electric railways, and in solar power generation.

This was Edison's worst nightmare – and he didn't take it very well at all. In fact, Edison waged an advertising temper-tantrum against AC power for many years, but that did not stop the advancement of this technology into what we have today. Edison himself, many years later finally admitted that he had always believed that AC power was superior to its DC counterpart (but business is business – right?).

Chapter 13: Riding on a Beam of Light

The fastest and most perplexing stuff in the universe

Light is a special form of electromagnetic radiation that is visible to humans and other life forms on planet Earth (typically with wavelengths between 380 and 780 nanometers, which is the region of sensitivity of the human eye). There's really nothing special in the universe about this subset of radiation except that we can see it without the need of any special instrumentation. In actuality, the universe is awash with all different types of electromagnetic 'light', all of it being equal in importance to visible 'light' – which is to say on a 'lighter' note*, that the Guru would never try to make 'light' of anything we may bring to 'light' with this 'lighthearted' introduction to, well, 'light'.*

Let's take the sun for example. Sunlight certainly illuminates things in our little region of space to the tune of roughly 100,000 lumens per square meter (lm/m^2) on Earth. This works out to about 445 watts per square meter (W/m^2) of visible light power hitting the ground at high noon during the summer months, averaged across the planet. This is quite bright as anyone will attest and has resulted in a booming market for an invention called sunglasses amongst humans.

But, what would happen if our eyes could not see visible light? Say we could only see the light just above visible, called *ultraviolet light*, or just below the visible, called *infrared light*. What would the world look like then?

In addition to the 445 W/m^2 of visible light, the Sun also produces 32 W/m^2 of ultraviolet light and a whopping 527 W/m^2 of infrared light as measured on the surface of Earth. Thus, if we could only see in the ultraviolet range we would conclude that Earth was a very dimly lit planet equivalent to the darkest of moonless nights. On the other hand, if we viewed the world from the infrared end of the spectrum there would be an even bigger market for sunglasses. So what are these different types of 'light'? Let's take a look at *electromagnetic wave propagation* and ultimately something called the *electromagnetic spectrum* to find out.

Electromagnetic Wave Propagation

As we can see from our previous discussion, the physical meaning of 'light' is quite relative and limited to an extremely small amount of *electromagnetic radiation* present in the universe. Electromagnetic radiation is nothing more than energy, which is transferred through space using (you guessed it) electromagnetic waves.

An electromagnetic wave is a natural consequence of what we have learned in the previous chapter on electricity and magnetism. Remember that we said that electricity and magnetism are just two sides of the same coin. You cannot have an electric field without a magnetic field because a changing electric field generates a magnetic field, and a changing magnetic field generates an electric field. We stated this without proof in Chapter 12, so let's develop the idea a little further here and see why this is so, and how it relates to this substance called light.

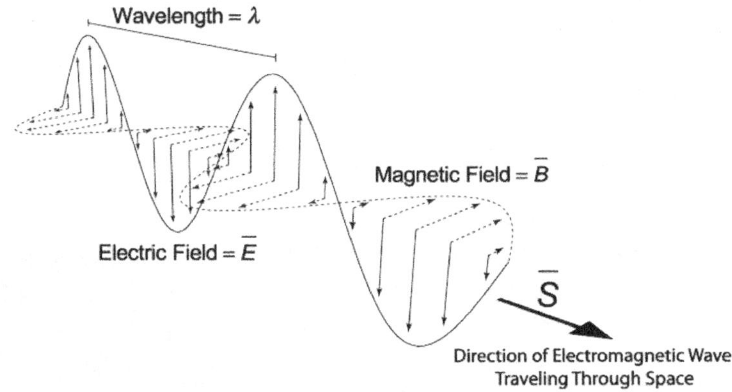

Figure 13.1: How an Electromagnetic Wave Travels Through Space

Take a look at Figure 13.1 where the Guru has plotted the measured results of an electric field ($\bar{E}$) versus its corresponding magnetic field ($\bar{B}$) propagating through space. Previously we noted that both are vector quantities and act at the same time. Thus, in this figure we could have either electron movement, or a magnetic field change, producing the effect shown. One of the interesting aspects of changing electric and magnetic fields is that they set up what is known as *self-propagating wave* motion away from, and independent of, the original disturbance that created it. This very concept created a lot of heartburn for physicists in the 19th century.

This phenomenon of electromagnetic radiation and wave propagation was first noted by the famous Scottish mathematician and physicist James Clerk Maxwell in 1862, when he derived a set of consistent mathematical expressions linking all of electricity and magnetism together – *called appropriately enough, Maxwell's Equations.*

The Guru Says...

Physicists believed in the 1800's that the universe (i.e., space) was permeated with a substance known as *aether*. This yet-to-be-detected or defined substance was the transference medium by which electric and magnetic fields propagated and electromagnetic waves traveled.

This bit of junk science dates back to Aristotle (340 B.C.), a great philosopher who unfortunately got more science wrong than right. For years physicists tried to detect aether, but to no avail. Some physicists got downright hostile at the notion of giving up aether, and resorted to more name-calling than scientific research. In the end, though, aether was relegated to the proverbial *ash heap of history*.

At least we do get an interesting English word out of this. When something is otherworldly or intangible, we refer to it as being *ethereal*.

Maxwell's equations link our previously discussed findings in Chapter 12 from Gauss, Faraday, and Ampere together into one nice neat mathematical package of integral vector calculus. The interested reader is encouraged to take a look at the development of these expressions, which are outside the scope of the Guru's guide, but interesting nonetheless. Unfortunately though, Maxwell never realized the full potential of his findings, since he refused to believe the self-propagating feature of electromagnetic waves in space.

Maxwell believed, like others working in modern physics at the time, that waves could not travel through the vacuum of space without some medium to transfer the energy. Waves just didn't work that way, and there had to be some sort of *'yet to be detected'* substance present in space – a mystical substance physicists referred to as *aether*. Even though physicists could detect electromagnetic radiation in a vacuum, there had to be something wrong.

Fortunately for science, a gentleman by the name of John Poynting discovered in 1884 that electromagnetic waves transfer power, and as a consequence of working with Maxwell's equations, also discovered that this power transfer occurred directly as a product of the electric field ($\overline{E}$) and the magnetic field ($\overline{B}$).

Thus, we define the *Poynting vector* ($\overline{S}$) as,

$$\overline{S} = \frac{1}{\mu_0}\overline{E} \otimes \overline{B}$$

Where $\overline{E}$ is the electric field vector, $\overline{B}$ is the magnetic field vector, and μ_0 is a physical constant called the permeability of free space and is given by $\mu_0 = 4\pi \times 10^{-7}\ (\mathrm{V \cdot s})/(\mathrm{A \cdot m})$ to make the units work out.

The Poynting vector, which sounds just like it's spelled, *'pointing'*, has units of Watts per square meter (W/m^2), and defines the relative power being transferred by an oscillating electrical and magnetic wave. More importantly though, it defines the experimentally observed pattern of electromagnetic propagation shown in Figure 13.1 and explains why electromagnetic waves travel through space without the need for any transfer medium.

Problem Solving Tip...

Since the pattern of one field generating another occurs automatically, the cross product of the result (the Poynting vector) also occurs automatically, and the wave self propagates through space without the need of any material like aether. This is a unique feature of electromagnetic waves that does not occur with other types of wave motion in the universe.

In computational work, we would normally use the scalar form of the Poynting vector, which would be given as follows,

$$S = \left|\frac{1}{\mu_0}\cdot\overline{E}\right|\cdot\left|\overline{B}\right|\sin\theta = \frac{1}{\mu_0}\cdot E\cdot B\sin\theta$$

Since we know that the electric and magnetic fields <u>always</u> occur at 90-degrees with respect to each other, we can simplify the expression further to yield,

$$\mu_0 \cdot S = E \cdot B$$

This is great, but our relationship between the electric and magnetic fields still has an intermediate term 'S' equal to the magnitude of the electromagnetic wave which is dependent on how accurately we can measure the permeability of space. What would be super cool is to figure out a way to directly relate the magnetic and electric fields and avoid the unnecessary computational step.

Let's rewrite our scalar expression for the Poynting vector such that we express it in terms of the ratio of the electric field to the magnetic field. Thus we'll divide both sides by B^2 (which doesn't really change anything) to get,

$$\frac{1}{B^2}(E \cdot B) = \frac{1}{B^2}(\mu_0 \cdot S)$$

$$\frac{1}{B^{\not 2}}(E \cdot \not B) = \frac{\mu_0 \cdot S}{B^2}$$

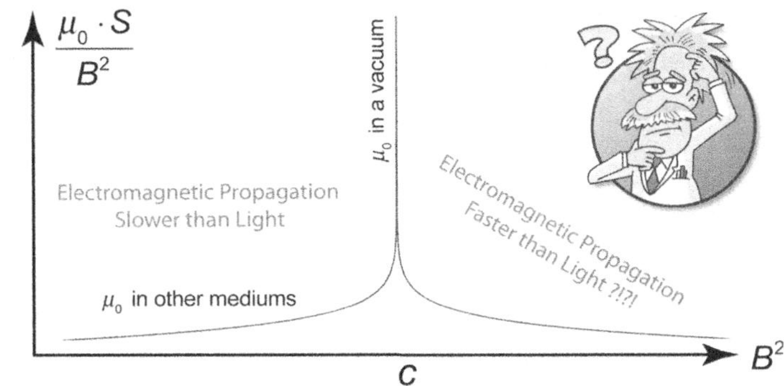

Figure 13.2: Plot Showing Asymptotic Solution of $\mu_0 \cdot S/B^2$ Always Approaching the *Speed of Light* 'c'

As luck would have it (well maybe it's not exactly luck, just a really good guess on the Guru's part, based upon a painful derivation on a piece of scratch paper), the quantity $\mu_0 \cdot S/B^2$, which is shown graphically in Figure 13.2, is equal to the speed of the electromagnetic wave propagation in meters-per-second, and this value approaches the same constant speed in a vacuum each and every time. We can therefore write our scalar expression for the Poynting vector as,

$$\frac{E}{B} = \frac{\mu_0 \cdot S}{B^2} = c; \quad E = c \cdot B$$

Thus the electric field is equal to the magnetic field times the speed of the electromagnetic wave propagation. This constant propagation speed is called the *speed of light* or 'c'. It is the topic of our next section, and the reason for the confused look on the Guru's face in Figure 13.2 .

The Speed of Light

There are lots of different ways to measure or calculate the speed of light in modern physics. Physicists have been performing various experiments for well over 100 years to figure out not only how fast light travels in a vacuum, but also how it behaves in other materials, such as glass or water, where the speed is noticeably slower than in the vacuum of space. Currently, the accepted *measured* speed of light in a vacuum is 2.99792458×10^8 m/s. Maxwell himself derived the *theoretical maximum* speed of an electromagnetic wave based solely upon the values of the permittivity and permeability of free space. This surprisingly simple expression is given as,

$$c = \frac{1}{\sqrt{\varepsilon_0 \cdot \mu_0}} = \frac{1}{\sqrt{\left(8.9x10^{-12}\ \frac{\text{C}^2}{\text{N} \cdot \text{m}^2}\right) \cdot \left(4\pi \times 10^{-7}\ \frac{\text{V} \cdot \text{s}}{\text{A} \cdot \text{m}}\right)}} = 3.0x10^8\ \frac{\text{m}}{\text{s}}$$

As you can see, the discrepancy between the measured speed of light versus the theoretical value is pretty darn small. For computational purposes in physics though, the above value of $3.0x10^8$ meters-per-second is the accepted value for the speed of light.

So then, why all the 'hubbub' about the speed of light and electromagnetic waves?

Well, the speed of light just happens to also be the fastest speed in the universe – *it is the 'universal speed limit'.* Nothing can travel at the speed of light, except light itself of course, and certainly nothing can travel faster than the speed of light, hence the Guru's puzzled look?

As we'll discover in Chapter 14, anything with mass cannot travel faster than the speed of light. Electromagnetic waves have no mass, they just move energy around, so they're not bound by this restriction. From a mathematical standpoint however, electromagnetic waves can theoretically move faster than 'the speed of light'; it's just that one can never cross that vertical boundary 'wall' defined by 'c' in a vacuum.

In modern physics this has led to the axiom, *"You can never get to there, from here"* meaning that anything traveling at sub-light speeds can never move beyond the 'light barrier', and anything moving faster than light can never reach sub-light speeds for the same reason.

Hey, this real physics stuff is starting to sound like science fiction – the only difference is that this 'fiction' is real, and is lurking out there in the darkness of space. Scared yet? You will be before we're done with this book, *because things are going to get a whole lot stranger from here on out*. Let's look at a couple of examples of things that we can do with our newfound knowledge of electromagnetic waves, and light in particular.

Example 13.1: The average distance from Earth to the Sun is 150 million km. How long does it take light from the Sun to reach Earth?

We can calculate the time of travel for a light wave from the Sun by directly applying the speed of light value from above. Thus,

$$t = \frac{150x10^6\ \cancel{km} \cdot \frac{1000\ \cancel{m}}{1\ \cancel{km}}}{3.0x10^8\ \frac{\cancel{m}}{s}} = 500\ \text{s} = \mathbf{8.3\ minutes}$$

It takes slightly over eight minutes for light from the Sun to reach Earth. Thus, we are technically not seeing the Sun as it currently looks right now, but as it looked eight minutes ago. The Sun could explode right now and we would not know for eight minutes.

Clearly there is a 'time delay' between an event and its observation in the universe due to the finite nature of the speed of light. *Does this mean that when we look at the night sky, we are technically looking back in time?*

Example 13.2: The star Betelgeuse (pronounced 'Beetle Juice') is a dying star and will eventually explode (supernova) in the not-too-distant future. If it is located 640 light years from Earth, and it exploded tomorrow, when would we see the explosion?

This 'trick' question is to see if you are paying attention. The reason astronomers measure extremely long distances in terms of light years is to make the math easy. If something were located at a distance of 640 light years away, it would be $6.0549x10^{10}$ m away, which really doesn't mean a whole lot unless you had to know the distance in meters. It's easier to reference everything in terms of the 'light yardstick' because the universe covers quite a bit of real estate.

So for our example, we would not see the star Betelgeuse go 'kaboom' for **640 years**. Thinking about it, it may have already exploded and we just don't know about it yet.

Example 13.3: NASA detects that its Mars Rover is on the verge of driving off of a cliff. It sends a signal to tell it to stop. How long must NASA wait to know if the stop command was successful?

(The average distance from Earth to the planet Mars is 228 million km)

The round-trip time for a radio signal from Earth to reach Mars and back can be determined to be the following.

$$t = 2 \cdot \frac{228x10^6 \ \cancel{km} \cdot \dfrac{1000 \ \cancel{m}}{1 \ \cancel{km}}}{3.0x10^8 \ \dfrac{\cancel{m}}{s}} = 2 \cdot 760 \text{ s} = \textbf{25.3 minutes}$$

So we better plan on roughly a half hour depending on where Earth and Mars are in their respective orbits. It is for this reason that mobile planetary rovers must have a certain degree of autonomy to prevent damage, since it takes a while to phone home for help.

Example 13.4: An amateur radio operator sends a Morse code signal having an electric field measured at 50 V/m. What is the magnitude of the magnetic field? What is the magnitude of the radiated electromagnetic wave (i.e., the magnitude of the Poynting vector)?

We can instantly solve for the magnitude of the magnetic field, since we know that the electric field equals the magnetic field times the speed of light or $E = c \cdot B$.

$$B = \frac{E}{c} = \frac{50\,\frac{\text{V}}{\text{m}}}{3.0x10^{8}\,\frac{\text{m}}{\text{s}}} = 1.67x10^{-7}\,\frac{\text{V}\cdot\text{s}}{\text{m}^2} = \mathbf{1.67x10^{-7}\ T}$$

Notice that we have no choice as to the magnitude of the magnetic field. Once we select one field, the other goes along for the ride. The magnitude of the resulting electromagnetic wave is given by $\mu_0 \cdot S = E \cdot B$ thus we can write,

$$S = \frac{E \cdot B}{\mu_0} = \frac{\left(50\,\frac{\cancel{\text{V}}}{\cancel{\text{m}}}\right)\cdot\left(1.67x10^{-7}\,\frac{\text{V}\cdot\cancel{\text{s}}}{\text{m}^2}\right)}{\left(4\pi x10^{-7}\,\frac{\cancel{\text{V}}\cdot\cancel{\text{s}}}{\text{A}\cdot\cancel{\text{m}}}\right)} = \frac{50\cdot 1.67\,\cancel{x10^{-7}}}{4\pi\,\cancel{x10^{-7}}}\,\frac{\text{V}\cdot A}{\text{m}^2} = \mathbf{6.65\,\frac{W}{m^2}}$$

The Electromagnetic Spectrum

Now that we know quite a bit about 'light' and how it is related to electricity and magnetism, it's time to take a quick look at how all this electromagnetic energy is spread out across the universe. This is a topic known as the *electromagnetic spectrum*.

A spectrum is a mapping of some type of physical property of interest. The electromagnetic spectrum, which is shown in Figure 13.3, is therefore a mapping from high frequency to low frequency of every imaginable form of electromagnetic radiation that occurs within the universe. It is an extremely large scale with each tick mark representing an order of magnitude (i.e., power of '10') increase in either the frequency or the wavelength.

Notice that we can interchangeably use the frequency (f) or the wavelength (λ) to refer to a portion of the electromagnetic spectrum. That is because they are related to each other by the familiar expression,

$$\lambda \cdot f = c$$

Or, the wavelength of an electromagnetic wave times its frequency is equal to the speed of light.

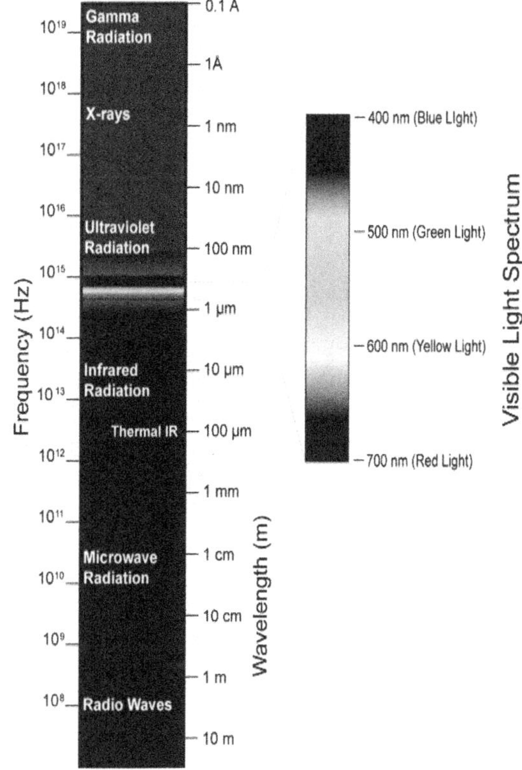

Figure 13.3: The Electromagnetic Spectrum for the Entire Universe

As we can instantly see, visible light is just a very small part of this large spectrum of electromagnetic radiation occurring between roughly 400 and 700 nm. Electromagnetic radiation above visible light is typically termed as *ionizing radiation* since it has the ability to easily excite electrons into a higher energy state and thus can cause injury, such as a sunburn if that electron just happens to be attached to you. Above that still, at a considerable distance along the spectrum, are higher energy waves producing what are collectively termed *x-rays,* so named by their discoverer Wilhelm Röntgen. At the top end of the spectrum is what is called *gamma radiation*. Electromagnetic waves of this high frequency (or short wavelength) are extremely energetic and dangerous as we'll find out shortly.

If we now continue down the electromagnetic spectrum from visible light, we see that just below this region is what is called the *infrared*. Infrared radiation is more 'felt' than seen by humans and right around 100 micrometers ($100\,\mu m$) is where normal radiant heat resides. Further below that is a region that we call *microwave radiation* and is the region in the spectrum in which microwave ovens and cell telephones operate. Everything below the microwave region is classified simply as *radio waves*. The radio wave section of the spectrum is the area in which all radio and television communication occurs, including amateur (ham) radio, AM, FM, and short wave. There's a lot of human activity packed into this region of the electromagnetic spectrum.

The Guru Says...

Why is it that you can see the food in a microwave oven, but cannot feel any of the microwave heat?

Well, light is a higher frequency (shorter wavelength) wave, thus it does not see the metal screen in a microwave oven's door. It passes right through without any problem. Microwaves, on the other hand, are much lower in frequency (look at Figure 13.3) and thus 'sees' the metal screen in the door as a solid object. The microwaves bounce back off the mesh unaware that there is an outside world.

As a final note, we should mention that the Doppler shift applies to light just as well as it applies to sound. In fact, as we mentioned before, Christian Doppler did his initial work on this topic with light and not sound (although his findings were correct for sound, and *not-quite-correct* for light). So without further ado, the Guru presents the correct findings below for the Doppler effect of light without proof (since the proof requires us to know something about special relativity, and we won't get to that topic until the next chapter).

$$f_{observed} = \left(\frac{1 - \frac{v_s}{c}}{\sqrt{1 - \left(\frac{v_s}{c} \right)^2}} \right) \cdot f_{source}$$

Where, v_s is the speed of the object, $f_{observed}$ is the observed frequency, f_{source} is the source frequency, and c is the speed of light (the special relativity part is in the denominator).

Example 13.5: What is the wavelength of an X-ray with a frequency of $1x10^{18}$ Hz?

Using our expression relating frequency to wavelength, we can see that the wavelength of an X-ray at the given frequency is,

$$\lambda = \frac{c}{f} = \frac{3x10^{8}\,\frac{m}{\not s}}{1x10^{18}\,\frac{1}{\not s}} = \mathbf{3x10^{-10}\ m = 300\ pm}$$

Example 13.6: Light from the Andromeda Galaxy is observed by a telescope on Earth to be 'blue shifted' by approximately 0.047%. An astronomer concludes that this galaxy is approaching us at a rate of 140 km/s. How does she arrive at this answer?

Performing a check on the astronomer's calculations, we can determine if the ratio of the observed to the actual light frequencies are as stated. Performing the substitution into the Doppler equation,

$$\frac{f_{observed}}{f_{source}} = \frac{1-\frac{v_s}{c}}{\sqrt{1-\left(\frac{v_s}{c}\right)^2}} = \frac{1-\frac{140\,\frac{\not{km}}{\not s}\cdot\frac{1000\,\not m}{\not{km}}}{3x10^{8}\,\frac{\not m}{\not s}}}{\sqrt{1-\left(\frac{140\,\frac{\not{km}}{\not s}\cdot\frac{1000\,\not m}{\not{km}}}{3x10^{8}\,\frac{\not m}{\not s}}\right)^2}} = 4.7x10^{-4} = \mathbf{0.047\%}$$

Problem Solving Tip...

In order to work the problem the other-way-around (i.e., figure out the speed from the observed light shift), we need to set it up as,

$$\frac{f_{observed}}{f_{source}} = \frac{1-\frac{v_s}{3.0x10^{8}}}{\sqrt{1-\left(\frac{v_s}{3.0x10^{8}}\right)^2}} = 4.7x10^{-4}$$

There is no direct algebraic solution to this equation, so we have to iterate (or guess) at solutions for the speed until we hit the correct one. Computers, of course, have no problem with a task like this.

It is curious to note that the Andromeda Galaxy is one of the only objects in the night sky that is actually traveling towards the Milky Way Galaxy. Almost everything else in the universe is red shifted towards the lower part of the electromagnetic spectrum and moving away from us. On a 'cosmically happy note', the Andromeda Galaxy is expected to collide with the Milky Way Galaxy in about 4.5 billion years. This is about the same amount of time remaining before our Sun dies – so *pick your poison.*

The Wave-Particle Conundrum of Light

Things would have been great in the world of physics if light just minded its business and remained acting as a wave. But *'noooooo'*, light has to act like an itsy-bitsy particle of massless matter under certain conditions, thereby screwing-up some really good-looking theories of the past. Just when you think you've got a pretty good idea on the rules controlling the universe, you find out that there's one more 'exception' to the rule.

It seems that at really high energies and frequencies like X-rays and gamma radiation, energy in the electromagnetic spectrum behaves quite well as a wave. The same holds true for really low frequency waves like radio waves. Somewhere however, right around the region where visible light and infrared heat likes to hang out, electromagnetic radiation also likes to behave like a particle, or a discrete 'chunk' of something, which can be quantified and counted, and imparts momentum to anything it comes in contact with – *something very, very un-wavelike indeed.*

The wave-particle conundrum is most pronounced within the thermal infrared (thermal IR) portion of the electromagnetic spectrum, where objects that are 'hot' do not produce visible light, but radiate electromagnetic energy in the form of heat. *Black body radiation*, as the effect is called, could manifest itself in something as simple as a hot piece of metal, so the phenomenon could not be easily ignored.

The Guru Says...

The expression for kinetic energy of light is vastly different than what we have seen previously (i.e., KE = ½mv^2) because of a simple reason – *light has no mass*. It doesn't weigh anything and cannot contribute to the mass of an object.

This might seem like a difficult concept at the start, to have kinetic energy without mass, but as the Guru has said, light is one of the most perplexing things in the universe.

Light also has momentum (i.e., $m \cdot v$) too, but only by virtue of it moving at the speed of light, of course.

In 1900, the German physicist Max Planck developed a theory for black body radiation that worked well with what was currently known about the electromagnetic spectrum. Planck called the little particles or 'packets' of energy that were radiated *quanta*, and unknowingly started an entire future subset of physics called *quantum mechanics*. Later, another physicist by the name of Albert Einstein (you knew we'd get to him eventually) proposed that these 'quanta of light' could be mathematically treated as real particles with a kinetic energy equivalent to,

$$KE = h \cdot f = \frac{h \cdot c}{\lambda}$$

Where 'f' is the frequency of the electromagnetic wave, 'λ' is the wavelength of the electromagnetic wave, and 'c' is the speed of light. The constant 'h' is called Planck's constant, and is equal to 6.626068x10^{-34} m^2·kg·s^{-1} and is yet another one of those 'fudge factors' keeping the equation in check.

Einstein also discovered that this 'quanta of light' also has a mechanically equivalent momentum 'M', which could also be described using Planck's constant as,

$$M = \frac{KE}{c} = \frac{h \cdot f}{c} = \frac{h}{\lambda}$$

The Guru Says...

Many years later, the term 'quanta of light' was replaced with a simpler name for a packet of light – the *photon*. The name of the photon was choosen to be similar sounding to other basic particles in physics like the *electron*, the *proton*, and the *neutron*.

Thus, we are able to describe electromagnetic wave motion not only as a wave with a distinct frequency and wavelength, but also as a massless particle having kinetic energy and momentum, and having the ability to do mechanical work on objects that have mass. Let's see how this works.

Example 13.7: A photon of visible red light from the Guru's laser strikes the glass prism as shown in the picture below. What is the kinetic energy and momentum of each photon if the wavelength of the emitted light is 655 nm?

Using our *Planck-Einstein relationships* from above to calculate the equivalent kinetic energy and momentum of a photon of light from the laser we find,

$$KE = \frac{h \cdot c}{\lambda} = \frac{6.626068\text{x}10^{-34}\ \frac{\text{m}^2 \cdot \text{kg}}{\text{s}} \cdot 3x10^8\ \frac{\cancel{\text{m}}}{\text{s}}}{655x10^{-9}\ \cancel{\text{m}}}$$

$$= \mathbf{3.0348x10^{-19}\ J\ /\ photon}$$

$$M = \frac{h}{\lambda} = \frac{6.626068\text{x}10^{-34}\ \frac{\text{m}^{\cancel{2}} \cdot \text{kg}}{\text{s}}}{655x10^{-9}\ \cancel{\text{m}}} = \mathbf{1.012x10^{-27}\ \frac{kg \cdot m}{s}}$$

Even though each photon's contribution is an extremely small number, it does contribute a finite 'push' on objects in space. This force is called the *solar wind* or *solar radiation pressure*.

Finally we should also note that the higher the frequency of the electromagnetic wave, the greater the energy or momentum we can obtain from the Planck-Einstein relationships. We can therefore conclude that higher frequency electromagnetic waves have more energy to impart than lower frequency ones, thus proving our earlier statement about the difference between ionizing and non-ionizing radiation (and why radio waves are harmless to humans, but gamma rays can kill).

Example 13.8: The Guru's laser in the previous example is rated at 1.0 watt of coherent light output at 655 nm. How many photons are produced by the laser each second? What force does this movement of photons produce? If the beam size is 1 cm^2, what pressure does the beam hitting an object produce?

This calculation is quite straightforward since we have already done the hard part in calculating the kinetic energy per photon at 655 nm. Since one watt is nothing more than one joule per second, we have,

$$\text{Photons per second} = \frac{\text{Radiated Power}}{\text{KE}_{\text{Photon}}} = \frac{1\frac{\not J}{s}}{3.0348x10^{-19}\frac{\not J}{\text{photon}}} = \frac{\mathbf{3.295x10^{18}}}{\mathbf{s}}$$

A photon isn't really a unit; so we express the final answer as just being in terms of 1/s. Clearly there are a lot of photons flying around in the universe at any given time.

The force produced by the beam can be found by multiplying the momentum of a photon at the indicated wavelength by the rate at which photons are being produced. This is nothing more than an application of Newton's second law, or the time rate of change of momentum.

Thus,

$$Force_{laser} = Rate_{photons} \cdot Momentum_{photons}$$

$$= \frac{3.295x10^{18}}{s} \cdot 1.012x10^{-27}\frac{kg \cdot m}{s} = 3.335x10^{-9}\frac{kg \cdot m}{s^2}$$

$$= \mathbf{3.335x10^{-9}\ N}$$

A small force, but a force nonetheless. For a one square centimeter beam size, the radiation pressure would be equal to **$3.335x10^{-5}$ Pa**.

We can conclude that in the world of *science fact*, force fields and tractor beams do indeed exist, but only the repulsive type. Go ahead and extend on the above example by letting the laser beam act for a long period of time and

The Guru Says...

Einstein used the particle nature of light to describe an observed physical phenomena called the *photoelectric effect*, whereby certain frequencies of light actually produced a voltage in certain types of metals.

Einstein won the 1921 Nobel Prize in physics based upon his work on the photoelectric effect. He never received a Nobel Prize for his Theory of Relativity, which was a much bigger deal and the cornerstone of modern physics.

figure out how long it would take for the beam to move a one kilogram mass a distance of one meter on a frictionless table.

Reflection and Refraction of Light Waves

Our final section on light is more down-to-Earth and focuses on two observable aspects of light propagation – that of *reflection* and *refraction*. Reflection is the bending or returning of light without it being absorbed, while refraction is the change in direction of light propagation with frequency.

We experience reflection and refraction of light all the time in our everyday experiences. In fact, we experience these two effects on light more than direct exposure to any light source. We look at ourselves in a mirror (reflection), look at the moon at night (both reflection and refraction), observe a rainbow (refraction), look though a magnifying glass or a pair of reading glasses (refraction), look at the shimmering of light on a pond (reflection), look at the sparkling of a diamond (both reflection and refraction)...

... and the biggest one of all – *all the light you see in the world around you is reflected light bouncing off of some solid object.*

After all of our somewhat painful electromagnetic equations, you'll be happy to know that the reflection and refraction of light waves are simple trigonometric relationships that can be summed up pretty quickly. Let's start with reflection.

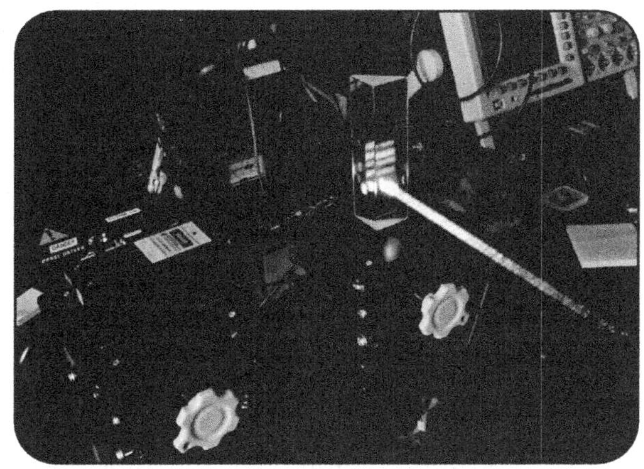

Figure 13.4: Example of Internal Laser Light Reflection Through a Glass Prism

Figure 13.4 shows the laser beam again passing through a glass prism and being completely reflected at 90 degrees with respect to the initial, or *incident*, beam position. Even though the light is passing through the glass of the prism, it is still being reflected, since the angle that it is hitting the rear face of the prism is the same as the angle at which the light is leaving the prism (i.e., it is entering at a 45-degree angle and is leaving at a 45-degree angle, thus the light is being reflected a total of 90-degrees as shown). Mathematically we can write this as,

$$\theta_i = \theta_R$$

Where we are saying that the angle of the original light beam (the incident light angle θ_i) equals the reflected light angle θ_R. So the angle of a light beam 'into' a surface equals the angle of a light beam 'out of' the surface. This is the basic physical law governing how the reflection of light occurs, and humans have known this simple fact about light waves for thousands of years.

Refraction, or the physical bending of light waves occurs because the speed of light slows down when it enters another material (it technically did in our prism photo, but the effect is negligible). Refraction produces bending of light waves as a function of frequency. The lower the measured frequency of the light, the more the light is bent when passing through, or around, a certain quantity of material.

This can be seen in Figure 13.5 where a coherent and incredibly parallel laser light beam (entering at the left) is being 'squeezed together' through diffractive bending associated with the curved interface of the acrylic wedge.

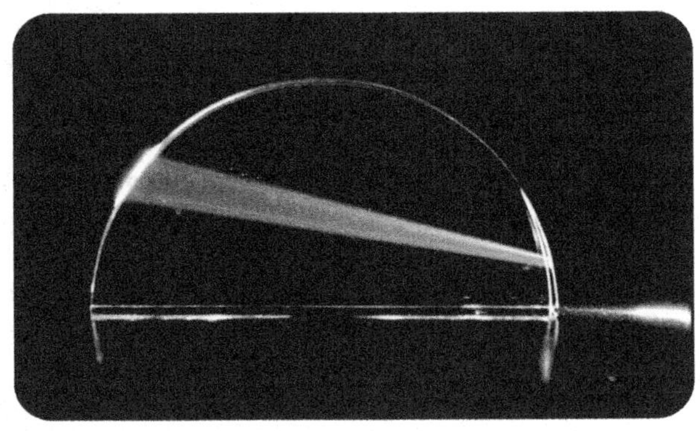

Figure 13.5: Example of Light Refraction Through an Acrylic Circular Wedge

Observations like this led Dutch astronomer Willebrord Snellius in 1621 to discover a relationship relating how light refracts through different materials as a function of the incident light angle (θ_i) and the material in question. This relationship, known as Snell's Law can be written as,

$$\sin\theta_i = \eta \cdot \sin\theta_r$$

Where θ_i is the incident light angle, θ_r is the refracted light angle, and η is a new physical parameter known as the *index of refraction* for the material in question.

The index of refraction is a material specific, as well as a frequency or wavelength specific parameter. By definition, the reference index of refraction is 1.0 in a vacuum. Just like our previous values for the magnetic permittivity constants, the refractive indices for air and a vacuum are essentially equal to one, so you can use a value of 1.0 without a significant loss of accuracy. A list of refractive indices for various materials at a wavelength of 589.29 nm (the scientifically agreed standardized wavelength of light) is shown in Table 13.1.

Table 13.1: Refractive Indices of Materials

MATERIAL OF INTEREST	REFRACTIVE INDEX η RELATIVE TO A VACUUM
Vacuum	1.00
Air	1.000277
Hydrogen	1.000132
Helium	1.000036
Water	1.3330
Ice	1.31
Glass	1.492
Sapphire	1.778
Faux Diamond (Cubic Zirconium)	2.18
Real Diamond	2.419

For more than one material interface that light passes through, we use the ratio of the indices of refraction for the two materials instead of a single index. In this case, Snell's Law becomes,

$$\sin\theta_i = \frac{\eta_i}{\eta_r} \cdot \sin\theta_r$$

Another Dutch astronomer by the name of Christiaan Huygens in 1678 explored the *then-new* concept of light waves and noted that the wavelength of refracted light was equal to the wavelength of the incident wave times the ratio of the change in speed of the wave through the different materials. He figured this out without the aid of lasers or computers, or even knowing what the speed of light was for that matter – *you now know more about these topics than he did*. His findings demonstrated that the index of refraction was nothing more than the ratio of the speeds of light in the different materials being observed.

Thus,

$$c_{vacuum} = \eta \cdot c_{material} \quad or \quad c_{material\ 1} = \frac{\eta_2}{\eta_1} \cdot c_{material\ 2}$$

Finally, we should note that there is a point at which refraction bends the light to an extreme point where the refracted light wave is parallel to the surface of the material doing the refracting. This condition is called *total internal reflection* and is the reason why fiber optic cables work. In a total internal reflection situation, the light just keeps bouncing inside the glass fiber cable until it reaches the other end.

Example 13.9: A light beam hits a diamond at a 30 degree angle with respect to a vertical plane. If the diamond is in a vacuum chamber, what angle does the reflected light bounce back? What angle does the light bend through inside the diamond?

We instantly know that the reflected light bounces back at the same angle as the incident angle. Thus **the reflected light wave would be 30 degrees from vertical on the other side of the diamond**.

The refracted light can be calculated using Snell's Law. From Table 13.1 we see that the index of refraction of diamond is $\eta = 2.419$, thus we have,

$$\sin 30^\circ = 2.419 \cdot \sin\theta_r$$

$$\sin\theta_r = \frac{\sin 30^\circ}{2.419} = \frac{0.5}{2.419} = 0.2067$$

$$\theta_r = \sin^{-1}(0.2067) = \mathbf{11.93^\circ}$$

So light would bend roughly 12 degrees inside a diamond. You now have a handy way to detect real diamonds from fakes.

Example 13.10: A child puts a hiking stick in a nearby pond and notices that it appears to bend underwater. If the stick is placed at a 45 degree angle with respect to the pond, what is the observed refracted angle underwater?

Applying Snell's Law for differing transmission mediums, we have for the case of a 45 degree incident angle with respect to a light-water interface, the answer to that time-tested question asked by children everywhere.

$$\sin\theta_i = \frac{\eta_{air}}{\eta_{water}} \cdot \sin\theta_r$$

$$\sin 45^\circ = \frac{1.000277}{1.3330} \cdot \sin\theta_r$$

$$\sin\theta_r = \frac{1.3330}{1.000277} \cdot 0.7071 = 0.9423$$

$$\theta_r = \sin^{-1}\left(0.9423\right) = \mathbf{70.4^\circ}$$

...and the stick appears to magically 'bend' by an additional 25.4 degrees.

Example 13.11: What is the speed of light inside a piece of sapphire?

Drawing upon Huygens' findings, we can write the expression for the speed of light inside the sapphire as,

$$c_{air} = \frac{\eta_{sapphire}}{\eta_{air}} \cdot c_{sapphire}$$

$$c_{sapphire} = \frac{\eta_{air}}{\eta_{sapphire}} \cdot c_{air} = \frac{1.000277}{1.778} \cdot c_{air} = \mathbf{0.5626 \cdot c_{air}}$$

Or close to half what the speed of light is in a vacuum. Given the peculiar behavior of light that we have seen, is any wonder why Einstein once pondered the following, *"What would it be like to ride a beam of light?"*

Chapter 14: The Relative Side of Everything

Welcome to the peculiar world of Professor Einstein

In our final section of the Guru's Guide, things start to get *really, really strange*. When we start looking at objects moving close to the speed of light, we find that all of the Newtonian mechanics that we've previously talked about starts to break down. Things start to occur in the universe that defy logic and common sense. Mass doesn't appear like mass anymore, and the notion of time itself starts to be altered and is no longer an immutable quantity. Everything is 'up for grabs' in a fast moving universe – except for light itself, which refuses to play by anybody's rules except its own.

This is not to say that Newton's laws should be abandoned for some other more complex theory – *quite to the contrary*. Newton's laws, and the motions described by Special and General Relativity, are exactly the same below roughly one-half the speed of light (or 0.5c) and only need some fine-tuning for things above that speed. Newton's work can be used to accurately describe, to a high level of precision and accuracy, the motion of any object in the universe, its force, momentum, work, energy, and so on. Since the vast majority of objects within the universe move at speeds considerably slower than 0.5c, we can be reassured that Newton's laws will be around for a very long time.

The real wrinkle between *Newtonian* and *Relativistic mechanics* is not Newton's laws themselves, but the fact that space and time gets a little 'goofy' at speeds close to light, and we need to rethink how we perform these types of fundamental geometric and temporal measurements. To solve this problem, one of the most remarkable physicists that ever lived, by the name of Albert Einstein, rethought how the universe must function on 'super fast forward', and he started with that pesky little non-conformist, the photon.

The Lorentz Transformation

In 1887, two American physicists by the names of Albert Michelson and Edward Morley set up an elaborate experiment to conclusively measure the relative motion of matter through the 'aether wind' of space (remember that worthless crud aether from Chapter 13). Their experiment, famously called the *'Michelson–Morley Experiment'* got negative results dozens upon dozens of times, and finally garnered attention that aether was merely an ethereal aberration of the mind. The Michelson–Morley Experiment ultimately meant the death of aether.

Meanwhile in 1892, the Dutch physicist Hendrik Lorentz developed a mathematical transformation describing how to adjust normal Euclidian motion to higher speeds approaching that of light in an attempt to – *yes, prove that aether exists and reconcile Maxwell's equations with the Michelson–Morley Experiment* (some physicists never say die). His equation, although most assuredly correct, still never did prove the existence of aether.

The *Lorentz Transformation*, which on the surface is a simple-looking geometric relationship between low speed Euclidian geometry and its higher speed relativistic counterpart, can be expressed in the following manner,

$$x_m = \frac{x_s - v \cdot t}{\sqrt{1-\left(\frac{v}{c}\right)^2}}; \quad y_s = y_m; \quad z_s = z_m$$

Where x_m is the position in the 'x' direction of the moving coordinate system, x_s is the position of the stationary coordinate system, 'v' is the speed of the moving coordinate system, 'c' is the speed of light, and 't' is how long you've been moving under this very fast condition. In the case where 'v' is very small, the equation reduces to our simple Euclidian form of distance equals velocity 'times' time.

Geometrically, the Lorentz transformation shown in Figure 14.1 is very similar to the inertial reference frame graphic the Guru presented in Figure 5.1, with the adjustment of the frame being dependent on the speed of light 'c'. Note also that it only needs to be represented in terms of one coordinate axis, the direction that you are moving.

So to reiterate, the Lorentz transformation <u>only applies to the coordinate axis in the direction of motion</u>. All other coordinate axes are unaffected. This is going to make for some pretty interesting physics as we move towards the speed of light, since the Lorentz transformation implies that things are warped (i.e., stretched or compressed) only in the direction in which they are moving.

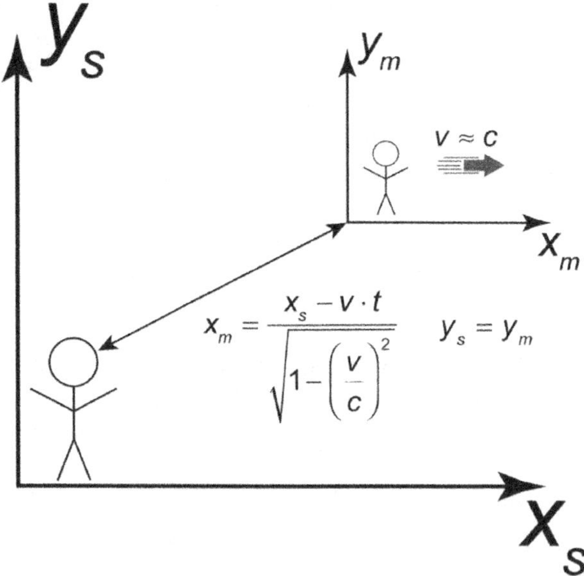

Figure 14.1: Our New Reference Frame Adjusted According to the Lorentz Transformation – *No Additional Assembly or Aether Required*

Solving the Lorentz transformation expression for time, we can arrive at a separate relationship describing the observed time as a function of the spatial distance traveled.

Thus,

$$t_m = \frac{t_s - \left(\frac{v}{c^2}\right) \cdot x_s}{\sqrt{1 - \left(\frac{v}{c}\right)^2}}$$

This led Irish physicist Joseph Larmor to conclude that time itself is not a fixed quantity in the universe and is 'relative' to how fast you are moving. A stationary elapsed time Δt_s will not be the same as a moving elapsed time Δt_m, and the two are related by the following expression,

$$\Delta t_s = \frac{\Delta t_m}{\sqrt{1 - \left(\frac{v}{c}\right)^2}}$$

The above equation represents what physicists call *time dilation*, which is a warping of the apparent time in two reference frames by virtue of an object moving at a high rate of speed. The effect is most noticeable at speeds near the speed of light, but applicable to all moving objects within the universe. Let's look at how clocks start to give different answers near the speed of light depending on where you are standing.

The Guru Says...

Clearly, time and space are 'bolted together' in the above expression – something referred to as *spacetime*.

Our expression states that the time measured by an individual in a moving reference frame is not the same as the time measured by an individual in a stationary frame for a given amount of distance traveled. This would imply that the time recorded on two identical clocks would be different depending on how fast the clocks are moving with respect to each other.

Example 14.1: In Figure 14.1, let's assume that we have two individuals with identically operating clocks. While the stationary observer counts one elapsed hour, the moving individual travels at 0.9c for the same period of time. How much elapsed time does the moving observer record?

Using our time dilation equation, we can write the expression for the moving reference frame in terms of the stationary frame as,

$$\Delta t_m = \Delta t_s \cdot \sqrt{1 - \left(\frac{v}{c}\right)^2}$$

We are told that the stationary clock records one hour ($\Delta t_s = 1$) while the moving clock (Δt_m) is buzzing around at 0.9c.

Substituting we get,

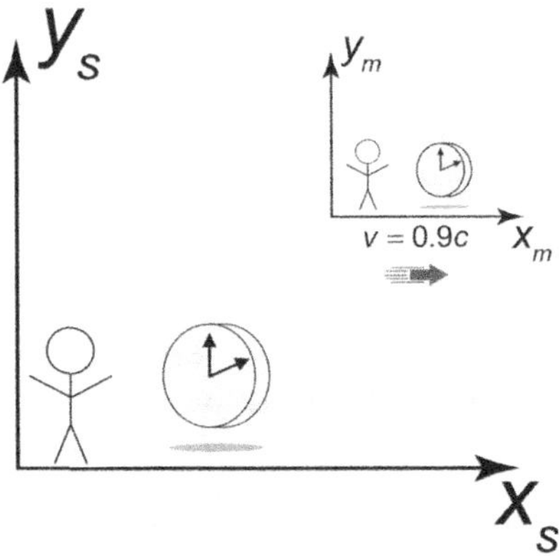

$$\Delta t_m = 1\,\text{hr} \cdot \sqrt{1-\left(\frac{0.9c}{c}\right)^2} = 1\,\text{hr} \cdot \sqrt{1-\left(\frac{0.9\not{c}}{\not{c}}\right)^2}$$

$$= 1\,\text{hr} \cdot \sqrt{1-0.81} = 1\,\text{hr} \cdot \sqrt{0.19} = 0.4359\,\text{hr}$$

$$= \mathbf{26.15\ min}$$

What takes an hour for the stationary observer, only takes slightly over 26 minutes to an observer moving at 0.9c. Both people arrive back at the same point in space with one clock showing roughly 34 minutes less elapsed time. How can this be so?

The faster *you* move towards the speed of light, the slower time literally 'ticks' for *you*. Your wristwatch ticks more slowly, onboard systems within your spaceship function more slowly, you age more slowly – *but you do not notice any changes*. In your reference frame your clock behaves normally, you just traveled through one hour of stationary spacetime in 26 minutes.

At a first glance, this could seem like there is something very wrong with the universe or that we are working in the world of science fiction, but this is reality. Space and time are really not separate entities. If someone wants to travel through a sizeable portion of space in a short period of time, expect time to 'take up the slack'. The faster you go, the weirder it gets. If our hypothetical traveler were moving for one hour (moving time) at 0.999c, he/she would return back to Earth to discover that an entire day has passed by. At the speed of light, time theoretically stops and an entire eternity could pass by in the blink of an eye.

Example 14.2: The Guru proposes a one-way time machine consisting simply of a device that let's you travel at 0.99999c for one minute? If he steps inside and pushes the 'go' button, how far in the future will he travel in that one minute? Does this device violate any physical laws?

The equation representing the Guru's time machine would be given by the Lorentz time transformation as,

$$\Delta t_{\text{real time}} = \frac{\Delta t_{\text{Guru time}}}{\sqrt{1-\left(\frac{v}{c}\right)^2}} = \frac{60\ \text{sec}}{\sqrt{1-\left(\frac{0.99999\not{c}}{\not{c}}\right)^2}} = \frac{60\ \text{sec}}{\sqrt{1-0.9801}} = 13416.4\ \text{sec} = \mathbf{3.73\ hr}$$

So by the Guru merely accelerating himself to 0.99999c, he can skip over 3.73 hours of normal spacetime and presumably arrive at the same point in space that he started from. In the Guru's time frame, he has only aged one minute, while everyone else has aged almost four hours. *Creepy huh?*

Even though the time machine proposed is possible and **does not violate the Second Law of Thermodynamics**, we'll see that the universe puts other restrictions on this sort of whimsical travel.

Let's look at an example of a not-so-fast moving object and see if the same rules apply.

Example 14.3: Let's say you fly from San Diego to Denver at a speed of 960 km/hr. If the flight takes 2.5 hours (or 9000 seconds) as shown on your watch, how much additional time have you added to your life?

A speed of 960 km/hr is equal to $8.88x10^{-7}c$. Using the equation from the previous example with a significantly reduced speed we find that your time is dilated for the trip by,

$$\Delta t_{\text{stationary time}} = \frac{\Delta t_{\text{your time}}}{\sqrt{1-\left(\frac{v}{c}\right)^2}} = \frac{9000 \text{ sec}}{\sqrt{1-\left(\frac{8.88x10^{-7}\cancel{c}}{\cancel{c}}\right)^2}} = \mathbf{9000.000000003548448 \text{ sec}}$$

Therefore, you've effectively 'stepped out of time' by 3.548 nanoseconds and, if you think about it, added this time to your life, over what you would have experienced by just standing still. Although, was the hassle of getting through the airport worth the additional 3.548 nanoseconds?

Problem Solving Tip...

You will need to watch your accuracy in calculations like these, since most calculators won't carry enough significant figures and will want to round the denominator to one.

The Guru wrote a little double-precision BASIC program to do the calculations, but you can find a super accurate calculator program on the internet pretty easily which will not truncate the answer.

We should note that when this finding was first proposed in the early 1900's, there was no way to test it. Really fast airplanes and atomic clocks capable of measuring a change like the one shown above wouldn't be invented for close to a half-century.

In 1971, two researchers by the names of Joseph Hafele and Richard Keating performed the famous *Hafele–Keating experiment,* whereby they flew several cesium atomic clocks in a commercial airplane around the world twice, both with, and against, Earth's rotation. These clocks were then compared against a set of identical clocks located at the United States Naval Observatory.

The results, which were published in 1972 demonstrated that time had indeed slowed down for the moving atomic clocks, by upwards of 273 nanoseconds compared to the stationary clocks. The Hafele–Keating experiment was a surprisingly simple way to verify one of the more peculiar aspects of modern physics.

Einstein's Theories of Special and General Relativity

Albert Einstein spent a lot of his early life fascinated with light and how it behaved. By the early 1900's he was well aware of the problems imposed by the behavior of light and the apparent conflicts with Newtonian mechanics. Although Maxwell's equations helped a lot to resolve some of the uncertainties of electromagnetic waves, the fact still remained that light did not want to play nice.

In 1905 Einstein published his Theory of Special Relativity, which explained some of the inconsistencies with the observed motion of light, as well as how the general laws of physics should manifest themselves at relativistic speeds.

Einstein's theory of 'Special Relativity' postulated that there are no absolutes in the universe. All frames of reference are relative and based upon the particular amount of spacetime you possessed (i.e., all inertial reference frames are relative). In order to resolve the problems observed with the behavior of light, he further postulated that the speed of light is always just the speed of light, the constant $3.0x10^8$ m/s.

The Guru Says...

Many people have heard of the Theories of Special and General Relativity, but few know the difference between the two.

The Theory of Special Relativity states that 'space' and 'time' are not two different entities, rather they are interrealted concepts – *one is exchanged for the other as you move.* Kind of like the similarities in work and energy we discussed earlier. Special Relativity is a kinematics exercise describing events that occur very close to the speed of light and, in the absence of gravitational effects, also states that the laws of physics are the same everywhere in the universe. In Special Relativity, space is 'flat' for the pusposes of analysis.

General Relativity, published much later in 1916, removed the restriction on 'no gravitational effects' on objects traveling close to the speed of light and further postulated that gravity is a phantom force, and is actually a consequence of matter warping space around it. Under General Relativity, space is 'curved' or 'warped' and objects move along the grooves of curved space.

This means that you can never add two relativistic velocities together and get a velocity greater than the speed of light. This is a very important aspect of special relativity that on the surface violates our basic understanding of how we add velocity vectors together. Thus, adding 0.9c plus 0.9c does not equal 1.8c, it still only equals something less than the speed of light.

Mathematically, Einstein said that the exact way to add velocities in the universe is through the following expression,

$$v_s = \frac{v_m + \Delta v}{1 + \left(\frac{v_m \cdot \Delta v}{c^2} \right)}$$

This expression, which again is consequence of the Lorentz transformation, states that the velocity of a moving object 'v_m' measured in a stationary frame 'v_s' is equal to its speed plus some velocity increment 'Δv'. For speeds much less than the speed of light, this expression reduces to $v_s = v_m + \Delta v$, which is our original expression for adding two velocities together. Let's look at two extreme examples of this concept to drive the point home.

Example 14.4: A person in a moving reference frame traveling at 100 m/s throws a ball in front of him at the same speed? How fast is the ball moving with respect to the stationary reference frame?

We know that the answer has to be 200 m/s just by adding the velocities together in a manner consistent with how Euclidian geometry works. Now, let's apply our new relativistic velocity addition expression to this low speed problem and see what the relativistic answer should actually be.

$$v_s = \frac{v_m + \Delta v}{1+\left(\frac{v_m \cdot \Delta v}{c^2}\right)} = \frac{100\,\frac{m}{s} + 100\,\frac{m}{s}}{1+\left(\frac{100\,\frac{m}{s}\cdot 100\,\frac{m}{s}}{\left(3x10^8\,\frac{m}{s}\right)^2}\right)}$$

$$= \frac{200\,\frac{m}{s}}{1+\left(\frac{1.0x10^4\,\frac{m^2}{s^2}}{9x10^{16}\,\frac{m^2}{s^2}}\right)} = \frac{200\,\frac{m}{s}}{1.00000000000011111111} = \mathbf{199.99999999977777778}\,\frac{\mathbf{m}}{\mathbf{s}}$$

Which we can conclude is still equal to 200 m/s when we round the answer to several decimal places. Thus, our relativistic velocity addition expression is consistent with the addition of velocity vectors at low speeds. Let's now redo the same example using a much higher speed.

Example 14.5: Another person in a moving reference frame traveling at 0.9c throws a ball in front of her at the same speed? How fast is the ball moving with respect to the stationary reference frame?

Einstein postulated that nothing could travel faster than the speed of light, thus we would expect our final answer to be closer to the speed of light than either individual speed, but never actually reaching the actual value of 1.0c.

Again applying our relativistic velocity addition formula we find,

$$v_s = \frac{v_m + \Delta v}{1 + \left(\frac{v_m \cdot \Delta v}{c^2}\right)} = \frac{0.9c + 0.9c}{1 + \left(\frac{0.9c \cdot 0.9c}{c^2}\right)}$$

$$= \frac{1.8c}{1 + \left(\frac{0.81\cancel{c^2}}{\cancel{c^2}}\right)} = \frac{1.8c}{1.81} = \mathbf{0.995c}$$

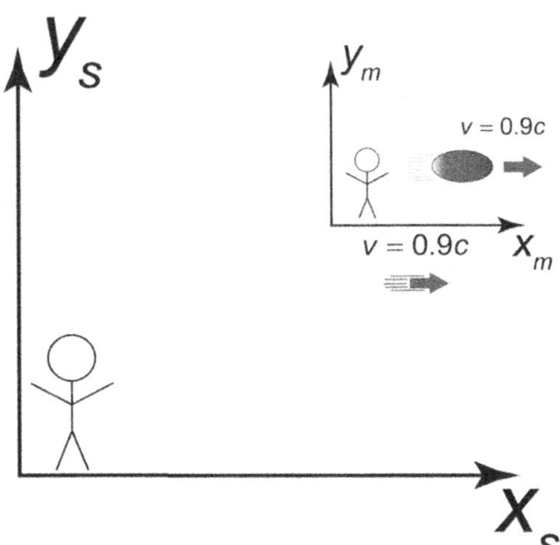

Thus we're getting close to the speed of light, but we'll never actually ever reach this speed no matter how many additional velocity increments we add.

Continuing with our discussion, Einstein additionally postulated that the laws of physics are the same in all reference frames, and that there are no 'special' or 'preferred' inertial reference frames anywhere in the universe. One reference frame is just as good as the next. Thus, we would expect that Newton's laws, impulse and momentum, work and energy, thermodynamics, wave motion, and electromagnetism would produce the same results in a lab that was standing absolutely still in space as it would in one that was traveling at 0.9c.

This was a remarkable statement, which could not be demonstrated in 1905 and took several decades to replicate and verify.

A short time after Einstein published his theory of relativity, the German mathematician Hermann Minkowski discovered that the Einstein's theories actually plotted a four-dimensional graph, called *Minkowski spacetime*, which is shown in Figure 14.2.

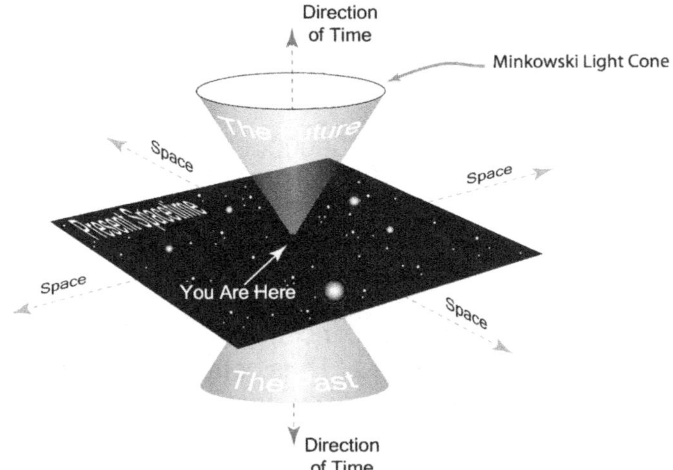

Figure 14.2: Example of Minkowski Spacetime as Realized by Special Relativity and the Lorentz Transformation

In this representation of the universe, we have a direction of time represented by the vertical axis and space represented by the other two axes. In the Minkowski spacetime model, our observable universe exists in the present spacetime. It continues to move forward along the time line with spacetime moving 'upward' as time progresses.

Since the speed of light is a finite constant independent on how fast you are personally moving, the Minkowski model defines a maximum observable universe defined by the limits on how far light can travel in a given period of time. This produces a 'light cone' thereby limiting anything you can ever do, or observe, at any given time. Also, the Minkowski model places no restrictions on the direction of travel along the time line, but we already know that little things like the Second Law of Thermodynamics prevents backwards motion and keeps would-be time travelers out of serious trouble.

Motion, Momentum and Energy in Einstein's Universe

Einstein also developed revised relativistic equations describing how mass, length, momentum and kinetic energy behave at speeds close to light. Given this, we can write the *relativistic mass* 'm_m' and relativistic length 'L_m' as,

$$m_m = \frac{m_s}{\sqrt{1-\left(\frac{v}{c}\right)^2}}; \qquad L_m = \frac{L_s}{\sqrt{1-\left(\frac{v}{c}\right)^2}}$$

The first expression states that the faster an object with mass (any mass – even a tiny speck of mass) travels towards the speed of light, the more mass it gains. It is for this reason that no object in the universe can ever achieve the speed of light, since at that point its mass would be infinite. Infinite mass requires infinite energy to move it so we can forget about any crazy light speed traveling notions.

The second expression states that an object will either appear longer or shorter along its direction of travel, depending on whether the object is traveling towards you (blue shifted) or away from you (red shifted). We touched upon this earlier when we discussed the Lorentz transformation, but the practical upshot of this is that we can't even depend on a ruler being accurate at high rates of speed since even basic spatial measurements become distorted.

Since momentum is equal to the mass of an object times its velocity, we can modify the above relativistic mass expression to show that the *relativistic momentum* is equal to,

$$Momentum_m = m_m \cdot v = \frac{m_s \cdot v}{\sqrt{1-\left(\frac{v}{c}\right)^2}}$$

As with the relativistic mass, the momentum of an object also approaches infinity as the object is accelerated towards the speed of light. So, maybe in Example 14.2, the Guru's time machine wasn't that good of an idea after all, since it would take a awful lot of energy to reach a momentum close to infinity.

As a result of the above two expressions, Einstein restated the kinetic energy of an object in a relativistic frame as,

$$KE_m = \frac{m_s \cdot c^2}{\sqrt{1-\left(\frac{v}{c}\right)^2}} - m_s \cdot c^2 = m_m \cdot c^2 - m_s \cdot c^2 = m \cdot c^2$$

Which gives us the classic relationship $E = m \cdot c^2$ that everyone knows is attributed to Einstein, but isn't sure why. The *mass-energy relationship* in special relativity states that mass and kinetic energy is the same thing under different circumstances. The Guru likes to think of matter as being nothing more than 'frozen kinetic energy'.

Bending Time Until It Breaks

One of the interesting side effects of special relativity is that the concept of two or more events occurring in a 'simultaneous' fashion has no meaning whatsoever. How you observe an event is really a function of how fast you are moving. Einstein explained this phenomenon using a railway car and two bolts of lightning hitting the rails in front of, and behind, the rail car. We'll do the same experiment using the Guru's head and a flying brick, and remembering that the relativistic length of space changes the faster we move, so two objects near each other in one reference frame need not be adjacent to each other in another.

Let's look at Figure 14.3 where we see three different scenarios involving the brick flying towards the poor Guru. When we are not moving, as time marches forward, so too does the brick and the Guru in the same spacetime. They are both in the right place at the right time. In this scenario, the Guru feels the kinetic energy transfer and we all know what the outcome of that experience is.

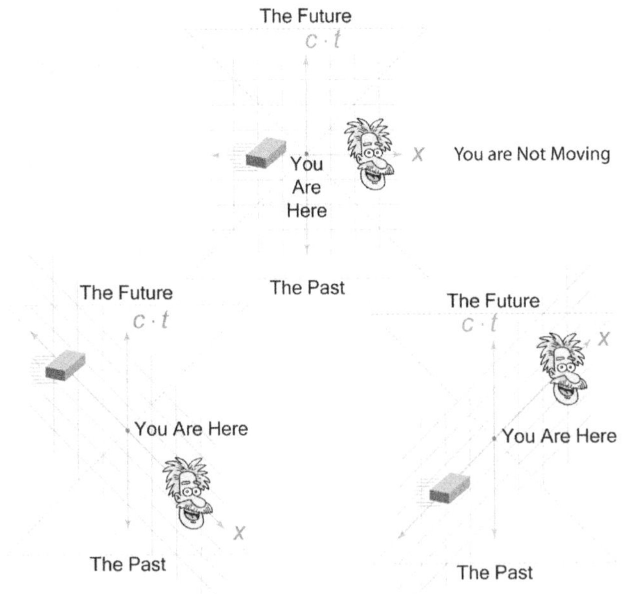

Figure 14.3: Why a Simultaneous Event is a Universal Falsehood – *When Exactly Does the Guru Get Hit with the Brick?*

In the next scenario, we are moving at 0.9c to the 'left', realizing that the concept of 'left', 'right', 'up', and 'down' is relative in space. In this case, we have traveled up the left side of the Minkowski light cone. As the Lorentz transformation predicts, spacetime is warped in that direction and the presence of the brick and the Guru do not occur simultaneously in the same spacetime. Thus, as time marches forward (up the diagram), we first encounter the Guru, then the brick. Since the brick and the Guru no longer share the same spacetime, the brick misses the Guru. Similarly, a shift to the 'right' would produce a similar phenomenon, but with opposite cause and effect. The brick would first appear in spacetime, followed by the Guru – again missing him.

So what's going on here? They can all be correct – can they?

Well the answer is yes. All three possible outcomes are available depending on what you are doing as an observer. We can either see the Guru and the brick hit each other at the same time, or see one or the other first, thereby observing the brick missing the Guru. Do you now believe that the universe can be a really strange place? Two different observers wouldn't even see the same event the same way. This puts a serious damper on what we humans consider 'reality' to be.

The Guru Says...

There's a little limerick that sums up why time travel is not a good idea. It goes something like...

"There once was a lady named Bright,
Who wanted to travel faster than light,
She set out one day, in a relativistic way,
...and returned on the previous night."

Time Now For Some Really Freaky Physics

Let's wrap up our journey of the relative side of the universe with some examples showing how truly warped the universe can get.

Example 14.6: The Guru accelerates a resting mass of 1 kg mass to 0.99c. How much would it weigh at that speed?

Our relativistic mass expression gives the final moving mass of the 1 kg object as follows,

$$m_m = \frac{m_s}{\sqrt{1-\left(\frac{v}{c}\right)^2}} = \frac{1\,\text{kg}}{\sqrt{1-\left(\frac{0.99\not{c}}{\not{c}}\right)^2}} = \frac{1\,\text{kg}}{0.1411} = 7.08\text{ kg}$$

$$\text{W}_\text{m} = m_m \cdot 9.81\frac{\text{m}}{\text{s}^2} = \mathbf{69.54\ N}$$

This expression works both ways. A massive object initially moving fast will lose mass as it slows down – *real mass, 'poof', gone...*

Example 14.7: Earth is moving along its orbit at a speed of approximately 100,000 km/hr (27,778 m/s). If you could stop the planet and jump off, how much mass would you lose?

Rewriting our relativistic mass equation in terms of the ratio of the stationary mass to the moving mass we can state,

$$\frac{m_s}{m_m} = \sqrt{1-\left(\frac{v}{c}\right)^2}$$

Or,

$$\frac{m_s}{m_m} = \sqrt{1-\left(\frac{27778 \, \cancel{m}/\cancel{s}}{3x10^8 \, \cancel{m}/\cancel{s}}\right)^2} = \mathbf{0.99999999571323730192}$$

You'd weigh a small fraction of a percentage less. The Guru does not advise this as a method of weight loss.

Example 14.8: A one meter long stick is traveling towards you, lengthwise, at a speed of 0.9c. What is its measured length as it approaches you? What is its measured length after it passes by you?

Our relativistic length expression gives the final blue shifted length of the 1 m long ruler as,

$$L_s = L_m\sqrt{1-\left(\frac{v}{c}\right)^2} = 1\text{m}\sqrt{1-\left(\frac{0.9\cancel{c}}{\cancel{c}}\right)^2} = \mathbf{0.4359\ m}$$

The red shifted length is just the inverse of this result, or 1/0.4359m or 2.2942 meters. Thus, <u>objects appear compressed as they approach you and stretched-out as they pass by</u>, basically the Doppler effect in overdrive. These objects will also literally appear 'bluish' or 'reddish' depending on the direction of travel, since the reflected light is also being shifted. This is not an illusion; it is a physical observation as real as anything you can measure in physics with the fanciest of instrumentation you could buy.

...and now for the equation everyone has been waiting for.

Example 14.9: A one kilogram mass of plutonium is converted into pure energy through radioactive decay. What is the kinetic energy released in joules?

Einstein's equation for the kinetic energy of an object relates rest mass to available energy content, thus we can write,

$$KE = m \cdot c^2 = 1\,\mathrm{kg} \cdot \left(3x10^8 \frac{\mathrm{m}}{\mathrm{s}}\right)^2 = 9x10^{16} \frac{\mathrm{kg} \cdot \mathrm{m}^2}{\mathrm{s}^2} = \mathbf{9x10^{16}\ J}$$

Thus, one mere kilogram of mass equals 90,000 Terajoules of energy. One can easily see why nuclear power (either fission or fusion) would provide an effectively unlimited amount of energy for our planet.

Concluding Thoughts

As promised, the Guru has taken you on a voyage of the universe and brought you back safely to the comfort of your reading chair. Throughout our journey we have covered all the basics (and then some) contained in an elementary physics course, as well as a course in vector mechanics, fluid mechanics, basic chemistry, astronomy, and some philosophy taking a sophisticated, but not too terribly complex, mathematical approach.

We have covered units and unit conversion, vector and scalar analysis, matter and the periodic table, Newton's laws, kinematics and kinetics, collisions, work and energy, rotational analysis, friction, fluid mechanics, thermodynamics, fundamental forces, electricity and magnetism, and the quantum nature of light and relativity – and a whole lot of scientific vocabulary and physical understanding of the world around you that you did not know before you started reading this book. *Yes indeed, we have covered a lot of territory and the Guru's fingers are tired from all the typing.*

At this point you should be able to pick up any college level physics book and understand the order in which topics are presented as well as the technical jargon. Don't be afraid to experiment with the examples provided in this book. Use them as a starting point to examine your own problems, or to figure out what a typical solution with the correct units looks like.

The Guru Says...

Want to contact the Guru with questions, comments, or suggestions? He can be reached by email at...

guru@gurusuniverse.com

Make sure to include the word 'book' somewhere in the subject line of the email (this helps him separate good email from junk). *...and don't forget to check out his website to see what's new in the universe.*

Concepts like momentum, energy, the Ampere, and moment of inertia should seem like an old friend instead of a scary new concept. When in doubt as to what the meaning of some new topic is, reach back for the Guru's Guide to understand the basics, since there is no magic in any of this. Physics follows a logical pattern of understanding which builds upon previous learned knowledge. The universe is a big place to explore. *Happy traveling...*

The Guru's Index of Key Concepts

F

G

H

I

K

L

M

N

O

P

Made in the USA
Las Vegas, NV
01 February 2026

40898634R00155